Human Exposure to Electromagnetic Fields

*To Elisabeth,*
*Amandine, Xavier and Maxime*

# Human Exposure to Electromagnetic Fields

*From Extremely Low Frequency (ELF) to Radiofrequency*

Patrick Staebler

iSTE

WILEY

First published 2017 in Great Britain and the United States by ISTE Ltd and John Wiley & Sons, Inc.



ISTE Ltd
27-37 St George's Road
London SW19 4EU
UK

www.iste.co.uk

John Wiley & Sons, Inc.
111 River Street
Hoboken, NJ 07030
USA

www.wiley.com



Library of Congress Control Number: 2017930986

British Library Cataloguing-in-Publication Data
A CIP record for this book is available from the British Library
ISBN 978-1-78630-121-5

# Contents

# Preface

*"Knowledge is the only treasure*
*that increases when shared".*

The laws of physics dictate that, whether we like it or not, we are all exposed to electric, magnetic and electromagnetic fields produced by the simple fact of using electricity. The number of electronic devices that radiate such fields has not stopped growing since the end of the 19th Century, when the first wireless communication took place and the spread of electricity began.

These devices have radically changed our lifestyle. They provide comfort, make tasks easier and offer opportunities for medical diagnosis and therapy. They have also deeply changed the way we communicate. Every proposed remote service must exist independently. In order to avoid interference between them, each one has its own frequency, so the radiation extends across the entire frequency spectrum.

This spectrum is a limited and fully regulated resource, the property of the states which grant its use. It will be increasingly busy around the clock. There will be no idle frequency, since future transmissions will dynamically manage the spectrum due to the arrival of software-defined radio providing ever more services. This means a higher data rate, and therefore more occupied frequencies.

Furthermore, the number of transmitters and receivers will continue to increase at a sustained rate. In addition to what are considered basic functions today (voice calls, internet and mobile video, contactless transactions, remote control), telephones will include Internet of Things (IoT) functions and others yet to be invented. The greater obstacles will be standardization and the protection of privacy, rather than technology.

We are eager for new technologies, despite the questions they raise for some people regarding the electromagnetic fields they radiate. These questions are logical because these fields are undetectable and intangible, so, for most of us, they remain a mystery.

Society's concerns with radiation reappear with each technological advance, including the wireless telegraph in the 1930s, the television in the 1960s and the mass usage of wireless telecommunications since the beginning of the 1990s. The debate about power lines appeared belatedly at the end of the 1970s.

Nevertheless, these fields can be dangerous to health and safety if they are of high intensity. The harmful effects have been universally recognized for decades, but the possible effects on health after long-term exposure to low levels continue to be the subject of controversial debates.

Exposure to electromagnetic fields is a complex domain since, on the one hand, the relevant physical quantities are abstract, and, on the other, it is a multidisciplinary topic, given the interactions with biological systems. Physics, biology, medicine and metrology coexist.

In daily life, the general population is exposed to rather weak electromagnetic fields. The fields are stronger in occupational environments. Electrochemical processes, magnetic induction, welding, electricity distribution and telecommunications all expose workers to high field levels when they are close to the sources of radiation.

To protect workers and the public, limits were established. They have become a part of legislation.

There is a lot of confusion between electric, magnetic, and electromagnetic fields. This confusion extends to biophysical, biological and health effects. Moreover, certain publications use these terms incorrectly. There is also a lack of knowledge about sources of radiation. Few people know how exposure limits are determined, what precautions should be taken to carry out radiation measurements and, even less so, how these measurements and limits should be compared. Forming your own opinion in this domain requires at least some curiosity and an interest in science.

In addition, legislation requires that employers protect their employees from strong electromagnetic fields and that manufacturers produce compliant products that only radiate a limited amount or else provide use restrictions.

## Intended audience

This book is intended for anyone who wishes to acquire a first level of knowledge about electromagnetic fields from 0 Hz to 300 GHz. Reading this book requires some scientific background. It contains several examples to facilitate understanding. The reader will be able to argue and discuss this controversial topic and analyze publications with a critical eye.

This book is also aimed at people who are responsible for the prevention of electromagnetic risk, such as engineers or technicians in charge of work conditions, health physicians, and, in particular, occupational doctors and trainers.

People in charge of ensuring compliance with regulations intended for the public and/or employees will have guidance on how to proceed and on metrology.

Manufacturers of household or professional materials will find elements to guide them in both the design (reduction and elimination of radiation at the source) and the production of an operating manual containing exposure data.

Finally, teachers can draw on several concrete applications to illustrate their lectures.

## Disclaimer

This work is a summary of directives, recommendations, guides, standards and other official or scientific publications related to human exposure to electromagnetic fields. Most of these documents are references. In the event that the proposed reformulations are misunderstood, it is best to refer back to the original documents.

We have gone to great lengths to ensure that the content of this work is exhaustive and thorough. However, if our readers notice any errors, we would be grateful if they could communicate them so that subsequent editions can be corrected.

## Organization and use of this book

This book summarizes all aspects of human exposure to electromagnetic fields from extremely low frequency (ELF) to radiofrequency or, to be more precise, from 0 Hz to 300 GHz. It allows the reader to form an opinion on the subject and adhere to regulations knowingly. This work is the result of several years of experience including presentations, standardizations, compliance assessments, measuring and designing equipment which radiates as low as reasonably achievable (www.exposition-cem.com).

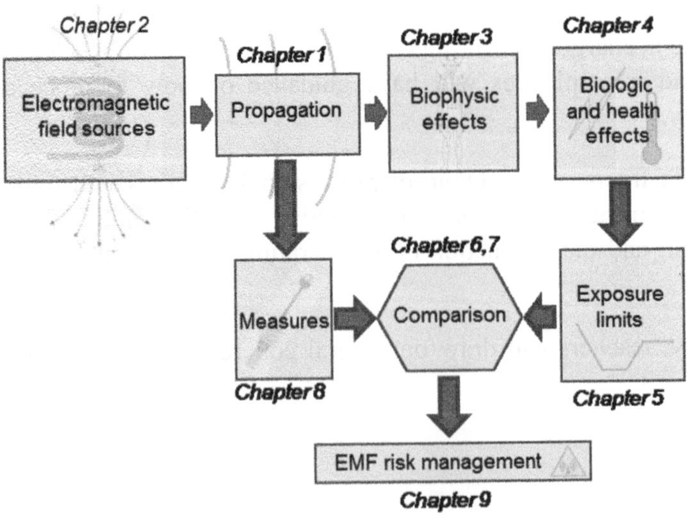


**Figure 1.** *Structure of the book*

The content is structured as presented in Figure 1. There are nine chapters and two appendices, although some of these topics deserve a completely separate book. The main terms introduced can be found in the glossary.

Chapter 1: This chapter aims to introduce electromagnetism. This chapter, like those that follow, focuses on "human exposure". Electric, magnetic and electromagnetic fields are explained by the laws of physics. They interact with organisms according to their nature and produce biological effects. It is essential to specify and clarify the vocabulary and definitions to avoid any misunderstanding. The relevant physical quantities are explained because they determine the effects and measurements. They also make it possible to define the exposure limit values. Although the frequencies extend infinitely, the "electromagnetic field" perimeter, as defined, is limited. This aspect has to be clarified. This chapter is fundamental because it is the gateway to grasping the human exposure concepts. For some, this chapter is a refresher for distant studies, but it should be approachable for those who have some gaps in their physical science backgrounds. Understanding this chapter is essential to take full advantage of this book.

Chapter 2: This chapter presents radiation sources encountered in daily life, others which are present in the workplace as well as others which are used for diagnostic or therapeutic purposes. It is impossible to be exhaustive because the sources are so numerous and heterogeneous.

Chapter 3: This chapter describes biophysical mechanisms. How do fields interact with material, in particular with biological tissue? This chapter explains why these tissues do not have the same electric properties as most materials used in electrical engineering. This chapter also takes a macroscopic approach. It describes how fields penetrate the human body and how they behave according to their nature and frequency.

Chapter 4: This chapter discusses the biological effects of fields, as well as the sensory and health effects. When the fields are intense, they can disturb certain senses or be imperceptibly harmful to health. This chapter presents the recognized acute effects as well as a summary of the studies on chronic (long term) effects. Direct effects are distinguished from indirect effects. The definitions are recalled to avoid confusion.

Chapter 5: Exposure limits (basic restrictions, reference level, etc.) logically follow biological effects. These limits were established to prevent adverse effects on health and well-being. This chapter also justifies the distinction between exposure of the general public and workers. It explains how these limits are determined. The values are summarized and illustrated in order to be quickly interpreted and compared. This chapter also considers people or workers at particular risk.

Chapter 6: This chapter explains the methods of comparison between measured or calculated fields and limit values, especially when these fields have complex waveforms. These methods are as significant as the limit values when assessing an exposure situation. In particular, the weighted peak method used to assess low-frequency exposure is introduced. Necessarily, this chapter includes many formulas. They may seem complex, but in reality most of them are comprehensible even to those readers who have some mathematical gaps.

Chapter 7: This chapter supports the previous one with examples of theoretical and real applications. Understanding them makes it possible to become familiar with the calculation methods presented. It can be read subsequently.

Chapter 8: This chapter introduces the basics of exposure assessment. Measuring techniques for electric, magnetic and electromagnetic fields are presented. Operating precautions and specific sources of uncertainty about the nature of these fields are explained. Measurement of contact and induced electric currents is also introduced.

The last section in this chapter succinctly approaches dosimetry, a technique used to precisely assess exposures when required or when measuring becomes impossible for physical reasons as indicated throughout this book.

Chapter 9: This chapter situates Directive 2013/35/EU (electromagnetic fields) in the European context and reformulates it by giving some practical guidance for its implementation. It refers to Appendix 1, which classifies equipment according to its radiation intensity. This chapter helps readers understand the interest of the exposure system limits presented in Chapter 5, which could seem obscure at the first reading.

Appendix 1 sorts equipment and workplaces according to the electromagnetic field strength to which the workers may be exposed. This appendix supports Chapter 9.

Appendix 2 presents the weighting functions implemented to assess exposure according to the calculation technique using the weighted peak method. This appendix complements Chapters 6 and 7.

Enjoy.

Patrick STAEBLER
March 2017

# Acknowledgments

The domain of human exposure to electromagnetic fields is very active, with many talented individuals working within it. This book was made possible because of lasting exchanges with:

– Dr. Martine Souques (health physician at the *Service des études médicales* [Medical studies department], EDF/France);

– Dr. Daniele Andreuccetti (researcher at the National Research Center "Assessment of human exposure to electromagnetic fields", IFAC-CNR/Italy);

– François Deschamps (scientist advisor in electromagnetic field exposure, RTE/France);

– Joe Wiart (Head of the chair, "Modeling, Characterization and Control of Electromagnetic Wave Exposure (C2M)" Telecom ParisTech/France);

who share their knowledge with such enthusiasm. I am grateful to all of them.

I express my sincere gratitude to Alain Azoulay (Honorary Professor at Supélec) who encouraged me to get into this project and who supported me in moments of doubt. Alain proofread this book before it was even a full draft.

I would also like to thank my friendly and generous colleagues who took the time to proofread chapters of this book related to their area of excellence: Dr. Martine Souques, François Deschamps, Thierry Letertre (teacher-researcher at the CentraleSupélec engineering school) and Péniamin Matossian (expert in electromagnetic compatibility).

I would also like to thank my friend Eric Lanzi, who volunteered to proofread this work. He provided many relevant comments driven by his own scientific curiosity and his love of new technologies.

Thank you to my wife, Elisabeth, who had the huge task of being my first reader, and to my son, Xavier, who verified the coherence of these chapters with a fresh outlook on the domain of electromagnetic fields.

Finally, I cannot end without thanking my colleagues at the *Société française de radioprotection, section Rayonnement non ionisant* (SFRP-RNI) [Radiation Protection Society of France, Non-Ionizing Radiation section] and my colleagues on standardization committees, who provided me with many helpful clarifications.

# 1

# Concepts of Electromagnetic Fields

Electromagnetic fields are produced from natural and artificial sources. We distinguish between electric, magnetic and electromagnetic fields. Whether static or variable in time, they each have physical properties that produce specific interactions with biological organisms: plant, animal and human.

To provide a better understanding of the interaction mechanisms, the concepts of electromagnetism and the associated terminology are presented in this chapter [FEY 15].

## 1.1. Concepts of fields

### 1.1.1. *Introduction*

In physics, a field can be defined as an area of influence. We are immersed in the Earth's gravitational field (the area within which the Earth attracts objects) and in electromagnetic fields (areas within which we can pick up television broadcasts and mobile phone signals, for example).

Gravity acts between two bodies that have a mass, while an electric field acts on positive or negative electric charges (electric charge is a fundamental property of matter, along with mass). These interactions are due to forces whose intensity is proportional to the mass of the objects or to the value of the electric charges. They decrease with the square of the distance that separates them and cancel each other out *ad infinitum*. The expression of these forces is identical. Their intensity in Newton (N) is given, respectively, by the law of gravitation and by Coulomb's law (in classical physics):

– law of gravitation between two masses $m_1$ and $m_2$:

$$F_{G_{1/2}} = G.\frac{m_1.m_2}{d^2} \quad \text{(N)} \qquad [1.1a]$$

– Coulomb's law between two electric charges $q_1$ and $q_2$:

$$F_{E_{1/2}} = k_C.\frac{q_1 \cdot q_2}{d^2} \quad \text{(N)} \qquad [1.1b]$$

where $d$ is the distance in meters between the two objects, $G$ is the universal gravitational constant, and:

$$k_C = \frac{1}{4.\pi.\varepsilon_0} = \frac{c^2}{10^{-7}}$$

Coulomb's constant ($N \cdot m^2 \cdot C^{-2}$), and "$c$" is the speed of light ($\approx 3 \times 10^{+8}\ m \cdot s^{-1}$). $\varepsilon_o$ is a constant that will be introduced later on. If we refer to the Earth, $F_{G1/2}$ represents the weight and [1.1a] becomes simply $P = m \cdot g$ with $g = 9.81\ m \cdot s^{-2}$.

Unlike the gravitational force, which always attracts objects, the electric force attracts opposite charges and repels them if they have the same sign. The force is directed according to the straight line between the two masses or charges.

Thus, the field is the area where forces can exert themselves remotely, without there necessarily being any propagation of matter or energy. Unlike electric fields, the Earth's gravity has the advantage of being directly perceived by living organisms.

The concept of the field enables us to introduce physical (vectorial)[1] quantities, values associated with a unit, to quantify these phenomena and put them into an equation for the purposes of analysis and prediction.

The physical quantity associated with electric charges is the electric field.

1 In this book, we will revisit the concept of vectors when necessary for understanding.

### 1.1.2. *Electric fields*

The electric field is introduced into Coulomb's law [1.1b] as follows:

$$F_{E_{1/2}} = q_2 \,.\, E_{1/2} \quad \text{with} \quad E_{1/2} = k_C \,.\, \frac{q_1}{d^2} \qquad [1.2]$$

$E_{1/2}$ is the electric field created by the charge $q_1$ in the place of $q_2$ (the symbol $E$ is reserved for the quantity that is the electric field). An electric field appears as soon as there is an electric charge. Generally, knowledge of $E$ makes it possible to calculate the force exerted on any charge.

In practice, an electric field is created by an inequality of charges between two distinct points. This inequality may be due to a difference in potential or voltage.

The most natural representation of an electric field consists of applying a voltage between two parallel planar electrodes, forming a planar capacitor (Figure 1.1). This set-up makes calculating the field strength simple.

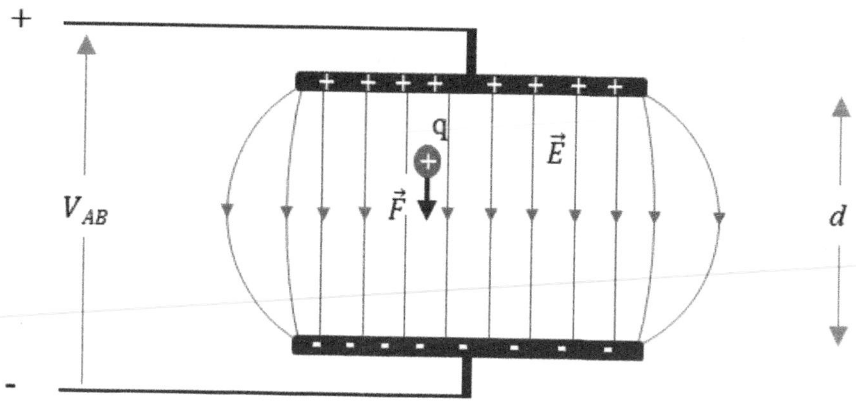


**Figure 1.1.** *Uniform electric field between two parallel planar electrodes*

At the center of this set-up, it can be demonstrated that the electric field is proportional to the voltage and inversely proportional to the distance:

$$E = \frac{V_{AB}}{d} \text{ (V/m)} \qquad [1.3]$$

where $V_{AB}$ represents the difference in potential or voltage (in volts) between the two electrodes and $d$ is the distance between them (in meters). This example introduces the electric field strength unit (V/m or $V{\cdot}m^{-1}$). A voltage of 1 volt between planes 1 m apart creates an electric field of 1 $V{\cdot}m^{-1}$.

The lines on Figure 1.1 represent the spatial distribution and orientation of the electric field. These are the force lines (force $F$ on charge $q$ illustrates [1.2]). They originate with the positive charge and end with the negative charge or *ad infinitum* in the absence of opposite sign charges. Field lines cannot intersect. There can be no electric field without electric charges.

When the voltage is constant or varies slowly, the field created is called electrostatic. It is limited to the space between the two electrodes.

The field is uniform when its strength and direction are constant, as shown in Figure 1.1. However, if its strength and direction depend on the position, the field is no longer homogeneous, as shown in Figure 1.2.

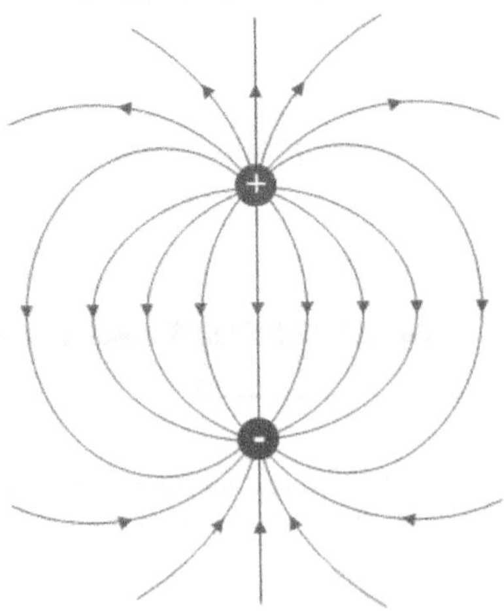

**Figure 1.2.** *Non-uniform electric field between two electric conductors of opposite polarity*

The electric particles subjected to a field are attracted by the electrode of opposite polarity and repelled by the electrode of the same polarity. They move if they are free, e.g. in a conductor or in the vacuum. When their movement is limited to a volume, e.g. in an insulated conductor, they gather on its surface (Figure 1.3).

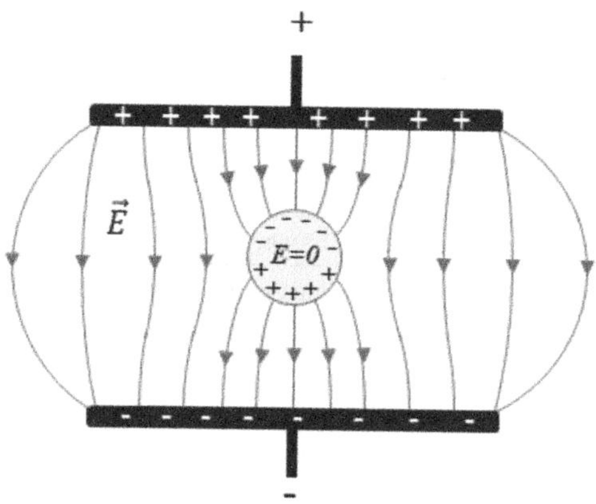


**Figure 1.3.** *Surface accumulation of charges on a conductor in an electric field*

The electric field induced within this conductor is of the opposite direction to the external field, in accordance with the principle that effects always counteract their cause (definition of the law of moderation in physics). The overall internal field, the sum of the external field and the induced field, is zero. This is the principle of the Faraday cage.

An alternating field will induce permanent movement of the free electric charges of a conductor. This movement creates an electric current.

Furthermore, the initial external field is altered by the conducting object itself, due to the new distribution of charges. This explains why an electric field is strongly influenced by any type of material, regardless of whether it is a good or poor conductor and even if it is insulated. The presence of a person and a measuring device (fieldmeter) in an electric field distorts its measurements if no precaution is taken. In return, this field can be easily mitigated.

More generally, all materials act on the electric field, because they all have electrical properties, even the most insulating among them. The resulting phenomena are discussed in more detail in Chapter 3.

| Key points of electric fields |
|---|
| The level of electric fields is expressed in volts per meter ($V{\cdot}m^{-1}$ or V/m). |
| An electric field appears in the presence of electric charges. These charges can be the result of a difference in potential (voltage).<br>That means a field exists as soon as an electric device is connected to the power supply, even if it is not working. |
| The field strength decreases with the distance from the source. |
| An electric field can make electric charges move. |
| The electric field alters the distribution of charges on the surface of a conducting object.<br>Reciprocally, an object that is slightly or highly conductive alters the field. |
| A distinction should be made between the ambient electric field and the electric field induced within an object during exposure. |
| Most construction materials protect against electric fields to some extent. |

**Table 1.1.** *Summary of electric fields*

### 1.1.3. *Magnetic fields*

A magnetic field is created when electric charges move to form a current. There can be no magnetic field without an electric current.

The force lines of the magnetic field form concentric circles along a conductor in which a current is flowing. These circles (forces) are perpendicular to the conductor, as shown in Figure 1.4.

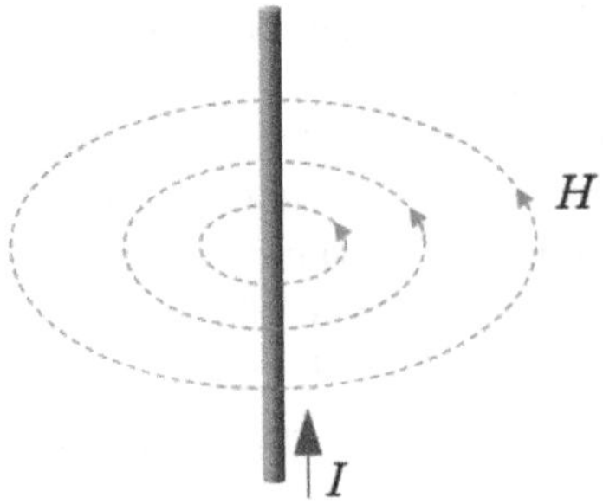


**Figure 1.4.** *Magnetic field line around a wire*

The magnetic field is characterized by its strength, represented by the symbol $H$ and measured in amperes per meter ($A \cdot m^{-1}$). This quantity is proportional to the intensity of the electric current and decreases with distance. For a spatially isolated, straight electric wire, the field is expressed by (Biot–Savart law):

$$H = \frac{I}{2.\pi.d} \text{ (A/m)} \qquad [1.4]$$

where $I$ represents the electrical intensity in amperes (A) and $d$ is the distance to the wire in meters. A current of 100 A intensity creates a magnetic field of 15.9 $A \cdot m^{-1}$ at a distance of 1 m.

A continuous current creates a static magnetic field, like a permanent magnet. An alternating current creates an alternating magnetic field.

Conversely, magnetic fields can induce electric currents and fields. This is the origin of many electrical phenomena. For example, a voltage appears in the presence of a conductive contour subjected to a variable magnetic field. When it is closed (loop), a current flows. This is the principle of induction heating, or that used by current generators.

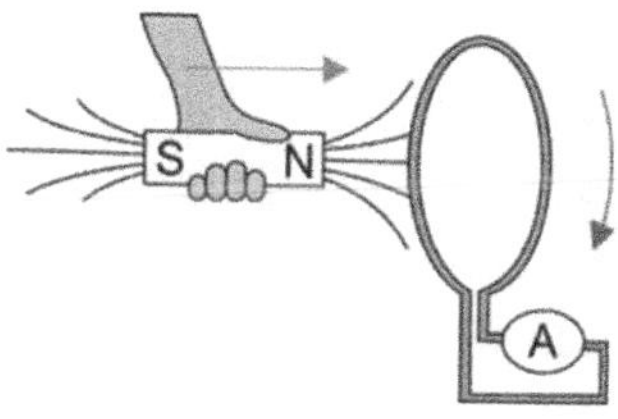


**Figure 1.5.** *Illustration of Faraday's law: surface subjected to a variable magnetic flux*

The electromotive force (open-circuit voltage) induced by a closed circuit is proportional to the surface of this circuit and to the time-varying magnetic flux. This is Faraday's law. When the field is uniform and perpendicular to a surface $S$, we have:

$$e = S.\frac{dB}{dt} \text{ (V)} \qquad [1.5]$$

where $B$ is the density of the magnetic flux, more commonly called magnetic induction. Its unit in the international system (SI) is the tesla (T). The microtesla (μT) is the most common unit in the domain of human exposure (1 μT = $10^{-6}$ T or 1 millionth of a tesla). The standard unit on the American continent is the gauss (10 mG = 1 μT).

A magnetic field also exerts a force on moving electric particles.

Biological tissues are conductors. Currents and electric fields are induced within them when they are subjected to a magnetic field that is variable in time according to the principle outlined above. This can trigger effects on biological organisms.

Magnetic fields and magnetic induction are connected by a constant that depends on the medium. In the vacuum or in the air, their relationship is:

$$B = \mu_0 . H \ \text{(T)} \qquad [1.6]$$

with $\mu_0$ (magnetic constant or vacuum permeability) = $4 \cdot \pi \cdot 10^{-7}$ H·m$^{-1}$ (henry per meter). We have the following relationships:

$$1 \ \text{A} \cdot \text{m}^{-1} \approx 1.25 \ \mu\text{T}$$

and

$$1 \ \mu\text{T} \approx 0.8 \ \text{A} \cdot \text{m}^{-1}$$

This makes it possible to simplify the notation of relationship [1.4]. Magnetic induction is therefore written as follows:

$$B = 0.2 . \frac{I}{d} \ (\mu\text{T}) \qquad [1.7]$$

where $I$ is the current intensity expressed in amperes and $d$ is the distance in meters. An electric current of 100 A produces a magnetic induction of 20 μT at 1 m around a single straight electric wire.

The medium crossed by the field alters the magnetic induction. Thus, equation [1.6] becomes:

$$B = \mu_r \cdot \mu_0 \cdot H \ \text{(T)} \qquad [1.8]$$

where $\mu_r$ is relative magnetic permeability of the medium and not necessarily a constant. It can depend on the frequency and magnetic field level. The relative permeability of biological tissues is 1, as for the vacuum or the air. It follows that magnetic induction is not altered by a human body presence (see Chapter 3).

Although not strict, the term "magnetic field" is often used for magnetic induction in the domain of exposure. This will be the case in this book.

Conveniently, the magnetic field $H$ is used when the field is created. However, magnetic induction $B$ should be considered when studying physical effects: the induction level makes it possible to calculate the force exerted on the moving charges and induced currents. $B$ can be measured.

Relationship [1.7] can be extended. For a given electrical set-up, the magnetic field is proportional to the intensity of the current and decreases with the distance. The more complex and compact the radiation source (electrotechnical systems, such as coils and motors), the faster the decrease. Thus, magnetic induction can generally be formulated as follows:

$$B = K \cdot \frac{1}{d^{\alpha}} \qquad [1.9]$$

$K$ is a constant dependent on the source and $d$ is the distance to the source in meters. The exponent $\alpha$, generally between 1 and 3, depends on the nature of the source (Figure 1.6). $\alpha = 1$ for a single straight electric wire [1.7]. It may vary depending on the direction. Induction can be calculated very precisely when this law is known. However, in many situations, the field can only be characterized through measuring or numerical simulation of the source.

Unlike the electric field, the low-frequency magnetic field is not altered by common materials ($\mu_r = 1$). It enters them with almost no distortion, which can pose a problem when it comes to protecting against it. However, this reduces the restrictions related to the human body presence while it is being measured.

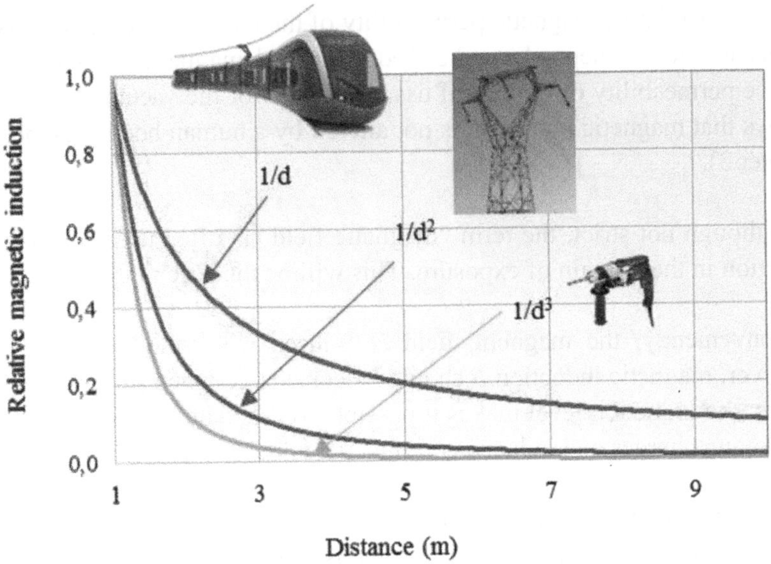


**Figure 1.6.** *Decrease in the magnetic field for various types of source*

| Key points of magnetic fields |
|---|
| Magnetic fields are produced by the circulation of electric currents. |
| The quantity used when studying the effects of magnetic fields is magnetic induction ($B$), often expressed in microtesla (μT) and sometimes in millitesla (mT). |
| The strength of the magnetic field diminishes with the distance from the source. |
| A time-varying magnetic field passing through a closed conductive surface creates an electric field as well as a current. |
| A magnetic field can exist without an electric field and vice versa. |
| At low frequencies, magnetic fields are not mitigated by most materials. |

**Table 1.2.** *Summary of magnetic fields*

### 1.1.4. *Introduction to electromagnetic fields*

How is an electromagnetic field created? It requires the simultaneous presence of an electric field created by electric charges and a magnetic field created by the flow of these charges.

When the two electrodes of the capacitor that was used before (Figure 1.1) are connected to an alternating voltage generator, a variable electric field appears. Positive and negative electric charges alternate on either side as the voltage varies.

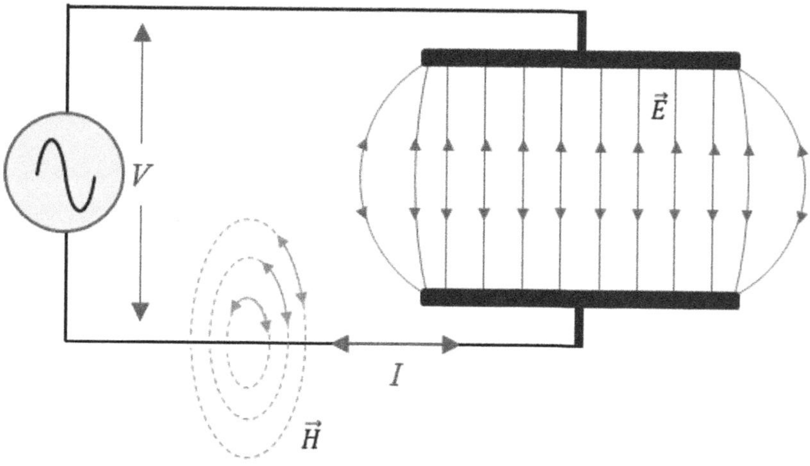


**Figure 1.7.** *Two electrodes under time-varying voltage create an electric and magnetic field*

The negative charges are electrons. Strictly speaking, the positive charges do not, circulate; they represent a deficit of electrons. The electron flow creates an electric current in the circuit and within the electrodes. This current produces the magnetic field.

By adapting the dimensions and shape of the conductors to the variation in current and voltage, we can create a radiofrequency antenna to make the electric field and the magnetic field radiate in the same direction (Figure 1.8). The association of these fields forms an electromagnetic field that will propagate over great distances. The antenna also aims to maximize the transfer of electric energy into electromagnetic energy.

The Scottish physicist James Clerk Maxwell successfully interrelated the electric field and the magnetic field in his famous equations. His equations form the basis of the theory of electromagnetism.

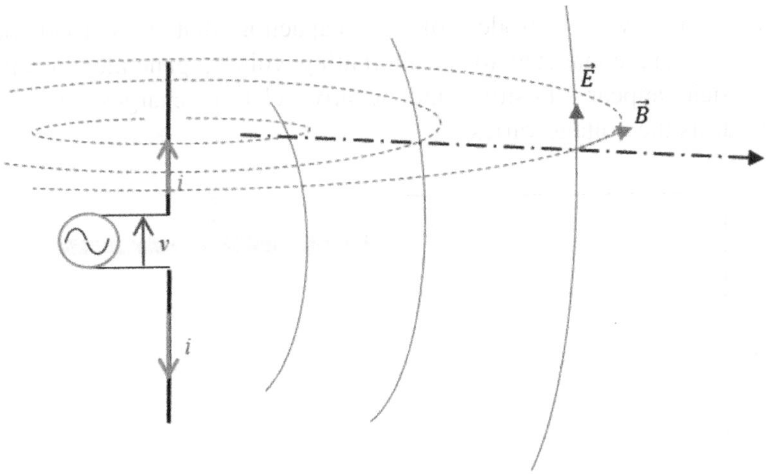


**Figure 1.8.** *Dipole antenna with electric field and magnetic field*

These equations describe the interactions between electric charges, electric currents, electric fields and magnetic fields. They indicate that a time-varying magnetic field creates an electric field that is also variable and vice versa, even in the absence of charges and current. It is, of course, necessary to initially have these charges and currents to generate the two fields, but once radiated, their existence no longer depends on it; one becomes the source of the other. This mutual generation enables them to propagate at the speed of light in the air or in the vacuum over an infinite distance from the source. The fields should therefore be considered as a single physical entity, the electromagnetic field. In this case, the variation of one is proportional to the variation of the other.

Maxwell's equations also indicate that in the presence of a direct current and voltage (steady state), these fields are not correlated, and one can exist without the other. There can be no propagation, only a limited area of influence.

## 1.2. Waves, frequencies and wavelengths

### 1.2.1. *Waves*

A pebble falling into a pond creates ripples, waves that spread away from the point of impact. Imagining a wave is easier than defining it.

A wave corresponds to the propagation of a disturbance that reversibly alters the physical properties of the environment in which it propagates. It transports energy without transporting matter. A wave exists only if there is a variation in disturbance.

Electromagnetic waves do not need a physical medium to propagate. They can propagate in the vacuum, unlike mechanical waves, which need a material medium (e.g. acoustic waves: sound propagates in the air, water, etc.).

The simplest waveform is the sinusoidal wave. This is the typical example of the voltage and current of the 50 Hz electricity grid. Complex waveforms are often used in telecommunications and radar systems.

### 1.2.2. *Frequencies and periods*

Frequency, represented by $f$, is used to characterize periodic signals over the duration. It indicates how many times an identical elementary event recurs per second. A static field, i.e. one that is constant in time, has a zero frequency.

The unit of frequency in the international system (SI) is the hertz (Hz). Its multiples are often used: kHz (kilohertz), MHz (megahertz) and GHz (gigahertz).

The period, represented by $T$, is the inverse of the frequency, the smallest interval of time after which an identical event recurs. The unit of the period is the second (s).

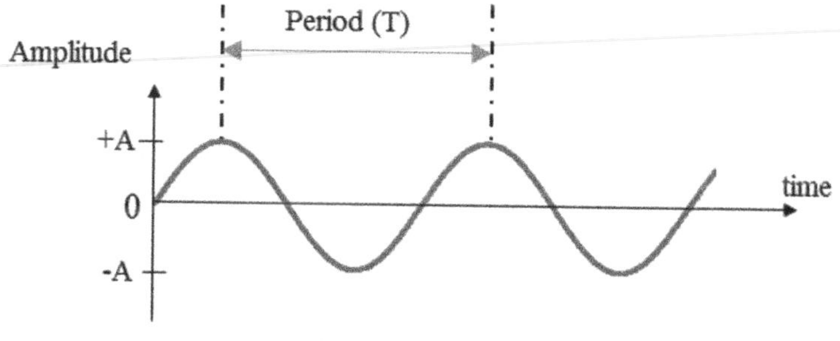


**Figure 1.9.** *Time profile of a sinusoidal wave*

Frequency is an important parameter of an electromagnetic field. It influences the design of electric and electronic circuits, the propagation and the measurement methods. Frequency is a physical quantity used in telecommunications and spectral analysis (spectrum occupancy, jamming, range of a wireless link, etc.).

The mechanisms of interaction with matter and the biological effects depend greatly on it. A field of equal strength can either have no effect or be dangerous to the health, depending on its frequency.

### 1.2.3. *Wavelengths*

Wavelength, generally represented by the Greek letter λ (*lambda*) and measured in meters (m), is the distance traveled by the wave during one period. It therefore depends on the frequency. It is also the shortest distance separating two maximum points of the wave at a given instant. It represents the spatial periodicity of the wave.

Wavelength (*m*) is calculated from the propagation speed of the wave in the medium and the frequency, according to the formula:

$$\lambda = \frac{v}{f} \text{ (m)} \qquad [1.10]$$

$v$ represents the propagation speed ($m \cdot s^{-1}$) and $f$ is the frequency of the wave in hertz. In the vacuum or in the air, $v = c$, meaning that electromagnetic waves propagate at the speed of light, which is usually rounded to $3 \times 10^8 \, m \cdot s^{-1}$.

The wavelength of a signal of frequency 100 MHz (FM radio band) is 3 m.

Wavelength is the reference value for calculating the dimensions of an antenna or characterizing the propagation close to a radiating source.

Wavelength also comes into play in the study of the interaction between electromagnetic fields and living organisms. This interaction depends heavily on the relationship between the wavelength and the size of the organism at certain frequencies.

## 1.3. Propagation of electromagnetic waves

The science of telecommunications (television broadcasting, telephony, radar, etc.) aims to optimize the propagation of electromagnetic waves (or electromagnetic fields), i.e. to ensure service coverage across a particular area, with a minimum quality. The spread of waves beyond a limited area remains an undesirable effect for applications outside of telecommunications.

The basic concepts of electromagnetic wave propagation are important for predicting the ambient electromagnetic field level, regardless of the application.

So that waves can propagate across significant distances in relation to wavelength, we have seen there must be simultaneous creation of an electric field $E$ and a magnetic field $H$. In the absence of one of the two elements, the field remains confined near to the source.

### 1.3.1. *Propagation in free space*

During propagation in free space, i.e. in a volume containing no obstacles, the energy transported by a wave is distributed across an increasingly large surface area. The law of conservation of energy states that energy propagating within a solid angle remains constant. As the surface crossed increases with the square of the distance, the energy must decrease in inverse proportion, i.e. the energy per unit surface area decreases with the square of the distance. It is divided by four when the distance doubles (if there is no absorption by the environment).

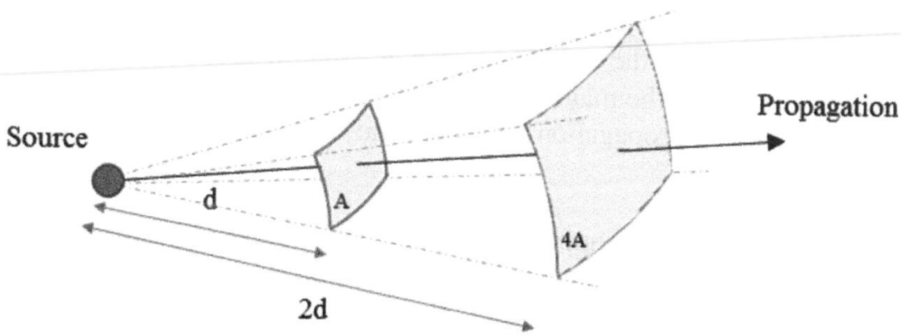


**Figure 1.10.** *Energy dispersal with distance*

Power density $S$ is defined as the quantity of energy per unit of time crossing a unit surface or as the power per surface unit. $S$ is expressed in watts per m$^2$ (W·m$^{-2}$). It is obtained by normalizing the power radiated by the surface of a sphere of radius $d$ according to the relationship:

$$S = \frac{P}{4 \cdot \pi \cdot d^2} \text{ (W·m}^{-2}\text{)} \qquad [1.11]$$

where $P$ represents the power radiated by the source.

This concept is important, because it makes it possible to characterize exposure at the highest frequencies (see section 5.4).

Initially, let us stand far from the source of an electromagnetic field, at a distance of many times its size. Even if the wave is spherical, it can be considered as locally planar.

**Figure 1.11.** *Exposure becomes uniform far away from the source*

In this set-up, the field power density is considered homogeneous over a large surface area. All the points are seen at equal distance from the source. The electric field and the magnetic field are perpendicular to one another and to the direction of propagation, the opposite direction from the source as shown in Figure 1.8.

The Poynting vector supplements field $E$ and field $H$ to form a right-angle trihedron. This vector is oriented in the same direction as the propagation. It represents the magnitude of power density $S$, which is also the product of the electric field $E$ and the magnetic field $H$:

$$S = E \cdot H \text{ (W·m}^{-2}\text{)} \qquad [1.12]$$

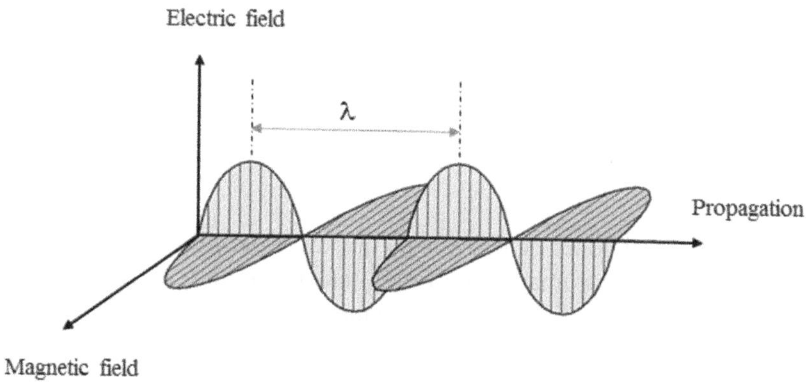


**Figure 1.12.** *Spatial representation of an electromagnetic wave in the far field*

It should be noted that the electric field (V·m$^{-1}$) to magnetic field (A·m$^{-1}$) ratio is consistent with electrical resistance and more generally with impedance. This relationship is called wave impedance in the medium. Its unit is the ohm (Ω):

$$Z = \frac{E}{H} \ (\Omega) \qquad [1.13]$$

Far from the source, the electric field and the magnetic field stay in phase and attenuate in the same way. In the vacuum, this relationship remains constant, as follows:

$$Zo = \frac{E}{H} \ (\Omega) \qquad [1.14]$$

*Zo* represents the wave impedance in the vacuum or the impedance in free space. Its value is 120·π Ω, i.e. around 377 Ω. This is also true for propagation in the air. It is also shown that the electric field is related to magnetic induction *B* by means of:

$$E = c \cdot B \qquad [1.15]$$

with $c$ the speed of light ($3 \times 10^8$ m·s$^{-1}$).

Formulas [1.12] and [1.14] make it possible to relate power density to the electric or magnetic field:

$$S = \frac{E^2}{Zo} = Zo \cdot H^2 \ (\text{W·m}^{-2}) \qquad [1.16]$$

Knowledge of field $E$, field $H$ (or $B$) or power density ($S$) is sufficient to characterize an electromagnetic field far from the source. At high frequencies, the electric field is generally measured.

An electric field of 30 V·m$^{-1}$ corresponds to a magnetic induction of 0.1 µT and a power density of around 2.4 W·m$^{-2}$ or 0.24 mW·cm$^{-2}$.

Combining [1.11] and [1.16] produces the electric field depending on the power radiated in watts and the distance in meters:

$$E = \frac{\sqrt{30 \cdot P}}{d} \ (\text{V·m}^{-1}) \qquad [1.17]$$

A radiated power of 10 W generates an electric field of 1.7 V·m$^{-1}$ at 10 m.

Power density diminishes regularly with the square of the distance, while the electric field and the magnetic field attenuate with the distance. This phenomenon is called *free space loss* (Figure 1.13).

During real propagation, loss with distance is generally made greater by the presence of hills and obstacles, such as plants or structures.

Telecommunications seeks to maximize the power radiated in relation to the power supplied. To do this, transmitters are equipped with antennae suitable for the wavelength of the signal to be transmitted. An antenna is a reversible electromechanical system: these properties remain identical in emission and reception, although the term radiation pattern is always used to characterize them.

An antenna can be designed to radiate over a large area or to concentrate energy in one preferred direction. By definition, the antenna gain ($G_i$) is the ratio of the power radiated in the direction of maximum radiation to the power that would be emitted by an ideal antenna radiating in all directions in the same way. This theoretical antenna is called an *isotropic antenna*.

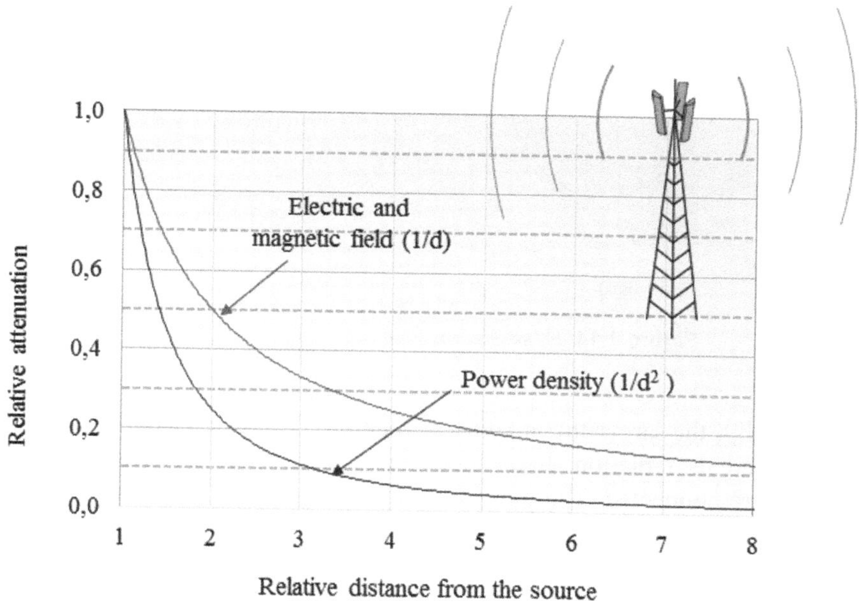


**Figure 1.13.** *Decrease of the electric field, the magnetic field and the power density with the distance from the source*

Gain is one of the key parameters relating to an antenna, along with which frequency band it uses. A gain is a dimensionless number (ratio between two identical quantities) generally expressed on a logarithmic scale in decibels (symbol dB). The following relationships make it possible to convert a linear scale into a logarithmic scale (dB) and vice versa:

$$G_{dBi} = 10 \,.\, \log_{10}(G_i) \quad \Leftrightarrow \quad G_i = 10^{\frac{G_{dBi}}{10}} \qquad [1.18]$$

The privileged coverage area, in which the maximum power is transmitted, is called the *main lobe*. Directions in which the power levels are lower are called *side lobes*. They can be sufficient to establish short-range communication. Among the side lobes, the back lobe can be distinguished, oriented in the opposite direction to the main lobe.

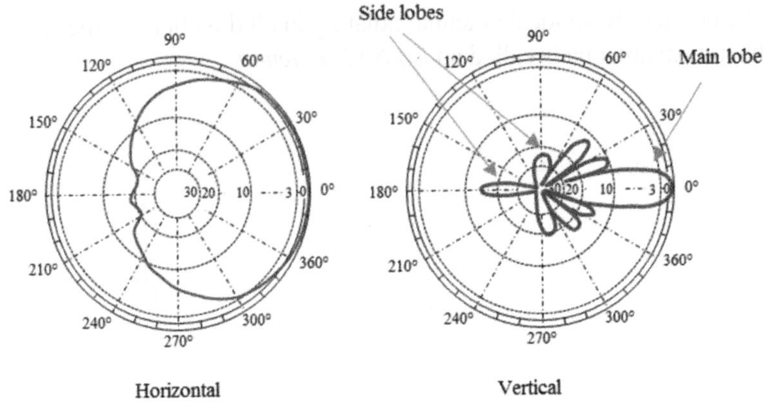


**Figure 1.14.** *Base station antenna radiation pattern*

Generally, the larger the antenna in relation to the wavelength, the more concentrated its radiation. It becomes more directional. Its aperture angle and gain are connected. The purpose of an antenna is to favor one direction or one propagation area.

The maximum power radiated by an antenna in its main axis is the product of its supply power ($P_c$) and its gain ($G_i$):

$$\text{EIRP} = G_i \cdot P_c \text{ (watt)} \qquad [1.19]$$

This power is called *Equivalent isotropically radiated power* (EIRP). The maximum available gain is used to determine the range of the transmission (distance within which the signal is usable).

Alongside EIRP, effective radiated power (ERP) is used, notably in a regulatory context. Power is no longer referenced in relation to an isotropic antenna, but in relation to a real reference antenna (dipole antenna). A constant factor makes it possible to connect these two powers.

Relationship [1.17] is formulated depending on the EIRP or the antenna gain and the power supplied to the antenna, $P_c$ in watts:

$$E = \frac{\sqrt{30 \cdot \text{EIRP}}}{d} = \frac{\sqrt{30 \cdot Gi \cdot P_c}}{d} \text{ (V·m}^{-1}\text{)} \qquad [1.20]$$

where $d$ is in meters. The electric field depends on the power of the transmitter, the gain of the antenna in the direction in question and the distance. It does not depend on the frequency. This comes into play when the signal is received or absorbed.

### 1.3.2. *Polarization of the wave*

In a far field, the concept of wave polarization is important in the study of electromagnetic field absorption by the body of the exposed person.

Polarization corresponds to the direction of the field around its propagation axis. By convention, the wave polarization angle is the angle between the electric field plan and the horizon. The wave is defined as vertically polarized when field $E$ retains a vertical direction as it propagates. Polarization is horizontal when field $E$ remains horizontal. These are called linear polarizations. Polarizations are called circular or elliptical when field $E$ rotates around the propagation axis ($H$ always remains orthogonal to $E$ and to the propagation direction). These polarizations can be treated as the sum of two linear polarizations.

### 1.3.3. *Near field/far field*

The properties of a field depend on the distance from the source. The near-field zone is distinguished from the far-field zone. This determines how the level of exposure to electromagnetic radiation is assessed. The border between these two zones relates to the dimensions of the radiating structure measured in wavelength. The relationships presented in section 1.3.1 are valid in the far field.

#### 1.3.3.1. *Far field*

The far-field zone extends indefinitely from a certain distance from the source. Waves propagate within this zone according to well-established properties: the wave is considered planar, electric and magnetic fields attenuate regularly with the distance, they remain orthogonal to each other and to the propagation direction. The power density $S$ and the field level at each point in free space can be predicted. Determining one of these quantities is sufficient (see equations [1.12], [1.16] and [1.20]). This zone is called the *Fraunhofer zone*.

#### 1.3.3.2. *Near field*

The near-field zone is located between the radiating source and the beginning of the far-field zone. The fields are reactive at immediate proximity ($d < \lambda/2\cdot\pi$), then become radiative. The electric field depends on the distribution of the electric circuit charges as well as the electric properties of the elements located in this space. The magnetic field is based on the current flow of the source. These fields decrease with the square or cube of the distance.

In the near zone, the fields are neither orthogonal nor in phase, their *maxima* and *minima* are not located at the same points *and their strengths can vary greatly over very small distances* ($d \ll \lambda$). The wave impedance in this zone is not equal to the wave impedance in free space ($Z \neq 377\ \Omega$). The electric field can diminish more rapidly than the magnetic field or vice versa, if the source radiates a predominantly electric field (capacitive sources, such as those used in high-frequency heating) or a predominantly magnetic field (inductive sources, such as those encountered in induction systems).

The behavior of fields in the reactive zone can only be studied through numerical simulation, because the complexity of their distributions makes them difficult, or even impossible, to put into equations. In the context of an experiment, specific measurements can be undertaken in the radiative zone, if many precautions are taken. The electric field must be measured separately from the magnetic field. The presence of an object, including a measuring probe, in this area can greatly alter the radiation, which adds a difficulty.

In the near-field zone, reactive fields always transform into radiation fields or radiative fields. Electric fields and magnetic fields feed on each other.

Applying the far-field formulae in the near field leads to overly approximate field and power density levels.

In determining at what distance the propagation zone begins in the far field, the greatest dimension of the radiating structure (single element or group of elements) must be taken into account. More precisely, the calculation is performed from the diameter of the sphere that delineates the structure.

In practice, when this dimension $D$ is small compared to the wavelength ($D < \lambda$), as is the case for omnidirectional sources or those with low directivity (mobile telephone, Wi-Fi, AM radio, HF industrial welding systems, etc.), the far-field zone can be reached at a distance of less than one wavelength.

When the size of the radiating element is large compared to the wavelength ($D >> \lambda$), as is the case for directional antennae (telephony base station antenna, satellite antenna, etc.), the far-field zone is reached only after several wavelengths. It begins at a distance equal to the square of this dimension divided by the wavelength.

The boundary between the zones is not strict; one zone gradually converts into the other. To quickly identify the boundary, the following practical relationship can be used, whatever the wavelength and dimensions of the antenna [IAR 13, BOL 01]:

$$d_L = \lambda + \frac{D^2}{\lambda} \qquad [1.21]$$

A body located beyond this limit is assumed to be exposed to a far field.

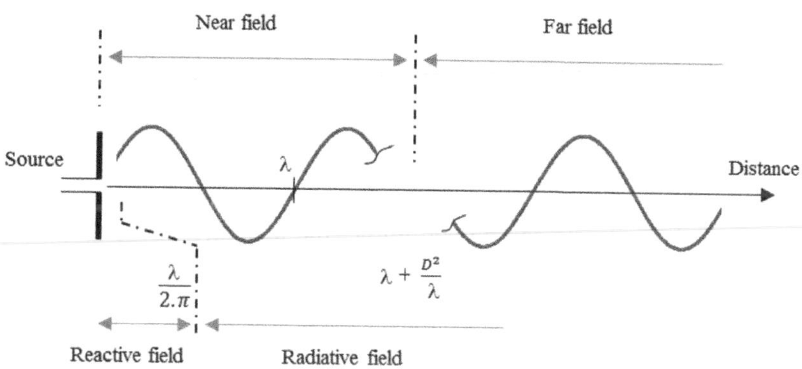


**Figure 1.15.** *Representation of near- and far-field zones*

Two examples are as follows:

– the far-field zone of a Wi-Fi antenna at 2.45 GHz ($\lambda$ = 12.2 cm), 3 cm in length, begins practically at 13 cm;

– the far-field zone of a GSM antenna of dimension 2 m emitting at 900 MHz ($\lambda = 33$ cm) is located beyond 12 m.

A small antenna mounted on a large structure leads to an intermediate distance (for example, an antenna built into a laptop or a tablet).

At low frequencies, an area where wavelengths can exceed several kilometers, exposure is always in the near field. As a reminder, the wavelength of a signal with a frequency of 1 MHz is equal to 300 m and the wavelength of a 50 Hz signal is 6,000 km in the air.

The near- and far-field limit proposed above is a convenient limit in the exposure area.

### 1.3.4. *Propagation in a real environment*

Electromagnetic waves are subject to the environment. They are partly reflected by the ground and by structures. The direct wave is combined with the reflected waves. They add up fully if they meet in phase and are subtracted if they combine in phase opposition. The signal level at one point can be above or far below that which would be received from a single direct wave. These phenomena, known as fading, can vary rapidly in space. *Maxima* and *minima* are encountered over a fraction of a wavelength. Rapid time variations occur if elements are moving in the propagation space. The fading can only be represented by statistical models. This complicates the measuring of electromagnetic fields.

However, in some environments, such as those relatively close to the source, the fading effects can be negligible; the intensity of the direct signal is much greater than the reflected signals.

It should also be noted that in the presence of obstacles, the higher the frequency of the waves, the straighter their line of propagation (weaker diffraction). The propagation of a signal at a few gigahertz tends to come close to that of optical radiation. An antenna positioned high up is of some interest for the range of emissions (e.g. FM radio or television). The most significant diffraction at AM radio frequencies (in the band of a few hundred kilohertz) makes it possible to position antennae directly on the ground.

Horizontal propagation directions are preferred in terrestrial telecommunications, to cover as extensive an area as possible. The power radiated toward space or toward the ground near the antenna is reduced (Figure 1.14).

Electromagnetic fields must be modulated to transport information. The waveform of the fields depends on the data to be transmitted and on the type of modulation. These are introduced in Chapter 2.

### 1.3.5. *Summary of electromagnetic waves*

| Key points of electromagnetic waves |
|---|
| An electromagnetic field consists of an electric field and a magnetic field that cannot be disassociated. |
| Electromagnetic fields propagate at the speed of light in the vacuum (or in the air). |
| Electromagnetic fields move away from the source. |
| At a great distance from the source, the electric field and the magnetic field are proportional. |
| The frequency and the wavelength characterize an electromagnetic field. These two quantities are inversely proportional: the higher the frequency, the smaller the wavelength. |
| Assessment methods for exposure distinguish the near-field zone from the far-field zone. |
| It is possible to precisely predict the value of the field far from the source when the space is free of any obstacles. |
| The electromagnetic field strength is obtained in the far field through measuring or calculating its electric field component ($E$) or its magnetic field component ($H$), or through the power density ($S$) measured in watts per square meter ($W \cdot m^{-2}$). |
| In close vicinity to the source, the electric field and the magnetic field must be considered independently. |

**Table 1.3.** *Summary of electromagnetic fields and their propagation*

## 1.4. Type of radiation

In physics, electromagnetic fields range from static fields to infinite frequency fields. These radiations are distributed into two large groups, depending on their fundamental effects on matter:

– ionizing radiations are located in the domain of nanometric wavelengths. They are able to break the atomic bonds of molecules or to eject one or several electrons from an atom to create positively charged particles (ions). These radiations have the highest frequencies (the modification of the structure of an atomic nucleus is part of nuclear physics);

– non-ionizing radiations cannot produce such effects.

### 1.4.1. *Ionizing radiations*

To understand the phenomenon of ionization, electromagnetic radiation must be imagined not as a wave that propagates (wave model), but as an emission of elementary particles called photons (particle model).

The wave model focuses on radiation power depending on the amplitude of the wave, while the particle model focuses on photon energy. Only photons with enough energy are able to ionize an atom. The amount of photons is not important. If they are not energetic enough, they will not succeed in modifying an atom. There is a threshold effect. The energy of a photon is proportional to the radiation frequency.

As the frequency is very high and the wavelength infinitesimal, like the size of an atom, physicists use the electronvolt (eV), a convenient energetic value, as the unit of measure. Energy and frequency are connected by the Planck constant $h$:

$$e = h\,.\,\upsilon \text{ (eV)} \qquad [1.22]$$

where $h = 6.63 \times 10^{-34}$ J·s or $4.14 \times 10^{-15}$ eV·s and $\upsilon$ is the symbol used in quantum physics for frequency (Hz).

Thus, for reasons related to physical phenomena, depending on the situation, frequency (Hz), wavelength (m) or energy per photon (eV) will be used to characterize the radiations.

For biologists, the threshold between ionizing and non-ionizing radiations corresponds to the energy required to break a water molecule or ionize a hydrogen atom, i.e. 13.6 eV. This divide occurs beyond optical radiation, within the far ultraviolet (UV). The regulatory limit is 0.1 μm ($3 \times 10^{+15}$ Hz), i.e. 12.4 eV [EC 08]. Visible light photons are therefore non-ionizing radiations, with energy of around 2 eV. The energy that would correspond to an elementary particle radiated by a microwave oven ($f$ = 2.45 GHz) is $10^{-5}$ eV.

X rays, γ rays and cosmic rays are ionizing radiations. They are able to dramatically alter matter. Their detrimental effects on health, even in small doses, are recognized. Non-ionizing radiations do not have these properties. However, they do interact with matter to create biophysical effects.

### 1.4.2. *Non-ionizing radiations*

Non-ionizing radiations begin at zero hertz, although strictly speaking, radiation and propagation do not occur, since the field remains confined to the immediate vicinity of the source. This radiation domain ends with far UV rays.

Non-ionizing radiations encompass two categories, usually called:

– *optical radiations:* these extend from infrared (which transmits heat), with a wavelength of less than 1 mm, to UV (which cause tanning), passing through visible light along the way. The visible spectrum covers wavelengths of a few hundred nanometers (1 nm = $10^{-9}$ m, i.e. a billionth of a meter). The corresponding frequencies are counted in petahertz (1 PHz = $10^{15}$ Hz);

– *electromagnetic fields (!):* this name is usually used for fields beginning at 0 Hz and ending at around 1 terahertz ($10^{12}$ Hz). This domain overlaps with that of optics. This name is reductive, because electromagnetic fields cover the whole frequency spectrum (from the physicist's point of view).

Electromagnetic fields are classified according to their applications, properties and effects. There are many classifications and sometimes a different definition for each sector. For example, designers developing tools powered by the 50 Hz network call frequencies above a few kilohertz *high frequencies*. Signals with frequencies lower than a few megahertz are called *low frequencies* or *video frequencies* by a radar operator who uses signals greater than several gigahertz.

### 1.4.3. *Electromagnetic spectrum*

The whole electromagnetic spectrum is shown in Figure 1.16. The wavelengths are given, as well as the usual name of the domains. Some thumbnail images illustrate the range of applications.

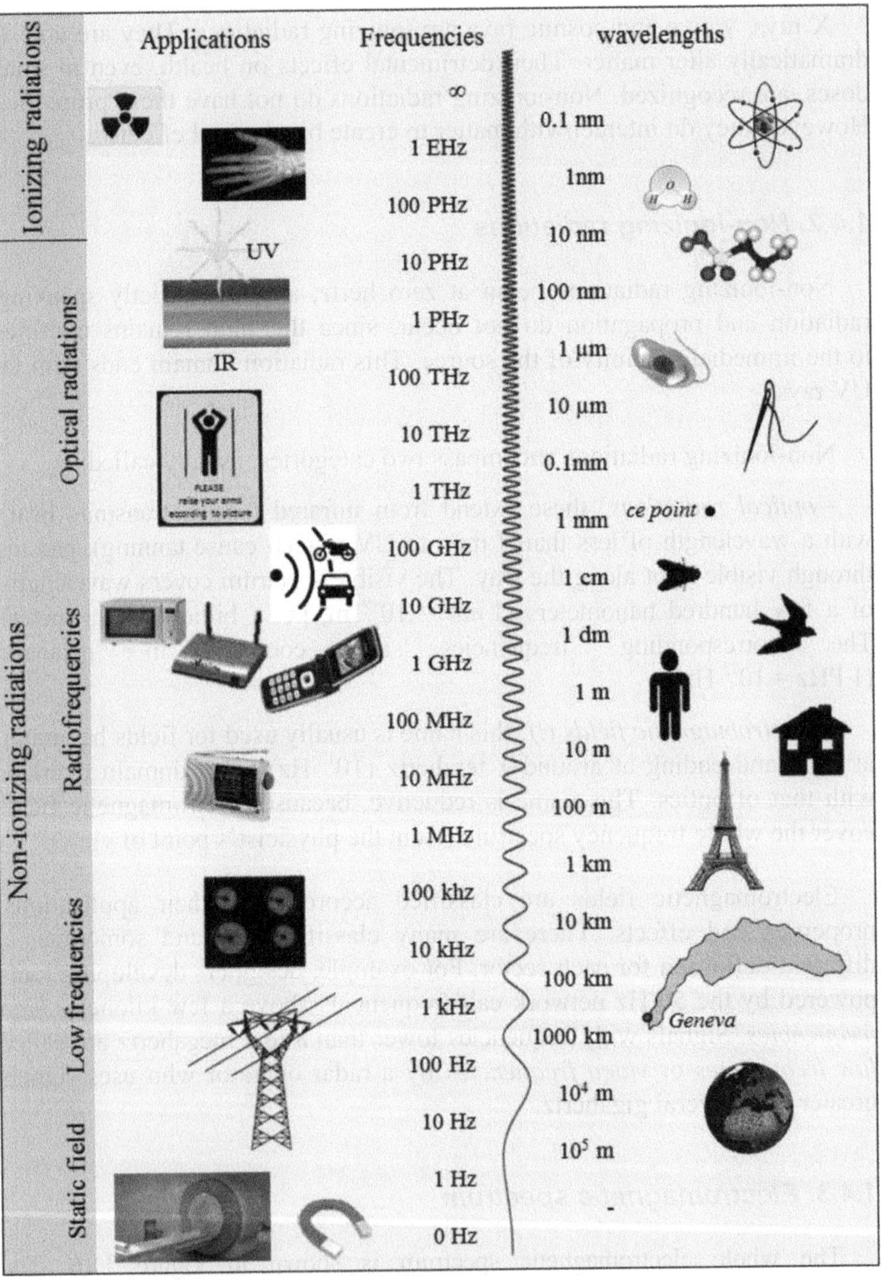


**Figure 1.16.** *Electromagnetic spectrum and applications*

### 1.4.4. *Frequency bands*

#### 1.4.4.1. *Classification of frequency bands*

Frequencies are classified by decade of wavelength on the basis of definitions from the International Telecommunication Union (ITU). Table 1.4 presents the main names associated with some types of application.

Breaking the applications down into subbands like this is arbitrary, because there is continuity in the physical properties of the electromagnetic radiations from one band to the next. Wavelength is sometimes used instead of frequency in classifying the spectrum: decametric waves for the HF band and millimetric waves for the EHF band, for example.

Initially, devices for measuring electric or magnetic field levels (extremely low frequency (ELF) fieldmeters) measured radiations of 50 or 60 Hz and had a bandwidth of 200 or 300 Hertz, like the first voltmeters. This made it possible to take into account complex signals containing a few additional frequencies (harmonics). As signals became increasingly rich in frequency, designers adapted the devices, without changing their name. Today, some use the acronym ELF for all frequencies lower than 10 kHz, or even 100 kHz. The terms audio frequency and video frequency would be more appropriate to describe these ranges.

Radiofrequencies are regulated from 8.3 kHz to 3,000 GHz, although there are hertzian transmissions below this band. Despite this, the term radiofrequencies often denotes frequencies from 30 kHz to 300 MHz (LF to VHF bands). The term "microwaves" is used for frequencies between 300 MHz and 300 GHz (UHF to EHF bands). Beyond this, the infrared spectrum starts, followed by the visible spectrum and then UV.

#### 1.4.4.2. *Industrial, scientific and medical band and unlicensed national information infrastructure band*

Industrial, scientific and medical (ISM) bands are frequency bands dedicated to industrial, scientific, medical and domestic applications [STD 17]. These bands are also used for short-range wireless links (Table 1.5). They do not include telecommunication applications at long distances, such as mobile telephones or radio broadcasting.

| Frequency range | Wavelength | Acronym* | ITU classification | Examples of application *** |
|---|---|---|---|---|
| Below 3 Hz | >100,000 km | TLF** | | Seismology, static electricity |
| 3–30 Hz | 100,000–10,000 km | ELF | Extremely low frequency | Some railways |
| 30–300 Hz | 10,000–1,000 km | SLF | Super low frequency | Electricity network, underwater telecommunications |
| 300 Hz to 3 kHz | 1000–100 km | ULF | Ultra low frequency | Underwater and underground telecommunications |
| 3–30 kHz | 100–10 km | VLF | Very low frequency | Maritime radionavigation geophysics |
| 30–300 kHz | 10–km | LF | Low frequency | AM radio, RFID, anti-theft system, radiosynchronization |
| 300 kHz to 3 MHz | 1 km–100 m | MF | Medium frequency | Shortwave radio, military communications, avalanche victim detector |
| 3–30 MHz | 100–10 m | HF | High frequency | RFID/NFC, CB, transhorizon transmission and radar |
| 30–300 MHz | 10–1 m | VHF | Very high frequency | FM radio, maritime and aeronautical radio |
| 300 MHz to 3 GHz | 1 m–1 dm | UHF | Ultra high frequency | DTT, mobile telephones, GPS, Wi-Fi, Bluetooth, RFID, radar |
| 3–30 GHz | 1 dm–1 cm | SHF | Super high frequency | Private networks, radar, satellite telecommunications |
| 30–300 GHz | 1 cm–1 mm | EHF | Extremely high frequency | Private networks, onboard automotive radar, very high-speed point-to-point radio connection, remote sensing, directed-energy weapon, security scanner gate |
| Over 300 GHz | <1 mm | THF** | | Remote sensing, radar, radiotherapy, spectroscopy |

* The international acronyms are more common.
** Unofficial name.
***Industrial applications and amateur radio operators have many dedicated bands.

**Table 1.4.** *Classification of frequencies*

A specific regulation [ECC 14] sets out usage requirements for these resources, such as radiated powers, bandwidths and air times. A license must be procured for any ISM application beyond this regulation (e.g. higher power or other frequency bands).

There are also additional bands and specific restrictions, even among states within the European Union.

| ISM bands (central frequencies) | Examples of applications |
|---|---|
| 6.78 MHz | – |
| 13.56 MHz | RFID/NFC, dielectric loss welding |
| 27.12 MHz | Dielectric loss welding, CB |
| 40.68 MHz | – |
| 433.92 MHz | Remote controls, baby monitors, toys |
| 2.45 GHz | Wi-Fi, Bluetooth, microwave ovens, RFID, webcam, computer mouse, home audio-video transmitters, diathermy, dielectric loss heating |
| 5.8 GHz | Wi-Fi, RFID, electronic toll collection |
| 24.125 GHz | Speed measurement |
| 61.25 GHz | Wi-Gig |
| 122.50 GHz | Satellite applications |
| 245.00 GHz | Radio astronomy |

**Table 1.5.** *ISM frequency bands*

Since ISM bands are authorized by the ITU for many applications, there is no guarantee for the quality of wireless links.

The 866 MHz frequency band is also used in Europe for ISM-type applications (remote controls, UHF RFID, wireless alarms, home automation). It is strictly regulated.

Many products also use the band centered around 915 MHz, an ISM band authorized across the whole American continent.

As well as ISM bands, there are unlicensed national information infrastructure (U-NII) or high-performance radio–local area network (Hyper-LAN) bands. These are employed by computer peripherals using wireless communication protocols from the 802.11 group (Wi-Fi, etc., see section 2.2.3.1.5.). These bands range between 5.15 and 5.85 GHz. They cover part of the 5.8 GHz ISM band. The frequency and radiated power of these bands are regulated according to the geographical continent.

### 1.4.5. *Area of exposure to electromagnetic fields*

The area of human exposure to electromagnetic fields encompasses all radiations whose frequency is lower than 300 GHz, i.e. the area where signals have a wavelength greater than 1 mm. Radiofrequencies cover a large extent of it.

In the rest of this book, the term "electromagnetic field" refers only to this "limited" domain.

A quick overview of applications is given here, demonstrating the range in frequency of electromagnetic fields. More precise examples are presented in Chapter 2.

The electromagnetic fields with the lowest frequencies usually encountered are those produced by the power network and by all the electrical appliances using it (electrical household devices, tooling, machines, etc.). Radiations are generally at 50 Hz in Europe (60 Hz in North America and Japan) and may be added to by multiple frequencies (harmonics).

Fields with frequencies between 20 and 100 kHz have appeared more recently, with induction heating systems (cooking hobs, ovens, etc.) and switched-mode power supplies.

Frequencies located between a hundred kilohertz and a megahertz have been used in radio broadcasting (AM) for more than a century, but these services are coming to an end. Other uses will be proposed.

At higher frequencies, between 80 MHz and 6 GHz, applications are numerous: FM radio, digital terrestrial television, telephony, Wi-Fi, microwave ovens, etc. They have become vital for many of us.

Electromagnetic fields are also encountered in the industrial environment, such as in the power production sector and the metallurgy sectors (induction heating, welding, etc.).

The medical sector also uses magnetic, electric and electromagnetic fields from 0 Hz to many gigahertz (MRI, diathermy, etc.).

### 1.4.6. *Summary of electromagnetic radiations*

| Key points of electromagnetic radiations |
|---|
| There is a clear distinction between ionizing and non-ionizing radiations. |
| Ionizing radiations, such as X rays and $\gamma$ rays, transport enough energy to break molecular bonds. |
| Non-ionizing radiations consist of optical radiations and electromagnetic fields. They transport little energy and are not able to act on a molecular level. |
| The domain of exposure to electromagnetic fields encompasses all frequencies lower than 300 GHz. It includes static fields. |
| In practice, the acronym ELF often covers all frequencies lower than a few kilohertz. |
| It is necessary to distinguish between electric, magnetic and electromagnetic fields. The first two should be considered separately very close to the sources (distance of less than a wavelength), which is always the case at low frequencies. |

**Table 1.6.** *Characteristics of electromagnetic radiation*

# 2

# Sources of Electromagnetic Fields

Mankind has always been exposed to electric, magnetic and electromagnetic fields. This chapter briefly outlines the existence of natural fields in our environment, before introducing some of the most common sources of artificial electromagnetic fields in our living spaces and also some present in our workplaces. These sources radiate because they involve electricity.

## 2.1. Natural fields

### 2.1.1. *Electric fields*

Terrestrial electric fields have existed since the atmosphere was created. There is a potential difference of 300 kV between the ground and the ionosphere – part of the atmosphere that lies above 50 km in altitude and is an electrical conductor due to the presence of free electrons.

This voltage produces a permanent electric field that can reach over 200 $V \cdot m^{-1}$ in a flat and clear area. The strength of this field depends on solar activity, the season, air humidity and weather conditions.

The field can exceed 20 $kV \cdot m^{-1}$ between an electrically charged cloud and the ground. It is strengthened by the presence of pointed items, such as lightning rods. When it reaches several hundred $kV \cdot m^{-1}$, lightning strikes occur.

A more modest static electric field may be produced naturally when an object is rubbed. This removes electrons from the object, polarizing it. Two experiments illustrate this phenomenon: a schoolboy's ruler attracts small pieces of paper after it has been rubbed on a woolen cloth, and a driver gets a small electric shock as he leaves his car after driving, i.e. rubbing in dry air. Although this shock is unpleasant, it presents no risk.

### 2.1.2. *Magnetic fields*

A huge magnetic field encircles the Earth, from the North Pole to the South Pole. The shape of its lines is identical to that of a single bar magnet. This field is locally distorted by the geology of the land and by man-made structures on a smaller scale. The Earth's magnetic induction is static on our time scale. It varies according to latitude, from 30 μT near the equator to 70 μT near both poles. It is around 47 μT in France. This field protects life on Earth from cosmic rays and helps migratory species find their bearings.

### 2.1.3. *Electromagnetic fields*

The Earth is also subject to natural electromagnetic fields:

– infrared rays radiated by all heat sources;

– light;

– ultraviolet rays;

– ionizing cosmic rays that are not intercepted by the Earth's magnetic belt;

– lightning: the electrical discharge produces an electromagnetic field pulse.

Optical rays are received primarily from the Sun. Light is the only source of electromagnetic radiation that humans can perceive.

## 2.2. Artificial fields

Aside from a few laboratory applications, the first artificial electric and magnetic fields originated in the production, transport and consumption of electrical energy, rapidly followed by telecommunication applications. Technological progress enables electromagnetic fields to use all frequencies up to a few hundred gigahertz.

Artificial fields are distributed across three frequency domains, depending on their interactions with matter and on their biophysical and biological effects (described in Chapters 3 and 4):

– static and quasi-static fields with frequencies lower than 1 Hz;

– low-frequency electric or magnetic fields between 1 Hz and 10 MHz, which may cause non-thermal effects;

– high-frequency (HF) electromagnetic fields between 100 kHz and 300 GHz, which may produce primarily thermal effects.

The last two categories overlap, because one effect may occur independently from the other.

Examples of radiating sources will be given for each domain hereafter. It is impossible to be exhaustive, especially as some are used for applications known only to their users. Furthermore, it is easy to imagine that a new source is being created everyday.

### 2.2.1. *Static and quasi-static fields*

Static magnetic fields are found in medicine (magnetic resonance imaging [MRI]), in some high technology domains (nuclear magnetic resonance [NMR] spectroscopy), in the transport of energy and people, in heavy-duty metallurgical applications (electrolysis, electroplating, smelting, etc.) and in the recycling and handling of ferrous and non-ferrous metals. Some research centers also use this type of field, for example in particle ace-lerators.

Artificial static fields are confined to the area around the source and decrease very rapidly with distance.

These fields either issue from permanent magnets or are generated from a coil supplied by a direct current (DC) from an accumulator, a battery or an alternating (AC) voltage, rectified and smoothed through filtering. A residual ripple may overlay this current.

Static electric and magnetic fields are totally independent. The former depend only on voltage, while the second depend only on current. One or the other may be zero. They must be considered separately.

The biological effects of a static magnetic field on a moving person and the effects of a slowly variable field on an immobile person are identical. This observation means that quasi-static fields, i.e. those with a frequency lower than 1 Hz, may be regarded as static fields.

It should be noted that effects due to time-varying fields, created by residual AC voltage or current, superimposed on the direct component, are often more significant than effects due to static fields (see Chapter 4).

#### 2.2.1.1. *Magnetic resonance imaging* (MRI)

MRI, based on the principles of NMR, is a relatively recent invention [FEL 03]. This painless technique is used first and foremost for medical diagnosis and has no known side effects, unlike, for example, computed tomography (CT) scans and X-ray scans, where the dose of ionizing radiation must be considered. Today, minimally invasive surgical operations are done using MRI, replacing "major" surgeries. The advantage for patients is obvious.

MRI produces a high-resolution 3D image of a body. This image is formed by layering up cross-sections (2D images). This technique is also used for studying other life forms (flora and fauna), as well as for scientific investigations. MRI is the only technique through which the central nervous system can be studied in great detail and cerebral activity is visualized in quasi-real time [MAR 10].

Without a doubt, MRI uses the strongest existing static magnetic field source. It exposes patients (public), university researchers and design and maintenance engineers to a strong field. Health professionals are also involved when they have to assist some patients, such as infants, or perform a surgical operation.

MRI equipment differ primarily according to magnetic field strength and the size of the tunnel, which may be used for examining a whole body or just a limb. It may be open or closed.

An MRI scanner generates three types of field simultaneously:

– a static magnetic field, Bo;

– three gradient magnetic fields;

– an electromagnetic field in the radiofrequency (RF) band.

#### 2.2.1.1.1. *Static magnetic field*

Static magnetic induction Bo is between 0.15 and 3 T for patients; it can exceed 10 T on some research equipment. The resolution or precision, or, to be more accurate, the signal to noise ratio, is proportional to the strength of this field. It must remain uniform and stable for the whole of the analysis.

This field is generated by a constant and intense current circulating in coils made from superconductive materials (with zero resistance). The consumption of the circuit is zero once a steady-state has been established, despite the presence of a current that may reach 1,000 A. Since the voltage is zero, there is no static electric field. This magnetic field can also be produced by permanent magnets for less powerful equipment.

A mechanism (counter coils) reduces this outward radiation to limit the exposure of professionals (surgeons, radiologists, nurses, operators).

The purpose of this static field is to orientate (polarize) the exposed atoms.

#### 2.2.1.1.2. *Gradient magnetic fields*

Three non-uniform magnetic fields ($B_X$, $B_Y$ and $B_Z$) are generated in the space [FEL 13, KAS 11]. Their gradient must be as constant as possible (linear variation) and may reach 100 mT per meter. This gradient also varies in time up to 200 $T{\cdot}m^{-1}.s^{-1.}$ These field gradients added to the static field Bo are produced by currents circulating in three pairs of orthogonal coils, one for each spatial orientation. The magnetomechanical stresses to which these coils are subjected (Lorentz forces) cause vibrations. These vibrations are the main source of noise from the scanner. Auxiliary coils (shims) are used to correct nonlinearities in the field.

The resulting field allows the sectional plan and its thickness to be defined, and the signals to be localized in the plan. The frequency variation possibility enables the analysis speed to be adjusted according to image quality requirements.

#### 2.2.1.1.3. *Electromagnetic field*

A pulsed RF electromagnetic field is also radiated uniformly in the tunnel by antennae adapted to the body part being diagnosed. Pulses stimulate the polarized atoms at their resonance frequencies, provided that they have a magnetic spin moment. Between two electromagnetic emissions, the atoms return to an idle state in a few hundred milliseconds, returning the energy stored during the emission. This relaxation time, which depends on the nature of the tissues, is sufficient for measuring the magnetic components of the atoms. The resonance frequency is proportional to the magnetic field Bo added to the field gradients. The frequency of the RF signals depends on the strength of the resulting magnetic field and the atoms to be analyzed. It is 42.6 MHz per tesla for hydrogen atoms. The RF radiation level rises to several kilowatts over very short periods of time, making it possible to limit the average power to a few watts. The 1 MHz to over 100 MHz RF band is used. This requires MRI equipment to be installed in an area shielded against HF radiation (Faraday cage) to avoid disturbing radio reception in the vicinity.

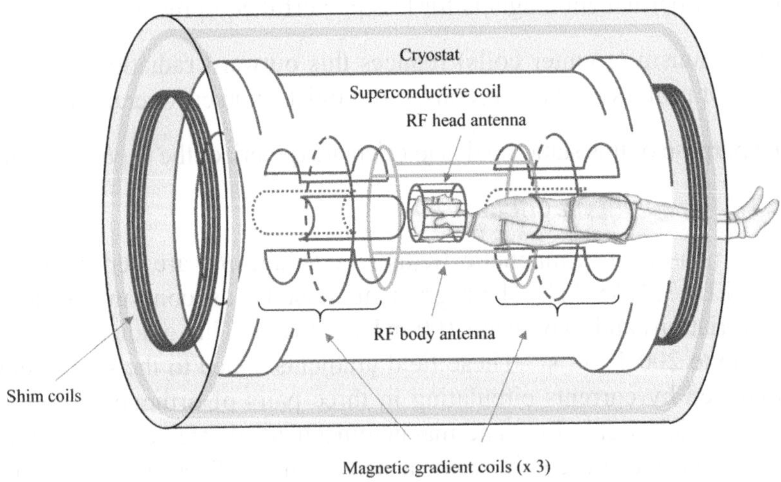


**Figure 2.1.** *Simplified representation of a conventional MRI scanner*

The announced arrival of metamaterials specially designed for MRI is promising, as they would make it possible to reduce the strength of the radiation or the length of the examination, while retaining the same image quality [SLO 15]. Patients and medical personnel would be less exposed.

MRI can also be classified in the sources of static magnetic fields, sources of time-varying fields and sources of HF fields categories.

### 2.2.1.2. *Transport*

More than a century ago, the first electric trains used DC voltage. Many metro and tram lines are still supplied in this way, creating a static magnetic field.

Conventional electric trains are generally powered either by a DC voltage (1.5 or 3 kV) or by a high-voltage AC voltage (15 kV, 16.7 Hz or 25 kV, 50 Hz), depending on the region or country. This choice is often dictated by the compatibility between old and new materials.

The electric power of a TGV (French high-speed train) is around 10,000 kW.

A third power rail (side or central rail) may be installed instead of an overhead line (catenary) for aesthetic reasons. To prevent the public (pedestrians, cyclists, etc.) from coming into contact with the conductor rail or possibly being exposed to radiation, it is divided into sections, which are powered gradually and then switched off as the carriage is moved. The return current passes through the running rails or a fourth dedicated rail. The maximum electric power of a tram is 1,000 kW.

The electric drivetrain of a tram is located under the carriage floor. Although magnetic field radiation can reach several tens of microteslas at certain locations in the ground, it decreases rapidly with distance, meaning that the exposure of travelers is very low at trunk (torso) height [EMF 16]. The electric field radiated near a catenary can reach a level equivalent to that radiated by a medium-voltage electric wire (~50 $V{\cdot}m^{-1}$).

Electric and hybrid vehicles use a battery to store all or part of their energy. The voltage is DC, from a few tens to a few hundred volts. The current is fast switched by electronic circuits (inverter) before powering the engine(s). It may exceed 200 A during full throttle phases. The magnetic field created is relatively weak and decreases rapidly due to the configuration of the electric wiring.

Figure 2.2 shows the magnetic induction on board a high-speed train (TGV) in the 1 Hz to 400 kHz band. It has been recorded every 250 ms over a duration of 20 minutes. The measurement probe was placed on a seat. The speed was obtained by a GPS receiver. There appears to be no correlation between induction level and speed.

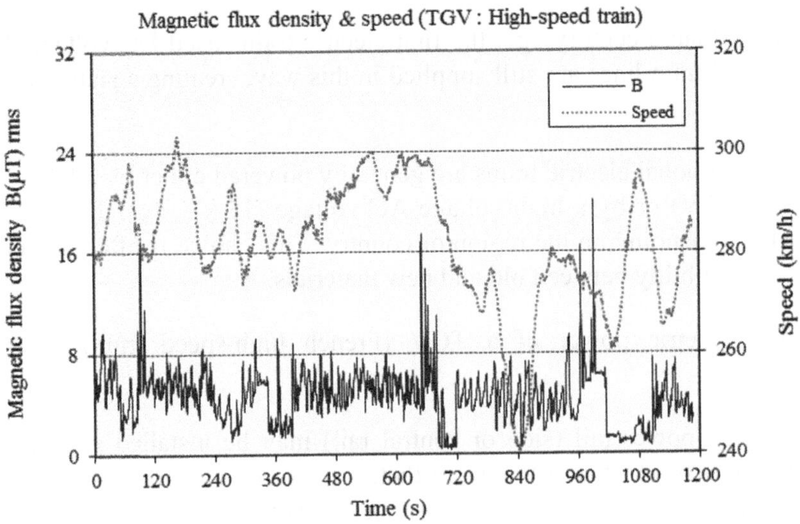


**Figure 2.2.** *Magnetic induction recorded on board a TGV*

This curve (solid line) does not distinguish between induction sources. These may be inside or outside the carriage. Indeed, the device can detect a sudden change in the ambient static field (passing a metallic structure at high speed or passing another train) or the electrical elements of the railway network.

### 2.2.1.3. *Manufacturing*

#### 2.2.1.3.1. Stirrers

Electromagnetic – or rather, magnetic – stirring is used in heavy industry to homogenize the temperature of molten metals and steel- or aluminum-based alloys. It also accelerates the mixing of elements and the purification of metals.

Molten metals subjected to a very low-frequency rotating magnetic field move due to the Lorentz law (a conducting element tends to move in a magnetic field). The quality of the materials when solidified is greatly improved. This type of stirrer does not require the movement of mechanical parts, thus increasing their reliability. Magnetic field coils are located outside pots or pipes of molten metal.

In addition, the intensity and orientation of magma flow can be controlled by a magnetic field. According to the location of the coils, the natural movements of the liquid can be slowed down, accelerated, or even reversed.

The frequencies used are a few fractions of a hertz at the most, adjusted according to the viscosity of the molten metal and the volume to be processed.

Fields of many teslas may be necessary. Fields in accessible places around equipment remain relatively weak (static or quasi-static field of a few tens of milliteslas). Fields must be suspended during infrequent operations near the inductors.

#### 2.2.1.3.2. Electrolysis

In manufacturing, aluminum is produced through electrolysis of alumina dissolved in a chemical bath. Electric circuits transport DC currents, which may exceed 400 kA over more than 100 meters, to the electrolysis cells. As well as the extraction of aluminum, the current keeps the bath at a constant temperature, by means of the Joule effect. A factory with many cells can consume half of the energy produced by one unit of a nuclear power plant (1,200 MW).

In this context, professionals may be exposed to static magnetic fields of several dozen milliteslas (application of equation [1.17]). Small ferromagnetic objects may be attracted to these fields. Also, rectified AC current residue ripples the DC current. The corresponding magnetic induction may reach a few milliteslas.

#### 2.2.1.3.3. Electroplating

Electroplating is a surface treatment that protects metal parts from corrosion through electrolysis. It is also used for decorative features. The parts are immersed in an electrolyte with an electrode made of nickel, zinc, chromium, gold, copper, etc. The presence of DC currents of a few kiloamperes may expose people to static magnetic fields of a few milliteslas.

#### 2.2.1.4. *Magnets*

Some applications of permanent magnets and electromagnets are briefly described here:

– static magnetic fields are common in daily life. We pay them no attention because they are localized and not very strong (<1 mT). These fields are generally created by permanent magnets that attract ferrous objects only when they are within a few centimeters. They are found in loudspeakers and at checkouts, for removing anti-theft devices. The manufacturing process of these magnets consists of placing suitable materials in a static magnetic field generated by a DC current;

– when a ferromagnetic object must be machined without leaving any trace of chucks, it can be held in place by a strong static magnetic field if efforts are contained. After this operation, the object retains a permanent residual magnetization, which is removed through demagnetization. To achieve this, the object is placed in a decreasing alternating magnetic field (see section 2.2.2.6.);

– the handling of ferrous metal objects with an electromagnetic lifting device is common, especially in recycling. A DC current supplies a coil that attracts the parts, which are released when it is switched off. For example, many tons of material can be lifted in this way to supply a furnace;

– the sorting of metals using a static magnetic field is also common. This field attracts scrap ferrous metal (iron, steel, cast iron, etc.) from a conveyor belt. The use of a variable field will eliminate non-ferrous metal, such as stainless steel, aluminum, copper and zinc, using eddy currents. These induce a magnetic field that opposes that of the source, creating a repulsive force (Lenz's law). This variable field may be generated by a permanent rotating magnet.

Sixty percent of the world's steel is produced from recovered scrap metal. The savings in energy and ore are significant.

Magnetic induction on contact with an industrial magnet can reach values between 50 mT and 2 T. This induction decreases with the square or cube of the distance.

### 2.2.2. *Low-frequency fields*

Low-frequency electric and magnetic fields are understood to cover the 1 Hz to 10 MHz domain (wavelength lower than 30 m). These fields must be considered independently, as they are not simply correlated. Exposure is generally assessed in an area where field strength is significant, in the close vicinity of the sources, i.e. in the near-field zone (section 1.3.3). The analysis of the magnetic field is sufficient in most situations, because the voltages implemented, and therefore the resulting electric fields, are low.

Exposure to electric and magnetic fields in the low-frequency domain is mainly caused by the use of devices directly connected to 50 or 60 Hz power, the transport of electric energy and manufacturing, with its tools, welding and induction heating equipment, etc.

Over 100 years ago, almost 95% of energy was used for lighting. Today, this figure has plummeted to less than 10% of the total consumption.

#### 2.2.2.1. *Electric energy*

The electric power grid was the first source of large-scale exposure. Large quantities of electricity are produced from turbines driven by water pressure (hydraulic dam), steam from nuclear reactors or fossil fuel plants. Wind and photovoltaic energy sources are more modest.

Each turbine drives an alternator, which produces electric power simultaneously under three AC voltages out of phase by 120°. They are carried to the users by means of transformer stations and overhead and underground power lines. The circuit is completed by systems for counting and protection against surges, overcurrent problems and grounding defects.

Losses in the power lines are greater when the current is significant (losses are proportional to the square of the current, according to the Joule effect principle: $P = R \cdot I^2$). The solution is to decrease the current, but since the power is the product of the voltage and the current ($P = U \cdot I$), the voltage must be increased in proportion. To achieve this, the voltage produced by a power station (27 kV) is first raised to 225 or 425 kV (very high voltage [VHV]) using transformers, before being transported over long distances. As the grid approaches the area which needs to be supplied, the voltage is decreased, first to 63 or 110 kV (high voltage [HV]), then to 20 kV (medium voltage [MV]), ultimately reaching the three-phase voltage of 400 V (low voltage [LV]). This voltage is delivered to craftsmen and small businesses. Individual consumers generally receive a single-phase voltage of 240 V (a voltage between one of the phases of three-phase voltage and neutral).

Large power consumers, such as railways and heavy industry, use HVs between 63 and 110 kV directly. They have their own transformer stations. Medium-sized enterprises are supplied with MV.

The values of HV and VHV depend on the country of the European Union, unlike service voltages (230/400 V), which remain the same, as does the frequency (50 Hz).

Dispatch centers distributed across a broad region ensure that the system is working well. All the power plants in a country are interconnected, and all European countries are interconnected, making it possible to smooth peaks in consumption and production.

The example of electricity transportation perfectly illustrates the influence of the topology of power lines on electric and magnetic fields.

#### 2.2.2.1.1. Power lines

*Electric field generated by power lines*

The electric field is proportional to the voltage on the line and decreases with its height from the ground. It depends on the geometric configuration of the cables (number, spacing and arrangement). This configuration is determined by the pylon type for overhead lines (Figure 2.3).

On the ground, the electric field is at its maximum where the wire dips in the middle, when the line is closest to the ground. It is at its minimum at the foot of the pylons, but reaches very significant values high up. It reduces as the distance from the line increases.

The presence of a ground wire (guard cable), which does not transport a current, decreases the electric field.

The electric field emanating from an overhead power line remains stable in time, because the voltage on the wires remains constant, regardless of the operating conditions of the network.

The electric field is influenced by the presence of objects, even if they are poor conductors, such as trees or buildings. The electric field, therefore, remains weak inside buildings close to the lines.

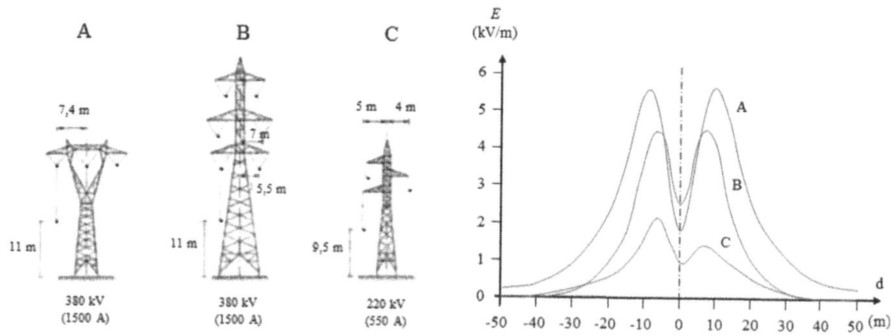


**Figure 2.3.** *Maximum transverse electric field depending on the configuration of the cables (determined by the pylons) [APA 06]*

Underground power lines are used to power city centers and commercial areas, for aesthetic, not wasting useful surface area and safety reasons. Their setting cost is much higher than for an equivalent overhead line.

The combined effect of a protective cable sleeve and the line being buried 1–2 m underground greatly reduces the electric field level on the ground.

## *Magnetic field generated by power lines*

The magnetic radiation generated by a line is proportional to the electricity consumption. It therefore varies considerably depending on the time, day and season. A peak and dip in strength is observed respectively in France at around 8 pm in winter and around 4 am in summer. A ratio of more than 5 is observed between the extreme values (Figure 2.4).

Like the electric field, the magnetic field diminishes with the distance from the line (height and deportation). Magnetic fields cancel each other out when the phases are balanced (the same current strength on each phase), especially when the cables are close together. However, a minimum gap is imposed by the safe distance due to the voltage, thermal expansion and movement caused by the wind.

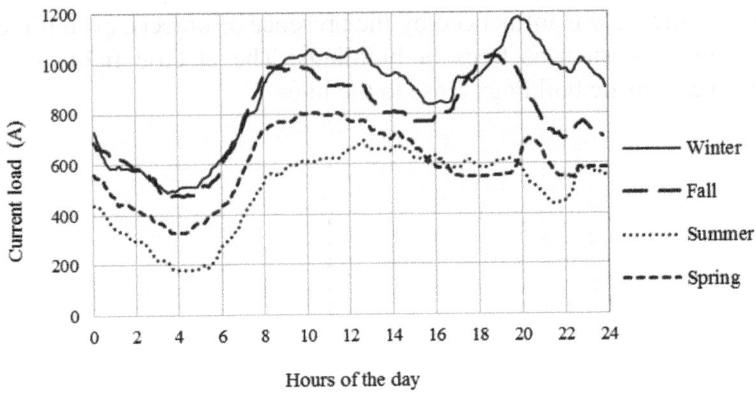


**Figure 2.4.** *Example of the daily current load of a 400 kV power line over the four seasons (source: RTE)*

Their arrangement, determined by the pylon type, is the least costly solution for reducing the radiation.

Finally, the presence of passive compensation loops decreases the strength of the magnetic field.

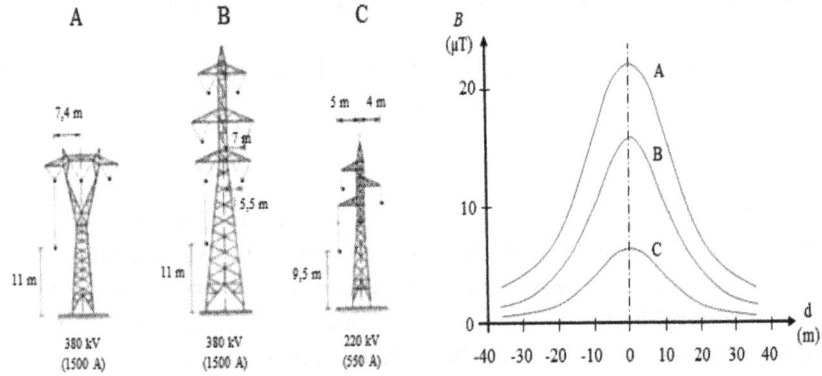


**Figure 2.5.** *Cross-section view of the magnetic field, depending on the type of pylon [APA 06]*

A single underground line radiates as much of a magnetic field as an overhead line at a very short distance, but this field diminishes more rapidly, because the buried cables are very close together (the radiation level becomes proportional to the gap between the cables beyond a distance equivalent to several gaps; to minimize the gap, the cables are laid out in trefoil or triangle formation, rather than on the same plane).

Laying active electrical compensation circuits further reduces the magnetic field level in a defined limited area.

#### 2.2.2.1.2. Electrical transformers

Electrical power transformers do not radiate much of a magnetic field, although they are composed of coils. The fields flow through the magnetic core and leakage is minimized to ensure that the device performs efficiently. Magnetic fields are greater around the connection cables carrying the lowest voltage (this corresponds to the secondary of a voltage-lowering transformer), because the current is stronger.

Transformers are installed in various types of station:

– primary stations transforming VHV (400 kV or more) into HV (60–110 kV);

– distribution stations reducing HV to MV (20 kV);

– substations converting MV into low service voltage (LV, 240/400V).

HV transformer stations are protected from risks of electrification and electrocution and from electric field risk (arc phenomena) by a safety barrier. The distance imposed is greater than that needed to protect people from magnetic fields.

MV/LV substations mounted at the top of power poles pose no radiation problems. Substations are also often placed in buildings, especially when the electrical inlet is underground.

It is difficult to use calculations to predict the electric and magnetic radiation levels around transformers or stations, because this depends greatly on the configuration of the sites.

#### 2.2.2.1.3. Fields generated by electrical household appliances

Magnetic fields are generated by all working electrical equipment. Appliances connected to the mains use a voltage that is certainly dangerous in the event of direct contact with a bare conductor, but it is low and radiates few electric fields.

The list of equipment and situations in Table 2.1 shows their electric field level.

| Situations/applications | Electric field |
|---|---|
| Razor | Negligible |
| Laptop | Negligible |
| Supply cables for a household electricity system | 0.1 V·m$^{-1}$ |
| Incandescent light bulb | 5 V·m$^{-1}$ |
| Household appliances: TV, blender, hairdryer, oven, refrigerator, etc. (30 cm) | <120 V·m$^{-1}$ |
| Electric blanket (1 cm) | 2,000 V·m$^{-1}$ |
| 230 V power line:<br>- under the line<br>- at 100 m | 9 V·m$^{-1}$<br>Negligible |
| 20 kV power line:<br>- under the line<br>- at 100 m | 250 V·m$^{-1}$<br>Negligible |
| 90 kV power line:<br>- under the line<br>- at 100 m | 1,000 V·m$^{-1}$<br>10 V·m$^{-1}$ |
| 400 kV power line:<br>- under the line<br>- at 100 m | 5,000 V·m$^{-1}$<br>200 V·m$^{-1}$ |

**Table 2.1.** *Measured electric field [EDF 05, WHO 07]*

Table 2.2 shows the magnetic field level radiated by various sources at 50 Hz.

The values given in Tables 2.1 and 2.2 are indicative. There are considerable differences between various models of the same appliance. The levels also depend on how they are used, set up and manufactured [LET 13]. Although the magnetic field level on contact with appliances can sometimes reach 1 mT (1,000 μT), it decreases very quickly with the distance, due to the complexity of the structure of the internal electric circuit [1.9].

| Situations/applications | Magnetic field |
|---|---|
| Electric razor with/without battery (1 cm) | Negligible/500 μT |
| Laptop (a few cm) | 1 μT |
| Household appliances without/with transformer (30 cm) | <1 μT/<5 μT |
| Television (30 cm) | <2 μT |
| Electric blanket (1 cm) | 3.6 μT |
| Major electrical consumer (radiator, cooker, etc.) (30 cm) | <10 μT |
| Appliances equipped with a motor (drill, etc.) (30 cm) | <20 μT |
| 230 V power line:<br>- under the line<br>- at 100 m | <br>0.4 μT<br>Negligible |
| 20 kV power line:<br>- under the line<br>- at 100 m | <br>6 μT<br>0.2 μT |
| 90 kV power line:<br>- under the line<br>- at 100 m | <br>10 μT<br>0.1 μT |
| 400 kV power line:<br>- under the line<br>- at 100 m | <br>30 μT<br>1.2 μT |

**Table 2.2.** *Measured magnetic field [EDF 05, WHO 07]*

Battery-operated appliances radiate much weaker electric and magnetic fields than appliances connected to the mains.

The average level found in a home is generally below 0.1 μT, provided that the home is not near a power line or another external source.

#### 2.2.2.2. *Electric motors*

Do people realize how many electric motors there are in our environment?

Motors, and more generally electrical machines, transform electrical energy into mechanical energy. Conversely, those that are mechanically powered produce a current. This is the case for generators, wind and hydroelectric turbines.

An electrical machine radiates the same field level, whether the current is consumed (motor) or produced (generator). The magnetic field is generated by the fixed coils of its stator and/or the mobile coils of its rotor. The stator or rotor may also be equipped with permanent magnets, but at least one of these two components must be coiled. The rotor may have neither coils nor magnets.

Squirrel-cage induction motors, which are supplied with single- or three-phase voltage, are the most commonly used in industry. These machines radiate the weakest magnetic field. Levels above 100 μT may be measured in the electrical connections of a fully loaded 900 kW motor. This decreases rapidly with the distance, reaching a few microteslas at 50 cm.

Synchronous motors are used when the rotation speed must be strictly controlled. The motor uses the power frequency for a fixed speed. Gears lower or raise this speed. An electronic power inverter with variable frequency can be used to precisely control the rotation speed over a large range.

The radiation frequency of the motors is proportional to the rotation speed. The number of pole pairs determines the proportionality coefficient. The greater this number, the slower a motor turns at a given frequency.

Hand-held power tools with motors, such as drills, grinders and saws, are widely used by all categories of people (laymen and professionals). These devices are equipped with universal motors when they are directly powered by the mains. A motor is called "universal" if it can be supplied either with a DC or an AC current. These motors radiate > 1 mT on the surface of the device.

Battery-operated devices are generally powered by a motor with a permanent magnet for reasons relating to operating life. They radiate much weaker magnetic fields (< 50 μT).

Eddy currents, induced by magnetic field leakage, pass through the metal frame of the motor. In turn, these currents induce a magnetic field that opposes the one that created it. The result is a reduced radiation level. The faster the rotation speed of the machine (i.e. the higher the frequency), the better the attenuation provided by the frame.

The waveform of the magnetic field involved in exposure level assessments depends on the electronic speed control circuit and its intensity depends on consumption, itself dependent on the resistive torque (load).

There are also linear motors, which produce a translation movement. They are used in machine tools to precisely control the position of the tool or the part being machined. The mobile part has several position-controlled coils. The fixed part is made up of permanent magnets aligned with alternating north–south polarities. Magnetic induction can exceed 500 mT at the contact of these magnets, fall beneath 0.5 mT a few centimeters away and no longer be measurable outside of the protective caps.

#### 2.2.2.3. *Industrial heating by means of magnetic induction*

The flexibility and precision provided by induction heating techniques have made this a popular process for transforming ferromagnetic materials. Examples include smelting, welding, brazing, bonding and thermal treatments [STD 23]. This type of heating is also used in the electronics industry for refining semiconducting blocks that, once sliced, will form wafers, base layers for the production of transistors, LEDs, microprocessors, etc.

The parts or materials are placed in the magnetic field created by a coil. This field induces eddy currents, which heat them up by means of the Joule effect. The current, and therefore the heat generated, depends on the frequency, the shape of the part, the quantity of material and its electric and magnetic characteristics.

Powers of several megawatts can be applied in furnaces using this principle. They use the 50 Hz voltage of the power. The field may widely exceed 1 T in the inductors, with an external residual level between 1 and 10 mT at a few cm and reduced by another order of magnitude at 1 m.

Frequencies from a few kilohertz to several megahertz are used, in particular to heat small parts, due to their superior magnetic performance and despite the electrical losses and equipment costs being higher.

When the frequency is relatively high, a HV must be applied to the coil connections to have a significant current. The radiated electric field level therefore increases significantly if no precaution (shielding) has been taken in the design of the system.

Inductors with flat coils (spirals) or cylindrical coils (solenoids) are often used. The field has axial symmetry when it is not distorted by the metallic environment. The field distribution may become complex when the inductor is adapted to the shape of the object. The magnetic field level may be higher near the supply cables of the industrial system than in the accessible vicinity around it.

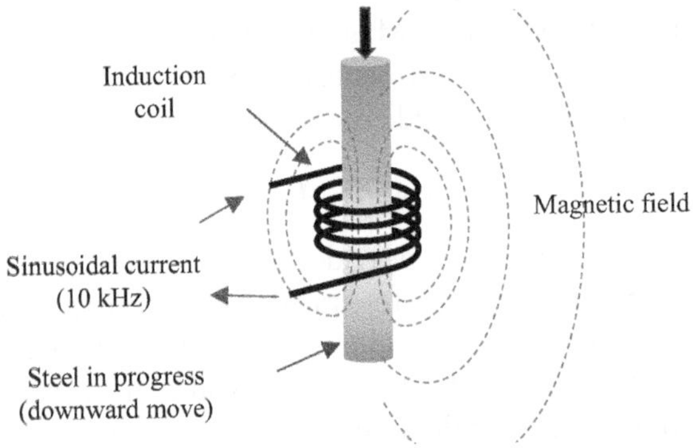


**Figure 2.6.** *Magnetic induction heating*

### 2.2.2.4. *Induction hobs*

Induction hobs for cooking food use the same principle as furnaces. The base and sides of a container made of a ferromagnetic material placed on an inductor (flat coil) are heated by induced currents (eddy currents). The contents heat through conduction (unlike microwaves, where the food heats the container). Frequencies between 20 and 100 kHz are used. The temperature is controlled either by reducing the supply current (by detuning the frequency) or by on/off switching.

When the inductors are not completely covered, because the container is too small or because it is not centered, the magnetic field leaks and the performance of the hob is affected. The radiation levels are higher. The control electronics of the hob interrupt the functioning, i.e. the radiation, in the absence of kitchen utensils.

### 2.2.2.5. *Welding*

Manual welding equipment is widespread in the metallurgy industry. The automotive industry and shipyards are among the major users of this assembly method. It is also used by many small and medium-sized enterprises (SMEs) for producing objects as varied as wheelbarrows and some kinds of pipe.

The operator uses his tools manually in two common processes: electric arc welding and resistive welding, also called spot welding. These processes can also be robotized.

#### 2.2.2.5.1. Arc welding

An electric arc with significant light intensity is created between an electrode (stick) and the objects to be welded. The passing of an electric current causes the metals to fuse together. The electrode contributes material. Welding can be performed under an inert gas (argon-based, for example). This gas envelops the molten metals, protecting them from oxidization.

The electric current between 50 and 1,000 A depends on the nature of the welding and the materials. It may be AC, DC or pulsed. This current flows from the generator to the electrode, passing through an electric wire and the electrode holder (torch). It then loops back through the pieces to be assembled and the chassis ground (Figure 2.7). A current in the opposite direction is possible. The magnetic field may exceed 10 mT at the cable contact and its decrease can be considered inversely proportional to the distance (unipolar or single-core wire [1.7]). Although the voltages used are sufficient to trigger the arc, they remain very low (<24 V) and the electric field is therefore negligible.

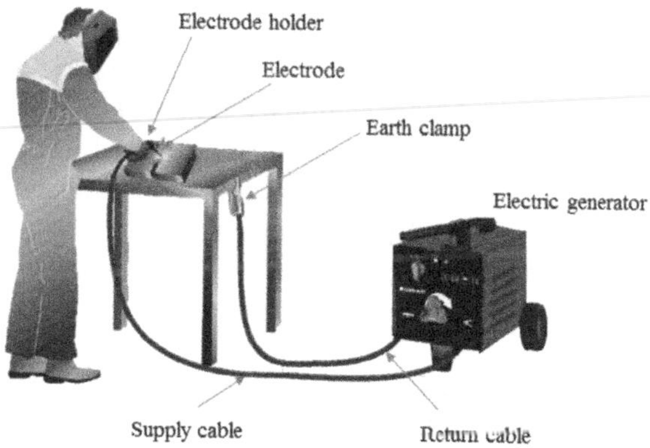


**Figure 2.7.** *Arc welding station*

### 2.2.2.5.2. Resistance welding

Resistance welding uses a gun equipped with electrodes that pinch several metal sheets together. An electric current of strong intensity flows in the copper electrodes, passing through the parts that are to be joined. They melt at the contact spot through the Joule effect and the pressure exerted by the electrodes welds them together (Figure 2.8).

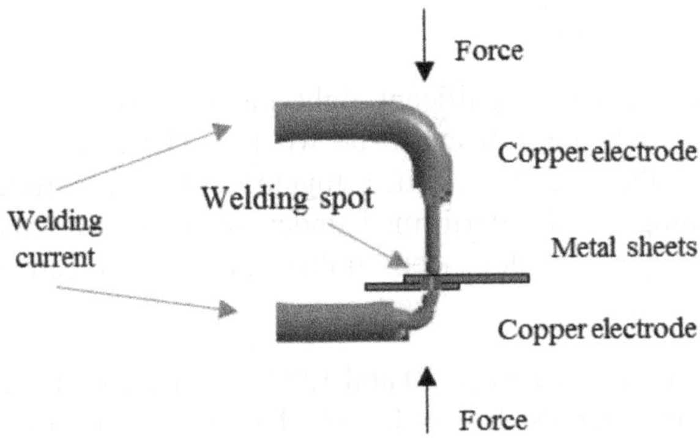


**Figure 2.8.** *Principle of resistance spot welding*

There are various welding gun technologies: guns with integrated transformer and guns with separate transformers supplied by a low (50 or 300 Hz), medium (1 or 2 kHz) or high (>6 kHz) frequency DC or AC current. Medium frequency and HF correspond to the ripple superimposed on the DC current. They are powered with single-phase, two-phase, or even three-phase voltage.

An operator puts the gun into position if it is portable, or puts in place parts to be assembled if the tool is stationary. The welding current is set to a typical value of 10 kA rms, lasting for several hundred milliseconds. These parameters are adjusted depending on the nature, thickness and surface condition of the metal sheets. The technology, size and adjustments to the guns influence the radiation. The size (reach and width) of the guns is dictated by the accessibility of the welding spot.

Portable welding guns are used when the elements to be assembled are substantial or of a significant mass. A typical example is the body of a vehicle. Using stationary welding guns has ergonomic advantages when the elements to be assembled are small and light. Once the elements have been put in place, the operator triggers the welding process. Currents of almost 100 kA are required when several spots must be welded simultaneously or when the surfaces to be joined are relatively wide.

**Figure 2.9.** *Portable spot welding gun and stationary spot welding gun*

The magnetic field is between a few hundred microtesla on the side of the gun body and around one hundred milliteslas in the center of the arms, in general. Radiation is more significant in the direction perpendicular to the arms. The exposure of the operator therefore depends on his position relative to the tool.

### 2.2.2.6. *Office equipment, household appliances*

Homes and offices are not free from electric and magnetic fields. Most devices are connected to 240 V and 50 Hz power and therefore radiate magnetic fields and weak electric fields (wireless communication devices are discussed in the next section).

Appliances with a small electric transformer radiate the highest magnetic fields. They remain modest, however, from a few tenths of a microtesla at 30 cm to a few tens of microtesla on contact (the levels of some models of radio alarm clock). Electric fields have a typical maximum value of 120 $V \cdot m^{-1}$, and more commonly a few tens of $V \cdot m^{-1}$ at 30 cm (Table 2.1).

The magnetic field level of many appliances has decreased, because they have become more energy efficient, leading to the miniaturization or even disappearance of transformers. Also, the replacement of cathode ray tube (CRT) and its deflection coils by LED flat screens (or, to be more accurate: LED backlit TFT screens) and OLED screens has caused magnetic field levels to decrease. Meanwhile, internal voltages have been reduced to a few tens of volts, making electric field levels negligible (the 25 kV necessary to make CRT operate are no longer in use).

The spread of laptops has also reduced the level of low-frequency radiation for the same reasons. The magnetic fields radiated by these appliances are lower than 100 nanoteslas (0.1 μT) beyond 30 cm and electric fields are no longer measurable in domestic environments.

#### 2.2.2.7. *Demagnetizers*

The demagnetization process removes any remanent (or residual) magnetic induction within ferromagnetic materials. These materials memorize the last magnetic induction to which they have been subjected.

The operation consists of immersing the material in a 50 Hz magnetic field radiated by a demagnetizer (coil), then reducing it gradually until the residual induction has been canceled out.

These devices are used in industry to remove the field memorized by objects during machining (flanged by a static field, handling by an electromagnet, magnetic particle inspection).

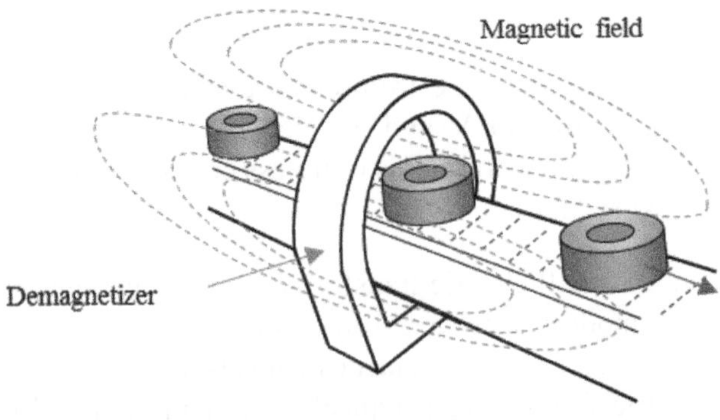


**Figure 2.10.** *The objects gradually leave the magnetic field, becoming demagnetized*

#### 2.2.2.8. *Magnetic particle inspection*

Magnetic particle inspection is a very widely used non-destructive inspection method in materials technology. Simple or complex ferromagnetic objects can be inspected after welding or smelting, or even after use (fatigue tests on safety critical parts).

This method makes it possible to observe discontinuities and surface or underlying cracks. This technique consists of creating an intense magnetic flux within the material to be tested. When there is a defect in its path, the magnetic flux is distorted, creating a magnetic leakage field. This field will attract the particles (colored or fluorescent) of a developer. Their distribution provides an indication characteristic of the defect.

In order to be revealed, the defects must be perpendicular to the field lines. There are many additional techniques for achieving this [GRA 14], chief among them are transverse magnetization through direct passage of a current in the object, longitudinal magnetization with magnetic heads (Figure 2.11) and multidirectional magnetization (rotating field of constant strength).

The field is applied for a maximum of a few seconds and the currents involved may exceed a few kiloamperes at 50 Hz. The magnetic field may exceed one millitesla on contact with the object. Demagnetization of objects may be required after inspection, in addition to cleaning.

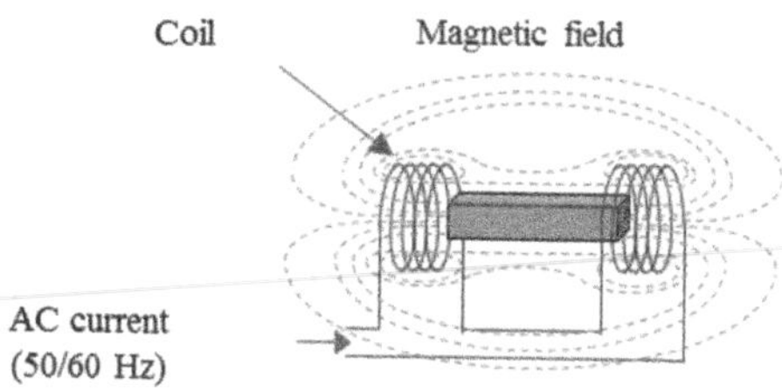


**Figure 2.11.** *Principle of magnetization through circulation of magnetic flux*

### 2.2.3. *HF fields*

When dealing with exposure to electromagnetic fields, the HF domain covers the 100 kHz (3 km wavelength) to 300 GHz (1 mm wavelength) band.

It is impossible to be exhaustive, because HF applications are so numerous and diverse. In general, this radiation is generated intentionally to spread information across great distances or transform electromagnetic energy into thermal energy.

The majority of these sources are involved in radio communications applications, including, for example, detection (Radar) and radiofrequency identification (RFID). There is a general trend toward augmenting frequencies. Transmissions above 10 GHz are growing in numbers.

HF electromagnetic radiation is also used in industrial, scientific and medical domains, for which frequency bands are reserved (ISM bands, section 1.4.4.2). They can be intense locally and are generally undesirable beyond a few centimeters or decimeters.

There can also be unintentional radiation due, for example, to malfunctions or insufficient shielding of an HF transmission chain.

#### 2.2.3.1. *Radio communications*

In 1888, Heinrich R. Hertz observed the electromagnetic waves that J.C. Maxwell had predicted a few years earlier. H.R Hertz did not believe that his discovery could have applications.

Historically, information was transmitted “digitally” at a few symbols per second (Morse code telegraphy = transmission of short and long pulses called dots and dashes).

In order to be propagated across long distances, a piece of information must be carried by a signal called a “carrier frequency” or simply “carrier”. The operation consisting of making parameters vary unambiguously depending on the content to be transmitted is called “modulation”. When the signal is received, the information is extracted through demodulation.

Modulation is called “analog” when the signal to be transmitted is continuous, such as that from a microphone, and “digital” when the input signal is a binary data stream.

Modulation is called amplitude modulation (AM), frequency modulation (FM) or phase modulation (PM) according to the parameter being modulated. These are either modulated separately – as in simple analog or digital modulations – or simultaneously – typically in high-speed digital transmissions. These allow wireless data transfers of constantly increasing data rates. Today, these data rates can reach several gigabytes per second. This increase has been made possible by the use of higher carrier frequencies with a broader bandwidth[1] and by increased spectral efficiency (quantity of information per hertz of bandwidth used).

Digital techniques are also used to encrypt information (e.g. paid TV channel or secure mobile telephony), compress data and detect and correct transmission errors.

Digital modulations have a remarkable spectral efficiency. They make it possible to broadcast a TV multiplex, combining four to eight channels in one frequency band identical to a single old analog TV channel, all with a superior or greatly superior service quality (high definition [HD]). The frequency domain allocated to TV stations has therefore been reduced. Since frequencies are resources that belong to states, the surplus bands, called the "digital dividend", have been given over to private telecommunication operators, so that they can deploy fourth-generation telephony (4G called also LTE), so on and so forth, in exchange for dividends.

Digital modulations make it possible to quickly transmit small data packets, thus leaving numerous time intervals free for other transmissions or for retransmission in the event of reception failure. This sharing of resources is called the Time Division Multiple Access (TDMA) method. When several frequencies (transmission channels) are attributed to one service, one of them can be temporarily allocated to a user. This technique is called "frequency-division multiple access" (FDMA).

The digital modulations and simple access techniques used by the GSM, for example, have given way to digital transmissions that modulate several thousands of subcarriers simultaneously and use sophisticated access techniques (orthogonal frequency-division multiplexing [OFDM]). They are used by digital terrestrial television (DTT), Wi-Fi and 4G.

---

1 The bandwidth corresponds to the frequency interval that can be used for an application (width of a transmission channel, measuring band, acoustic band, etc.). Depending on its relative width (bandwidth to central frequency ratio), the band can be narrow or broad. The bandwidth allocated to an application, which determines the data rate for a given modulation, must not be confused with the carrier frequency, which defines the propagation conditions and, in part, the range. In information technology (IT), the term "bandwidth" denotes the data rate, often expressed in a number of bits or bytes per second.

These techniques make it possible to optimize the use of the frequency spectrum, in addition to modern digital modulations.

The most recent transmissions also use multiple input multi output (MIMO) technology, which involves at least two transmitter antennae and two receiver antennae (Wi-Fi, etc.). With two physically independent links, the transmission can be made more robust or the data rate doubled, depending on the propagation conditions.

The signal can therefore be modulated simultaneously in amplitude, frequency and phase, use different pathways and be radiated several times (redundancy), making digital transmission particularly reliable. The energy received by the unit of data (bit) determines the quality of the link.

Essentially, a transmitting station consists of a transmitter and an antenna. The transmitter is composed of a modulator, which processes the signal to be transmitted, a carrier generator and an RF power amplifier. The antenna transfers the energy produced by the transmitter into space. Its electromechanical characteristics and orientation determine the way in which this energy is radiated. The properties of the antennae have evolved less, unlike the modulation techniques and electronic systems.

The configuration of a receiver is symmetrical to that of a transmitter (Figure 2.12). The antenna picks up the radiation, an amplifier increases the signal level, filters isolate the useful signal and a demodulator extracts the information from the carrier.

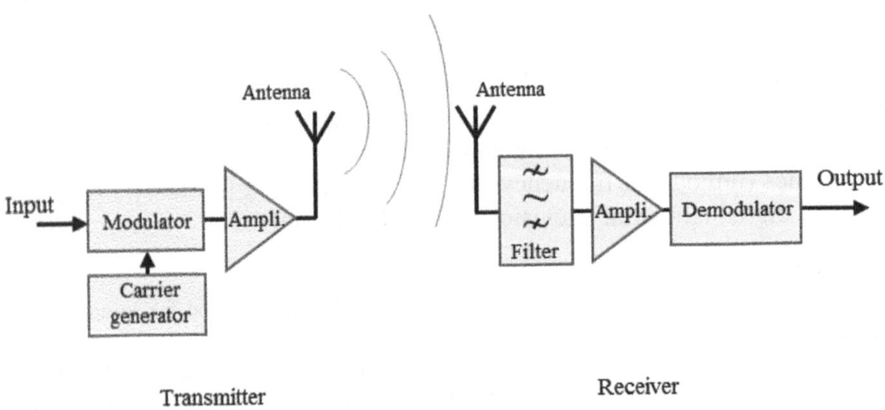


**Figure 2.12.** *Principle of a telecommunication link*

Transmission systems are briefly described in the following sections by order of appearance or increasing complexity.

Analog transmissions are disappearing, giving way to digital transmissions, which enable rare frequency resources to be shared and provide a high level service quality.

### 2.2.3.1.1. Radio and television broadcasting

Radio and television broadcasting occupy almost identical frequency bands in every country in the world. Transmissions are one-way, from one transmitter to several million receivers (down link).

#### *AM radio*

AM radio uses "long wave" (LW, kilometric waves), "medium wave" (MW, hectometric waves) and "short wave" (SW, decametric waves) bands, covering the 143 kHz/284 kHz, 526 kHz/1606 kHz and 3/26 MHz bands, respectively. These frequencies are classed as low frequencies, medium frequencies and HFs, as defined by the International Telecommunication Union (Table 1.14).

The sound quality of this broadcast is reduced, because the bandwidth of the audio signal allocated to this service is limited to 4.5 kHz in Europe. The bandwidth required for the transmission (channel) is 9 kHz.

In LWs and MWs, the power radiated is between 100 W and 10 kW for local coverage. A few megawatts (MW) are required to guarantee national coverage. Antennae are a considerable size. They are made up of pylons positioned tens of meters apart, powered by a transmitter, and passive pylons. These are non-active earthed elements, acting as reflectors to prioritize one transmission direction. LW and MW services are to be stopped, giving way to new digital radio services (according to the country).

In SWs, the power radiated exceeds 100 kW and can reach 2,000 kW. Directive transmitting antennae (narrow beam) provide reception far beyond the horizon and even at the opposite end of the world. Despite the roundness of the Earth, SWs propagate without relays over several thousand kilometers, because of reflections between the ionosphere and the surface of the Earth (these two media are considered to be conductors and therefore reflect the waves).

The electric field can exceed 100 $V{\cdot}m^{-1}$ inside the pylons and at their base. It decreases very rapidly up to a distance equal to around the height of the antenna. It then increases slightly before decreasing continuously. It is difficult to predict the radiation near these antennae, because of the reflections and currents induced in the metal structures [IAR 13, ICN 09b].

*FM radio*

FM radio uses FM, a technique that consists of slightly modifying the frequency of the HF carrier around its nominal value to the rhythm of the low-frequency audio signal. The bandwidth allocated to an audio channel is 15 kHz. A transmission bandwidth of less than 200 kHz is required for a stereophonic transmission (two audio channels), providing a quality close to that of high-fidelity (Hi-Fi) reproduction.

In Europe, FM radio broadcasting uses the frequency band between 87.5 and 108 MHz, corresponding to the wavelength domain around 3 m. The output power of an FM transmitter between a few watts and a few hundred watts makes it possible to radiate a power above 100 kW EIRP [1.19]. Coverage can reach 50 km, provided that the antenna is fixed to a pylon or tower, which may be over 300 m tall (e.g. the Eiffel Tower). The main characteristics of French radio transmitters (radiated power, frequency and antenna pattern) and their localization can be found on the website www.annuaireradio.fr.

The higher the frequency rises above 30 MHz (VHF band), the more the topography and obstacles between the transmitter and the receiver attenuate the signal. When necessary, a local low-power retransmitter positioned on a hill relays the broadcast to get a valley out of a radioelectric shadow.

The electric field in a pylon and in its close vicinity remains difficult to predict, because the size of the structures (crossbars, braces and gaps) are similar to the wavelength. The fields reach 900 $V{\cdot}m^{-1}$ near the antennae. Inside the pylons, average levels of 300 $V{\cdot}m^{-1}$ have been measured [ICN 09b]. Currents are induced in them, creating strong, localized electric fields that are very sensitive to changes in the environment (e.g. presence of a person). Access to these areas is restricted, reserved for professionals who are trained and equipped with protection systems (electromagnetic protective clothing, field detectors, etc.).

Field strength varies rapidly with distance and position, depending on the configuration of the antennae and the direction of their sidelobes (Figure 2.14 shows a field decreasing with distance). Close to an antenna, the fields on the ground are always much weaker than those present in the main lobe. When the main lobe reaches the ground, the levels are weak, even when multiple reflections are taken into account.

##### 2.2.3.1.2. Television broadcasting

Although television studios have used digital cameras for a few decades for archival quality and the optimization of recording volumes purposes, the transmission of signals remained analog and digital transmission techniques were being trialed. This delay enabled individuals to keep their CRT sets.

Finally, analog programs were replaced and have ceased to be broadcast in Europe. This change was achieved by preserving the (UHF) frequency bands, allowing individuals to keep their antennae.

Today, the transmission chain is entirely digital from the camera to television screen pixels. This technology has improved picture quality and removed the ghost pictures caused by echoes and other interference, all while retaining a constant quality (without adjusting devices). The services have been improved: more channels, HD, multi-language features, text, etc., while reducing the transmission power per channel. A power of around 10 kW makes it possible to transmit a DTT six-channel multiplex, whereas double the power was required for a single channel of analog TV.

Transmission carried out according to the Digital Video Broadcasting-Terrestrial standards in Europe is based on OFDM techniques. These techniques make it possible to distribute information between thousands of subcarriers, permitting communication that is robust with regard to jamming and other propagation hazards. Subcarriers are shared between several programs, forming a TV package or multiplex. These modulations are often used in hertzian television, underlying DTT and satellite digital television.

Constant advances in coding, modulation and compression techniques (MPEG2, MPEG4, etc.) make it possible to propose ultra HD (4k) and 3D, without requiring additional spectrum resources for the transmission channel.

Radio and TV receivers are passive. They do not require a link with a base station. Therefore they emit no HF signal.

TV transmission antennae in the HF band share sites (pylons, towers, etc.) with FM radio broadcasting antennae (VHF band). They are generally mounted at higher levels, because UHF waves are propagated more as optical rays. Electric fields of 600 V·m$^{-1}$ have been measured near antennae [ICN 09b].

Table 2.3 summarizes the main technical characteristics of radio and TV services.

| Services | | Frequency bands | Bands | Transmission powers (PAR) | Maximum ranges (km) |
|---|---|---|---|---|---|
| Radio | LW | 143 kHz<br>284 kHz | LF | 50–2,000 kW | ~1,000 |
| | MW | 526 kHz<br>1,606 kHz | MF | a few kW to 300 kW | ~200 |
| | SW | 3 MHz<br>26 MHz | HF | ~500 kW | ~10,000 |
| | MF | 87.5 MHz<br>108 MHz | VHF | 10 W–150 kW | ~50 |
| TV | DTT | 470 MHz<br>690 MHz | UHF | 20 W–100 kW | ~100 |

**Table 2.3.** *Radio and television services*

FM transmissions are disappearing and being replaced by digital terrestrial radio (DTR).

#### 2.2.3.1.3. Mobile telephony

The second-generation (2G) GSM mobile telephony was the first telecommunication service to be commercialized on a very large scale at the beginning of the 1990s. It became widespread due to drops in prices for service and portable devices.

2G uses 900 and 1,800 MHz bands. The wavelengths are 33 and 16 cm, respectively.

The base station antennae or relays are typically fixed to masts 20–30 m tall or placed on natural or artificial elevations (water towers in rural areas, tall buildings in urban areas, etc.) to guarantee optimal propagation conditions. The area covered by a base station is called a "cell". The cells overlap to form a cellular network. When the user moves, his mobile switches from one cell to another with no perceptible interruption (intercellular transfer or handover). He can continue his conversation, receive a call instantly, make a call or exchange data. The power and operating life of a mobile, determined by the performances of its battery, have partly caused its range, and therefore the size of the cells, to be limited.

Furthermore, the connection capacity of a base station is limited to a few dozen mobiles. Stations are necessarily denser when the number of users is high (e.g. in business centers and malls). The network is therefore made up of base stations with a transmission power of over 100 W, forming macrocells with a radius of several km, and stations with a less power, up to less than 1 W. There are therefore microcells, guaranteeing a typical local range of 2 km to cover a neighborhood, picocells, which cover a mall or station, and femtocells, which give access to the cellular network within a radius of 10 m (accommodation, railway carriages, etc.).

External antennae can be mounted at heights of between a few meters and a few dozen meters. The exposure level at base stations is lower than a few $V{\cdot}m^{-1}$ on the ground beyond the safety perimeters of the antennae. In comparison, the reception threshold of a mobile telephone (sensitivity) is around a few tens of $\mu V{\cdot}m^{-1}$.

A mobile telephone operator must deploy several thousands of stations to cover a whole country, such as France. These stations are connected to the central network system by point-to-point microwave links, wired networks, or fiber optics.

The link between the base station and the mobile is two-way. To achieve this, a frequency band is reserved for the down link, the link from the base station to the mobiles, and another band is attributed to the up link, the link from the mobile to the base station (see Table 2.4, mobile telephony services and frequencies).

2G telephony uses relatively simple carrier modulation techniques. Transmission from and to a mobile is not continuous. It uses the TDMA technique, whereby a time slot of an eighth of the time is attributed to a user every 4.6 ms (217 Hz). This slicing is sometimes audible when a mobile is placed against a radio receiver or an insufficiently shielded audio amplifier.

The initial data rate of 9 kilobits per second (kbps), enough for a good quality conversation, was increased to transmit more and more digital data. The improvement of spectral efficiency, the attribution of additional time slots during data transmission and the modification of communication protocols (transmission in packets) have contributed to the rise in new mobile telephony services and to the advent of smartphones. 2.5G (GPRS), then 2.9G (EDGE) were quickly deployed, due to software adaptations on the existing material bases of base stations and mobile telephones.

The need for more and more data rates, related to new services (mobile Internet, mailbox synchronization, online games, etc.), led to the third generation or 3G, also known as UMTS or IMT 2000. Each frequency is used by various users, each with their own encoding. It is rather as if everyone is speaking at the same time, but in different languages in the same place. You simply need to be compatible with your interlocutor to understand one another, despite the other conversations continuing with an equivalent sound level. This access technique is called Wideband Code Division Multiple Access (WCDMA).

3G base stations have a transmission power of several tens of watts. 3G facilitates data rate of 144 kbps, even up to 2 Mbps in very good conditions.

Improvements, essentially to modulation techniques, have been made to 3G, producing 3.5G and its derivations (HSPA, etc.).

The need for service in mobility (high data rate exchange on the internet, video downloading, TV reception) constantly requires a greater data rate in the down link, which 4G delivers. The theoretical rate exceeds 100 Mbps per second, and fifth generation (5G), with its gigabits per second (Gbps), is planned. To guarantee such performances, the transmission power must be slightly above that of 3G.

The modulation technique becomes complex, optimizing the frequency spectrum while guaranteeing good transmission quality in an urban environment. To achieve this, the information is distributed over several hundred or even several thousand frequencies. It is also spread over time (interlacing) so that the loss of several frequencies or a mute in transmission for a short lapse of time does not alter the overall service quality. Each subcarrier can be digitally modulated on 16, 64 or more amplitude and phase levels. This technique, also used in digital television, for ADSL and Wi-Fi, is called OFDM- Quadrature Amplitude Modulation (QAM). An OFDM channel can be shared between various users. The 2G and 4G cellular telephony networks are similarly structured.

### *Base station antennae*

A base station antenna is a vertical assembly of radiating elements equal in size to half a wavelength (dipole of 15 cm at 1 GHz), regularly spaced and protected by a radome transparent to waves. This set-up thus reaches a height of between 1 and 2 m and a width of around 25 cm. This layout makes it possible to obtain a narrow beam (main lobe) in the vertical plane (between 5° and 15°) and a wide one in the horizontal plane (between 60° and 120°) (see Figure 1.14).

The beam tilts slightly downwards, by 3°–10°, mechanically (during the mounting of the antenna) or electrically (through dephasing of the elements). With a gain of 7–20 dBi and supplied by a power of 10 W, these antennae produce an electric field of less than a few volts per meter at 50 m in the main lobe axis. This field is weaker by an order of magnitude in the direction of sidelobes if there is no obstacle on the wave path. This is shown in Figures 2.13 and 2.14.

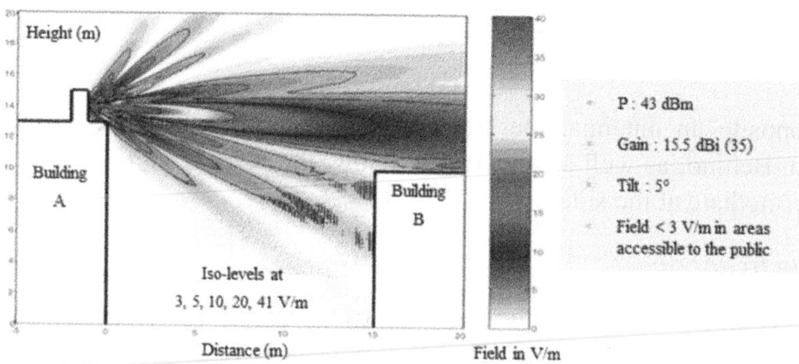


**Figure 2.13.** *Electric field around a GSM base station mounted on the roof of a block of flats (source: ANFR)*

In the axis of the antenna, the electric field decreases with the distance[2] (the power decreases with the square of this distance). Furthermore, the proximity of an antenna does not mean a high field level. It is difficult to

2 Numerical implementation of [1.20] with data from Figure 2.14: the electric field is $\frac{\sqrt{30.10.63}}{150} = 0.9\ \mathrm{V \cdot m^{-1}}$ at 150 m.

predict an exact level underneath the antenna, due to radiation from the sidelobes (Figure 1.14), but it does not exceed 1 V·m$^{-1}$ and undergoes significant variations with the distance. In all areas open to the general public, including the floors of a block of flats facing an antenna, the field typically remains under 3 V·m$^{-1}$. It is much weaker underneath the antenna [COP 13].

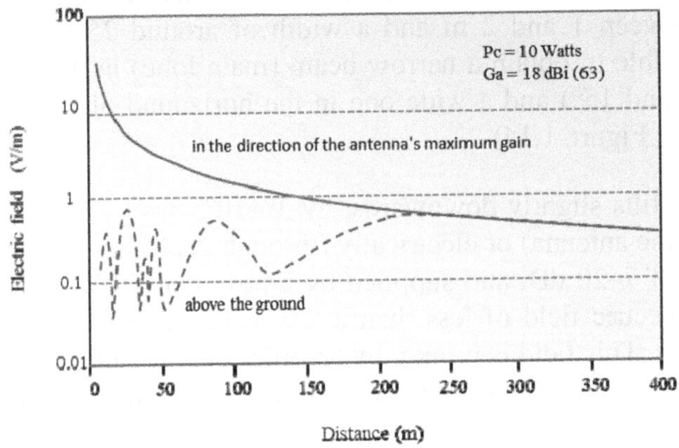


**Figure 2.14.** *Electric fields in the axis of an antenna and on the ground*

Opposite an antenna, the field can exceed 50 V·m$^{-1}$ at a distance of 50 cm. Behind, as well as above and below, the level is 10 times weaker. It is intermediate at the sides.

### *Mobile telephone*

A 2G GSM mobile telephone radiates a maximum power of 2 W peak or 250 mW averaged at 900 MHz and of 1 W peak at 1800 MHz. The ratio of 8 between the peak power and the averaged power is due to time-sharing of the radiation frequency between several users (TDMA technique). This power is required to guarantee a quality link even when it is located in a RF slight shadow area, such as within buildings (an elevator is the worst case, since it forms a Faraday cage, which can greatly attenuate electromagnetic waves). The transmission power of the mobile reduces automatically to exactly what is required when it approaches a base station to preserve its battery life (adaptive power management). This function can considerably reduce the exposure of the user to his mobile. The transmission power is correlated to the received signal strength indicator displayed on the telephone. Its radiation power increases when the reception quality declines.

Although the transmission power of base stations is greater by a factor of 10–100 than the transmission power of a mobile telephone, a user is exposed to levels 10,000 times higher from his mobile telephone due to contact between his or her head and the mobile [WIA 13].

The accumulation of exposure from the mobile and the base stations must also be taken into account when determining the overall exposure of a user. With this in mind, the Low EMF Exposure Networks (LEXNET) project, supported by the European Commission, aims to propose technical recommendations to reduce the overall exposure of the general public by 50%, without compromising service quality (www.lexnet-project.eu).

Furthermore, a mobile is constantly trying to remain connected to the cellular network whenever it is switched on. To achieve this, it exchanges short messages with the base stations. It may send one message every hour if it remains inside a cell, or several messages per minute to many stations in a situation of rapid mobility (e.g. traveling in a car). In this second case, its operating life is substantially reduced. Therefore, even if it the user is not having a conversation, he or she is unwittingly exposed to radiation from his or her mobile. However, these functions can be deactivated (airplane mode, etc.). Furthermore, smartphones are constantly exchanging with the network to synchronize mailboxes, diaries and other services.

The frequency bands allocated to mobile telephony services are outlined in Table 2.4. These bands bring together the up link and down link channel frequencies.

| Services | Frequencies |
|---|---|
| LTE 700 | 703–788 MHz |
| LTE 800 | 791–862 MHz |
| GSM 900 | 880–960 MHz |
| GSM 1800<br>LTE 1800 | 1,710–1,880 MHz |
| UMTS 2100 | 1,920–2,170 MHz |
| LTE 2600 | 2,500–2,690 MHz |

**Table 2.4.** *Mobile telephone services and frequencies in Europe*

#### 2.2.3.1.4. Digital radio

Digital radio uses the same multicarrier modulation techniques as television and mobile telephony (OFDM), with the advantages that have already been mentioned: transmission quality and reduced bandwidth without requiring more power than authorized for AM or FM analog modulations.

The sound is encoded and compressed according to the MPEG2 or MP3 principle. It can be recorded without difficulty on any digital support. Data such as the artist's name, his news and his album cover are transmitted along with the sound.

There are two broadcasting standards in Europe or in the world for DTR:

– *Digital Audio Broadcasting (DAB)*: DAB is a European standard with two variants: DAB+ and terrestrial digital multimedia broadcasting (T-DMB). It is radiated in VHF (from 174 to 240 MHz), and in the UHF band between 1.45 and 1.5 GHz (also called L band). An SHF band is reserved for it. The transmission power radiated may exceed 10 kW and the antennae are mounted on masts several tens of meters tall, just like for television and FM radio. DAB is being used tentatively in France in large urban areas, after a few attempts and some encouragement. Broadcasting occurs simultaneously with FM radio. This double broadcasting enables a car radio to switch automatically to the best quality service without the listener being aware of the shift;

– *Digital Radio Mondiale (DRM)*: DRM is a worldwide broadcasting standard for LW, MW and SW (<30 MHz). These waves therefore propagate in the same way as they do for AM radio, with the advantage of a more extensive range and superior sound quality, because multicarrier modulation is also used. The transmission power is the same as that of AM. DRM is an advantageous radio broadcasting method in very large countries (with a range of several hundred kilometers).

Radio transmissions can also be received digitally by satellites, just like TV broadcasts. Frequencies of around 10 GHz are used. The power densities of electromagnetic fields reaching the ground are around 1 $pW{\cdot}m^{-2}$, i.e. $10^{-12}$ $W{\cdot}m^{-2}$ (equivalent to an electric field of a few tens of $\mu V{\cdot}m^{-1}$). They only can be picked up by parabolic antennae with a very strong gain, i.e. greater than 1,000 or 30 dB, although they measure a few decimeters.

### 2.2.3.1.5. Local wireless networks

Local wireless networks have flooded our daily life since the beginning of the 2000s. Except within the confines of a basement, it is rare to find an enclosed space where a Wi-Fi connection (often with free access) is not available. These wireless networks use communication protocols from the IEEE 802 family, usually issued from the IT domain.

These systems use the ISM frequency band at 2.45 GHz and the "free" U-NII bands centered around 5.4 GHz (section 1.4.4.2). These systems must verify that there is no transmission on a frequency before radiating to avoid interrupting other transmissions. If necessary, they wait for it to be free or use a different frequency. This technique of sharing or access is called listen before talk (LBT).

The peak powers radiated are between 1 mW, to cover a radius of a few meters, and 1 W EIRP, for a range of more than 100 m (in a clear area) [ICN 08].

Wi-Fi belongs to the wireless networks family called Wireless Local Area Nework (WLAN). Transmissions can be performed at speeds of a few tens of Mbps to Gbps on several channels centered in the ISM bands. European regulations restrict the maximum power radiated (EIRP) by Wi-Fi routers (modem) to 100 mW peak at 2.45 GHz and 200 mW at 5.4 GHz indoors. The maximum power authorized for outdoor use is 1 W. This level facilitates a wireless connection over a distance of up to 100 m.

Wi-Fi transmission works through data packets. It is not continuous, even when large files are downloaded. The average power radiated, the value used during exposure assessment, is obtained from the product of the maximum radiation power (peak value) by the relative radiation time given in percentage. Table 2.5 summarizes the maximum transmission time values (duty cycle [DC] = $\frac{t_{\text{emission}}}{t_{\text{total}}}$ in %) measured for various applications [WOU 13].

These values must be multiplied by the number of devices connected within the limit of the maximum router data rate (obtained during file transfer). The maximum strength of the electric field is around 1 $V \cdot m^{-1}$ at a distance of 1 m from the terminal.

| Applications | Duty cycle (95%) |
|---|---|
| Surfing the web | 0.62% |
| Voice on Skype (VoIP) | 1.3% |
| Video on Skype | 2.0% |
| YouTube video (360 p) | 2.1% |
| YouTube video (HD) | 64.5% |
| Downloading files | 65.2% |

**Table 2.5.** *Relative radiation times measured for various Wi-Fi applications (IEEE 802.11g protocol – OFDM 54 Mbps)*

The power radiated by mobile appliances (laptop, tablet, etc.) is often reduced in order to not affect operating life. Furthermore, they often receive much more information than they transmit; their averaged radiated power is reduced to a few per thousand (‰) of the available power.

Exposure level is not necessarily proportional to radiation power. Transmission protocols function by means of acknowledgments of receipt, which involves retransmission of corrupted packets received. A reduction of maximum radiation power can lead to repetitions, which extend the radiation time. This can increase the averaged radiated power, which was not initially the desired aim.

Bluetooth is one of the Wireless Personal Area Network (WPAN) transmissions and also uses the IEEE 820.15.1 family protocols in the 2.45 GHz ISM band. The peak power levels authorized are distributed over three classes: 1 mW, 2.5 mW and 100 mW [ICN 08]. Bluetooth transmission durations are short, except for transfers of large files. Power of 1 mW is sufficient for a range of a few meters; it is also the first choice for handheld devices (headset, cellphone to car radio link, etc.).

The power absorbed by a head from a Bluetooth headset (1 mW) can be compared with the 250 mW that a mobile phone can radiate.

The maximum peak strength of the electric field at a distance of 50 cm is around 0.4 $V \cdot m^{-1}$ for a radiation power of 1 mW, a value that should be divided by 2 at a distance of 1 m. These values must be weighted by the Duty Cycle to determine the exposure level.

The electric field strengths for radiations at 100 mW are identical to those of Wi-Fi.

### 2.2.3.1.6. Radiofrequency identification

RFID is a technology that was used at the end of the Second World War to recognize airplanes by means of radar (IFF system). Electromagnetic fields make it possible to remotely identify people or objects equipped with a transponder or an electronic tag. Each tag contains a unique manufacturer code, an ID registered to the user and possibly free memory space. The format of these data is standardized internationally. Unlike bar codes, a tag does not need to be visible.

The application domains are numerous: logistics, tracking, access control, anti-counterfeiting, payment, etc. The agri-food, industrial, textiles, luxury, health, goods transport and passenger transport sectors use this technology.

RFID operates over a band at 125 kHz or the ISM bands at 13.56 MHz, 433 MHz, 866 MHz, 2.45 GHz and 5.8 GHz. The two technologies supplanting the others in terms of volume are Near Field Communication (NFC) at 13.56 MHz and RFID UHF (866 MHz in Europe). Europe is planning to allocate an additional UHF band centered around 918 MHz (American band) to deal with the arrival of Internet of things (IoT) applications based on RFID.

Applications such as antitheft, keyless access systems (including car remote controls) and electronic toll collection are also based on the RFID principle.

An RFID system is composed of an interrogator, also called a reader, equipped with an antenna, and one or several tags fixed on the objects to be identified (Figure 2.15).

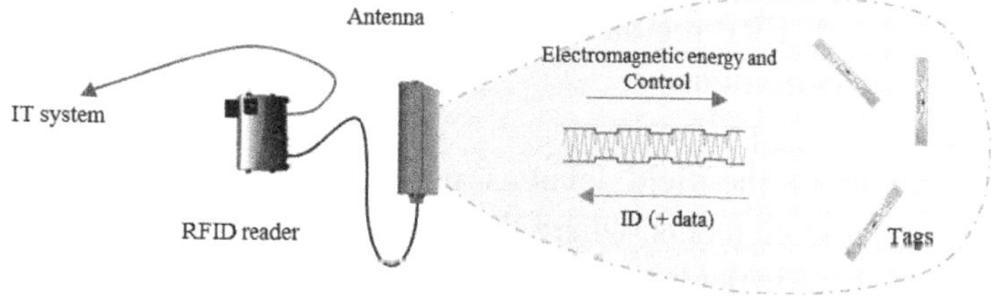


**Figure 2.15.** *UHF RFID system*

The reader antenna radiates an HF field, according to a protocol that is generally standardized worldwide. This is picked up by the tag, which sends back its identification and any data relating to the application. Tags may be active, i.e. containing a power supply to retransmit information, but are usually passive. In this case, they must pick up their energy from the ambient electromagnetic field. As they do not have enough of it to perform a true transmission, they more or less reflect the incident field (backscattering technique). To do this, they modulate the reflection coefficient of their own antenna (through modification of their impedance matching), depending on the data to be exchanged. The reader picks up this (partly) reflected field and extracts data from it. Passive tags are less expensive and have a long lifetime. There are also battery-assisted passive tags, which need a battery for functions other than transmission (e.g. powering a sensor).

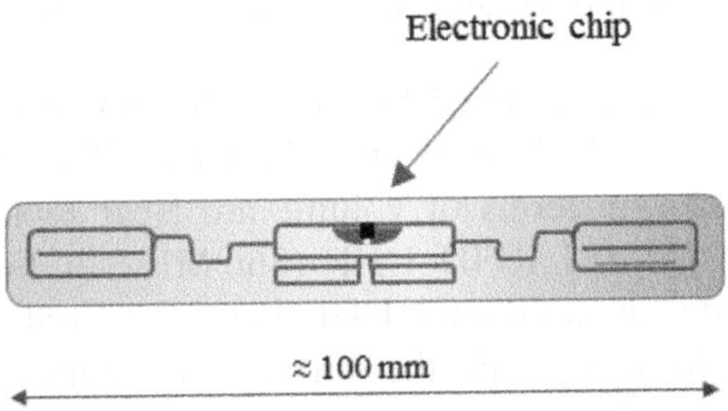


**Figure 2.16.** *UHF electronic passive tag*

UHF RFID technology mainly uses tags with dipole antennae for economic reasons (Figure 2.16). The minimum field level required for them to work is around 1 $V \cdot m^{-1}$. One portion of the energy captured powers an electronic chip and another is (more or less) reflected in all directions. The field reflected in one direction is therefore 100 times weaker than the ambient field generated by the reader, whose radiation power is limited in Europe by the 2 W ERP regulations (see section 1.3.1). A range of 10 m is possible in good conditions.

The decrease in the field level is not continuous in an environment containing metallic elements, where the field is at once attenuated, reflected and diffracted. It is difficult to predict which areas will be covered, i.e. those with high enough levels to activate a tag. The movement of a tag in a difficult electromagnetic environment greatly facilitates its detection and reading.

NFC technology enables data exchange through inductive coupling at 13.56 MHz. The tag and reader are equipped with a loop antenna. Tags in the form of a credit card are common (Figure 2.17). The magnetic field decreases very quickly with the distance. The range is limited to around 10 cm in the best cases. The tag can become a reader and vice versa, according to the application. Typical applications are contactless payment, a function that is integrated into mobile telephones, and public transport access control (e.g. Navigo pass, Oyster Card, etc.).

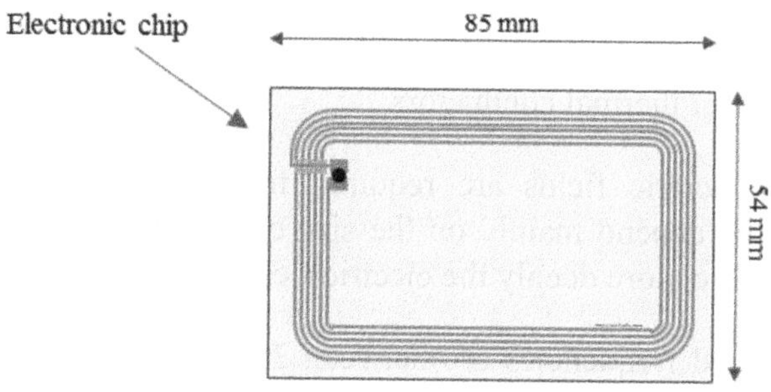


**Figure 2.17.** *NFC electronic tag (13.56 MHz)*

Other RFID applications:

– active tags at 2.4 or 5.8 GHz. These tags can communicate across distances of several tens of meters. The battery must be changed if the lifetime of the product is significant or the tag is reused. Active tags radiate a signal lower than a few hundredths of a milliwatt;

– *localization*: passive tags can be localized in real time across a broad area (warehouse, car park, etc.) by real-time locating systems (RTLS). The radiation level of the antennae covering the area is limited by the general regulations applicable to RFID systems (2 W ERP in UHF in Europe);

– antitheft gates detect passive security tags with a tuned antenna. A distortion of propagation is detected when they penetrate the magnetic field (e.g. 125 kHz) radiated by a gate. The field may reach 300 μT locally. There are also electromagnetic field systems, which usually operate at 8.2 MHz.

### 2.2.3.2. *Industrial and medical applications*

#### 2.2.3.2.1. Dielectric loss heating

Dielectric loss heating uses HF electric fields and is an electrothermic technique. The electric field is used to heat insulators and materials that do not conduct electricity well, such as organic materials, on which magnetic fields cannot act by induction.

An electric current cannot flow through an insulating material (movement of electrons). An electric field must be applied to stimulate its molecules.

Through their resistance to change, the molecules will raise the temperature, depending on the frequency and power (this is discussed in Chapter 3). There are no other techniques for heating these materials quickly, because they are often poor thermal conductors.

Significant electric fields are required for this type of heating. The frequencies used depend mainly on the size of the objects in question. The lower they are, the more deeply the electric field penetrates them.

The authorized frequencies are between 3 and 30 MHz for HF heating and between 300 MHz and 3 GHz for microwave systems.

This heating technique is used in application domains as varied as chemistry, plastic material processing, such as polyvinyl chloride (PVC) and polyurethane (PU), food processing, the glass industry, wood drying and shaping, textiles and pharmaceutical products.

Dielectric loss heating equipment is mainly installed in confined spaces and at chest height.

*HF heating*

The electric field is absorbed by insulating materials, such as wood, plastics, latex, etc., and transformed into heat through dielectric losses. This technique can be used to dry, to accelerate a bonding process and to thermoform plastics.

These materials are generally placed between two planar electrodes, forming a dielectric capacitor, according to the principle shown in Figure 1.1. Their surface area can reach several $m^2$ with a gap of more than 1 m.

Frequencies are chosen from the bands dedicated to industrial, scientific and medical applications (Table 1.5), depending on the size of the electrodes and the homogeneity of the required heat.

The efficiency of the thermal process increases with the frequency and with the square of the electric field strength. Powers of up to 10 kW are applied per $dm^3$. The electric field level is limited by the dielectric rigidity, which represents the maximum field that triggers an electric arc (breakdown) in the material.

When the system is working (presence of a strong electric field), the manufacturing processes do not require the intervention of an operator. The radiation of these systems presents a definite risk in the event of radioelectric leakage. Also, the wires, connectors and shielding must be regularly checked.

Conducting elements located near the electrodes must be earthed to prevent them from charging through capacitive coupling (without electrical contact) and avoid any indirect effects (see section 4.2.2).

The electric fields on the operator vary between 10 and 300 $V{\cdot}m^{-1}$. They may reach 1 $kV{\cdot}m^{-1}$ in front of some equipment.

### *HF welding*

The heat produced by the electromagnetic field melts the materials, which will join together under pressure, which will be released after cooling. The energy that is not converted into thermal energy radiates. The power and the duration of the electromagnetic field, the pressure and the cooling time depend on the materials and determine the quality of the welding.

Plastic sheets, swimming pool liners, inflatable products, tanks, etc. are assembled in this way to guarantee that they are completely air and watertight.

The electric field is established between two electrodes, which enclose the materials. Their shape depends on the elements to be assembled and on the power required. Some electrodes have a surface area of a few $cm^2$ and are adapted to point welding. Disc-shaped electrodes may be used to weld long lengths of materials.

To avoid disturbing the environment and minimize the exposure of the operators, an electromagnetic shield, protection dome or chassis ground is required. In some cases, the electrodes are placed in shielded rooms of several $m^3$ in volume.

Powers exceed 100 kW in the 13.56 MHz or 27 MHz ISM band. The field may reach 1,000 $V \cdot m^{-1}$ near the electrodes and decreases quickly with the distance, reaching localized peak values of 100 $V \cdot m^{-1}$ at 50 cm.

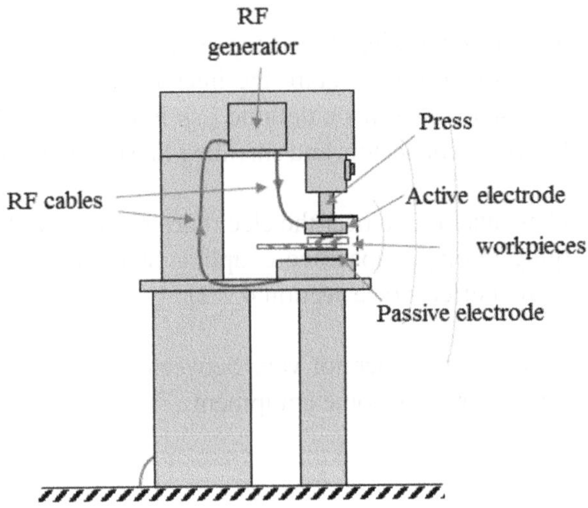


**Figure 2.18.** *Dielectric loss welding press*

### *Microwave ovens*

Microwave ovens use two main frequencies (915 MHz and 2.45 GHz). As for HF heating, these ovens have the advantage of heating the interior of the materials directly, which conduction, convection and infrared heating techniques cannot provide. Surface overheating is avoided. Another advantage of the technique is more homogeneous and rapid temperature rise.

Applications include heating, liquefaction, polymerization, sterilization, pasteurization, drying, dehydration, etc. (non-exhaustive list) Various industrial sectors use this technique: printing, pharmacy, cosmetics, chemistry and food management, material synthesis from electronic components, etc.

Microwave ovens are widely used by the public and cooking professionals. They generate electromagnetic fields at 2.45 GHz (ISM band), absorbed by water molecules. Foods that contain a lot of water will absorb proportionally more waves than dry food. These ovens are first and foremost Faraday cages shielded against electromagnetic radiation for reasons of efficiency, protection of people and electromagnetic compatibility (electromagnetic pollution). The power level cannot be homogeneous in the oven because of its reflecting walls. In the absence of a turning plate, a fan stirs the waves to facilitate a uniform rise in temperature in the food.

Safety systems switch off the radiation when the door is opened. The most vulnerable security element is the joint, which acts as a seal between the door and the oven cavity. This joint or, failing that, the oven, must be replaced in the event of the slightest weakness. The power of kitchen ovens is relatively modest, a maximum of around 1 kW. The maximum authorized leakage level must not exceed 5 mW/cm$^2$ measured at 5 cm [STD 27].

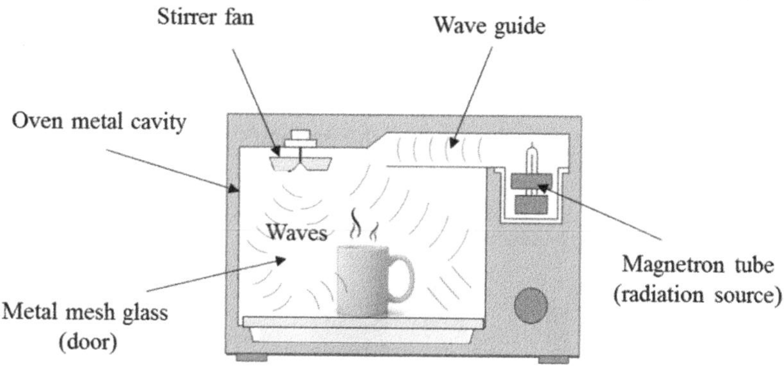


**Figure 2.19.** *Kitchen microwave oven*

Industrial microwave ovens at 2.45 GHz exceed 30 kW of power. The 915 MHz frequency is chosen when the volumes are significant, because the electromagnetic field penetrates more deeply into the matter (skin effect, see section 3.2.3). Some ovens warm up 2 tons of food in less than 1 h (temperature raised to just under 0° to avoid problems relating to defrosting). Rapid heating prevents bacteria from developing and foods from oxidizing.

Low-power appliances are used in research laboratories. There are also vacuum, pressure and controlled atmosphere microwave ovens.

Despite the very high powers of certain ovens – levels with which we could establish radio communication beyond the Moon – the leakage level of the electromagnetic fields remains low. The cavity that forms the Faraday cage attenuates the signal by a factor greater than $10^6$ (i.e. 60 dB).

Very high power microwave heating techniques are reserved for high value added products.

#### 2.2.3.2.2. Medical applications

Electromagnetic radiation is used in the health sector for diagnosis, therapy and surgery. Very high radiation levels are sought for the benefit of the patient, but they may present a hazard to health and safety, including for medical personnel [AND 13b]. The lack of specific regulations for prevention and protection in this domain, other than for MRI and NMR, should be noted.

*Magnet therapy*

Magnet therapy is used to soothe joint or muscular pain and diminish stress. It is an alternative medicine. Low-frequency (ELF) magnetic fields generated by coils are applied to part of the patient's body.

**Figure 2.20.** *Magnet therapy treatment [AND 13b]*

These fields generally have a sinusoidal waveform. The frequencies used are between 1 and 300 Hz; the typical frequency is 100 Hz. The maximum field strength inside the coil is generally between 1 and 10 mT.

### *Transcranial magnetic stimulation*

Transcranial magnetic stimulation (TMS) is a non-invasive technique that stimulates or inhibits the cerebral cortex. The technique consists of applying an intense and rapidly varying magnetic field to induce an electric field in the brain tissues, creating stimulation.

The field is generated by a coil placed on the patient's scalp, through which a current of several kiloamperes flows. Stimulation is created by an isolated pulse or series of pulses, with frequency lower than 1 Hz or lower than 20 Hz. Each pulse lasts less than 1 ms.

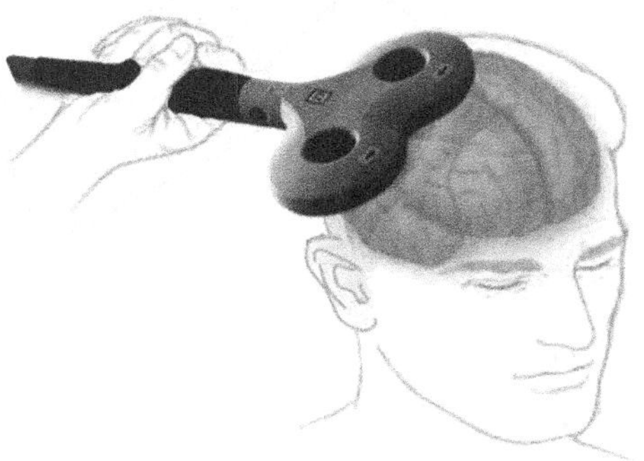

**Figure 2.21.** *Transcranial magnetic stimulation*

The type of applicator depends on the strength and distribution of the desired magnetic field. Magnetic induction may exceed one tesla on contact with the inductor.

### *Diathermy*

Diathermy uses HF electromagnetic fields to heat tissues for therapeutic purposes (muscular trauma, joint injury, etc.). Short waves with 27.12 MHz frequency and microwaves at 915 MHz or 2.45 GHz are usually used.

SW applicators are formed of two electrodes between which an electric field is established (Figure 2.22(a)). Radiation can be continuous, with power of around 300 W, or pulsed, at 1,500 peaks watt. Locally, the field reaches several hundred volts per meter. The patient is typically exposed for 10–20 min. A single applicator with an HF antenna is sufficient for infusing microwaves into tissues (Figure 2.22(b)). The power is limited to around 100 W.

The radiation parameters depend on the body part to be treated. Microwaves can reach only superficial tissues, while SWs cross through them to treat deep joints or muscles (e.g. the hip). Precautions must be taken to avoid the indirect effects of electromagnetic fields (e.g. localized burning due to the presence of metal objects, such as piercings and passive or active implanted medical devices).

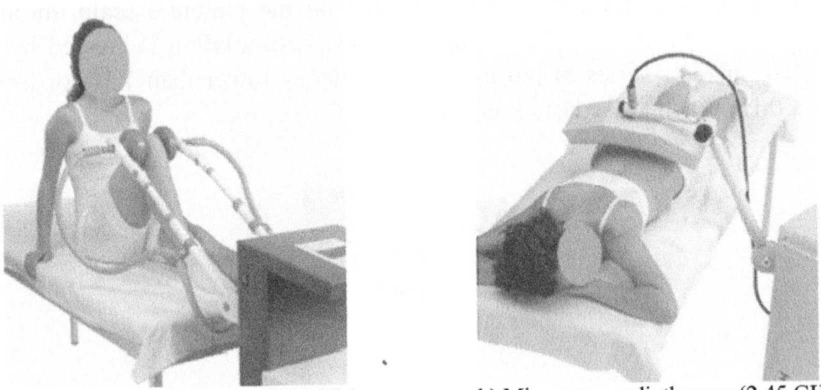

a) Short-wave radiotherapy (27 MHz) b) Microwave radiotherapy (2.45 GHz)

**Figure 2.22.** *Diathermy treatment*

### *Electrosurgery*

Electrosurgery involves applying an HF electric current to the biological tissues of a patient. The current flow between two electrodes creates a temperature elevation proportional to the square of its intensity and to the duration of the application. Coagulation (hemostasis) requires a temperature lower than 100 °C, an incision requires a higher temperature and cauterization requires a temperature of around 500 °C. The current parameters must therefore be adjusted to the surgical operation.

Electrosurgery is practiced in many medical specialties (dermatology, gastroenterology, etc.).

HFs, ranging from 200 kHz to more than 2 MHz, are chosen to avoid stimulating the nervous and muscular tissues and prevent electrolytic processes.

The device can deliver an average power of 300 W and a peak voltage of 4 kV. The voltage depends on the gap between the electrodes as well as the conductivity of the tissues. The current of a few amperes is pulsed (a series of bursts). The electromagnetic radiation level depends on the type of device:

– monopolar devices use two separate electrodes (scalpel and return contact);

– bipolar devices have two electrodes fixed together in the shape of pliers.

Radiation is greater for the monopolar versions (Figure 2.23) due to, on the one hand, the length of the electric circuit and, on the other hand, the large gap between the electrodes and between the wires.

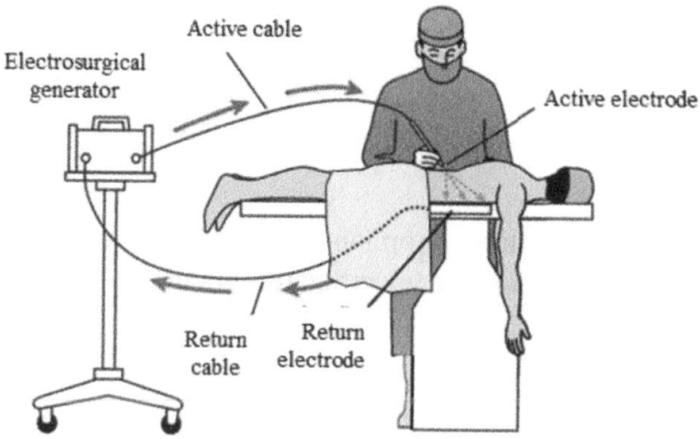


**Figure 2.23.** *Electrosurgery: monopolar scalpel*

The use of a twisted pair (two wires rolled into a helix) or, better still, of a coaxial cable and close electrodes, reduces the radiation (bipolar devices, Figure 2.24).

The risks caused by the electromagnetic field exist for the patient outside the surgical area and for the surgeon. Adverse direct effects may occur, just as they can near all sources that use these frequency ranges (Chapter 5). A working method must be respected to avoid all risks.

Indirect effects are also possible. An induced current may circulate through a capacitive effect between the device, the earth and the patient if the latter is not correctly insulated, or through other instruments. Metal objects present in areas where there are strong fields may charge and create contact currents. The appearance of electrical arcs according to the device may ignite volatile products, such as disinfectants.

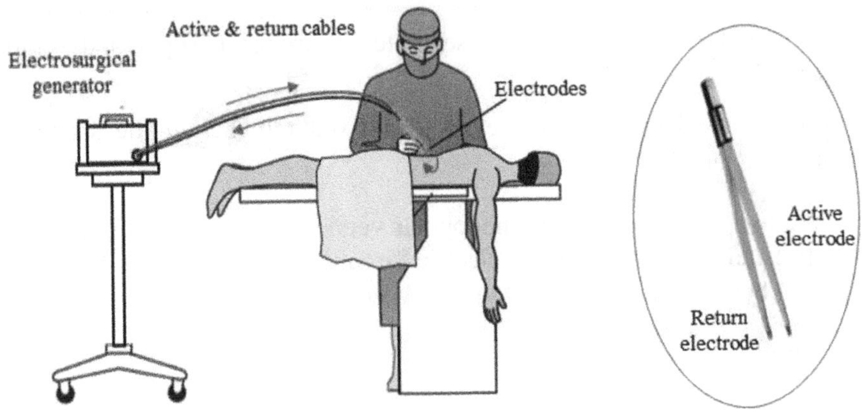


**Figure 2.24.** *Electrosurgery: bipolar scalpel.*

Furthermore, electrosurgery appliances may disturb other equipment that is sensitive to electric fields (e.g. electrocardiograph).

### 2.2.4. *Summary of sources*

The main sources of radiation to which the public is exposed are the power grid, the appliances that connect to it and telecommunication devices. These also include radio and television transmitters and mobile telephony base stations. Although the radiation power of television channels has been reduced by the shift to digital (DTT), the power radiated by the transmitter remains the same overall due to the increased number of stations. The powers radiated by mobile telephony base stations are weaker, but they are numerous and close together.

The introduction of new telephone networks, such as 4G, does not significantly increase the average level of electromagnetic fields in the environment.

The overall trend since radiofrequencies were discovered has been to use higher and higher frequencies.

The number of sources near our bodies is constantly increasing. Wi-Fi and Bluetooth functions are built into mobile telephones. These low-power transmitters work intermittently. The ways these sources are used is also changing, since a telephone is also used as a messaging device, a games console and a television. The exposed area of the body is therefore also changing.

Generally, the radiation from household appliances, whether they are HF or low frequency, is weak at a few centimeters' distance.

The strongest sources encountered by the public are those used for diagnosis and therapy. Examples include MRI, transcranial magnetic stimulation, radiotherapy and electrosurgery. Health personnel are also exposed to these sources.

Overall, people are exposed to modest field levels. Workers may be exposed to specific sources radiating high fields in areas that are normally inaccessible to the general public. Some examples of these sources are industrial heating equipment (induction oven, microwave oven, etc.), welding equipment (arc, resistive or HF), metal treatment and telecommunication equipment (at close proximity to telecommunication antennae, radars, etc.).

A list of equipment radiating a weak field can be found in Appendix 1 (Table A1.1). Tables A1.2, A1.3 and A1.4 present a list of equipment and situations where fields are potentially high.

# 3

# Biophysical Mechanisms

This chapter introduces the interaction mechanisms between electromagnetic fields and biological systems, providing an explanation for biological effects. These help justify the limits of human exposure to electromagnetic fields, as per the guidelines, standards and recommendations.

Describing these mechanisms and their effects is complex, because it involves several scientific disciplines, such as physics, biology and medicine. Electric, magnetic and electromagnetic fields differ, since they act according to different physical phenomena on exposed biological tissues and organisms.

The first part of this chapter deals with the interaction between fields and matter in general. The second part focuses on the specific interaction of fields with biological tissues. Both occur at the microscopic level and reading about them will facilitate understanding of macroscopic coupling phenomena, which are introduced in the third part of the chapter.

Biological effects, including those on health, are discussed in Chapter 4.

When electromagnetic fields penetrate tissues and when their energy is transformed into heat, the significant biological effects which may occur are called "thermal effects". Effects produced through other mechanisms are called "non-thermal effects". These effects can have consequences for health well-being, depending on the frequency.

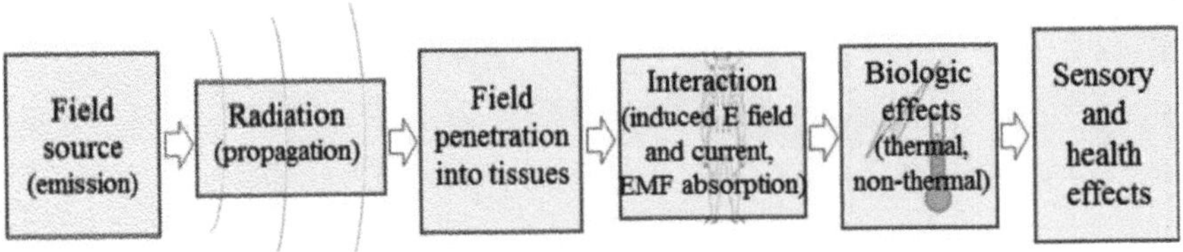


**Figure 3.1.** *Complete process of the mechanisms that may lead to effects on the health and well-being*

Figure 3.1 summarizes the complete process of the various mechanisms.

The main recognized phenomena are described in linear behavior for frequencies up to 300 GHz. This excludes interactions due to optical and ionizing radiation (section 1.4) and irreversible effects.

## 3.1. Interactions with matter

Matter consists entirely of electrically charged particles. The electron is the smallest charged particle, which is negative. The atomic nucleus, made up of protons and neutrons, is a positively charged element. In its natural state, matter is generally neutral and at rest, because the charges are balanced. In reality, this is true in appearance only, as on a microscopic level particles move randomly in all directions across very small distances, under the effect of thermal agitation. The sum of these movements is zero.

Materials with many free charges are called "conductors". Those that do not have enough of these charges are called "dielectrics" or, more commonly, "insulators". They cannot conduct electricity. Biological tissues have both properties, making them different from the materials usually used in electrical engineering.

When an electromagnetic field is applied, matter is subjected to three phenomena:

- interactions due to low-frequency electric fields;
- interactions due to low-frequency magnetic fields;
- energy absorption from high-frequency electromagnetic fields.

The electrical properties of matter, as well as the concepts of conduction currents, displacement currents and dielectric losses, are discussed here, with reference to conductive and dielectric materials. These concepts are necessary for understanding the complexity of the properties of tissues and their interaction with fields, as explained further.

### 3.1.1. *Matter exposed to an electric field*

Electric fields exert a force on the electrically charged particles present in matter. This force (statistically) moves free particles and attracts bonded particles, although it cannot move them beyond the atomic level. When these particles are unbalanced, they produce their own electric and/or magnetic field. This phenomenon is reversible, i.e. in the absence of any field, the matter is once more at rest, in a neutral electric state.

Thus, matter is subject to two principal phenomena in the presence of an electric field:

– drift of free charges, i.e. the conduction current;

– polarization and/or orientation of bonded charges. This phenomenon creates a current called a displacement current, although there are no charge migrations, and dielectric losses.

#### 3.1.1.1. *Conduction current*

The application of an electric field on a medium exerts a force on the charged particles. Under its action, the free particles permanently subjected to thermal agitation move statistically (or drift) in the direction of the field, macroscopically forming an electric current. This current is due to electrons in good conductors and to ions in liquid electrolytics (solutions of molecules separated into oppositely charged ions).

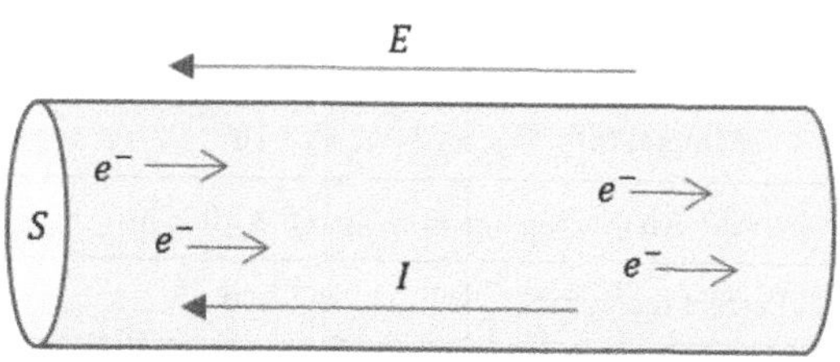


**Figure 3.2.** *Electric current in a conductor when an electric field is applied*

$E$ is the electric field in the conductor, $e^-$ represents an electron and $I$ is the current (proportional to the electron flow). The electric quantities are vectorial, i.e. orientated in space.

Electrical conductivity, symbolized by the letter σ (*sigma*), is the ability of a medium to allow a certain number of electric charges to move freely. Its unit is the siemens per meter ($S \cdot m^{-1}$). The more numerous and mobile the charges, the better the conductivity. The conductivity of metals is high: greater than $10^{+4}$. It is lower in electrolytes and almost zero in the usual dielectrics ($<10^{-10}$). The examples in Table 3.1 show that this physical quantity has one of the widest ranges at 20°C (68°F).

| Media | Conductivity ($S \cdot m^{-1}$) |
|---|---|
| Perfect conductor | $\infty$ |
| Silver | $62.5 \times 10^6$ |
| Copper | $58 \times 10^6$ |
| Aluminum | $36 \times 10^6$ |
| Carbon | $25 \times 10^3$ |
| Electrolytes | 1 to $10^2$ |
| Biological tissues | $10^{-4}$ to $10^{+2}$ |
| Seawater | 5 |
| Drinking water | $10^{-3}$ |
| Semi-conductor | $10^{-4}$ to 1 |
| Distilled water | $10^{-6}$ to $5 \times 10^{-4}$ |
| Glass | $10^{-14}$ to $10^{-10}$ |
| Mineral oil | $10^{-12}$ |
| Air | $5 \times 10^{-15}$ |
| Perfect dielectric | 0 |

**Table 3.1.** *Conductivity of various media*

Conductivity σ enables current density $J_C$ (current crossing a unit surface: $J_C = \frac{I}{S}$; see Figure 3.2) to be linked to the electric field $E$ through Ohm's law:

$$J_C = \sigma . E \ (\text{A·m}^{-2}) \qquad [3.1]$$

$E$ is in V·m$^{-1}$. Conductivity is the inverse of resistivity. The flow of an electric current in a resistive medium produces heat. This is the Joule effect.

### 3.1.1.2. *Displacement current, bonded charges polarization*

A dielectric medium cannot, by definition, conduct a current. However, electric fields penetrate it and interact with the bonded charges. These interactions give it electrical properties.

Electric charges move slightly away from their equilibrium position when a field is applied. These limited movements allow electric dipoles to appear. Positive charges move in the direction of the field and negative charges in the opposite direction. The medium becomes polarized. There are three main microscopic polarization mechanisms and one macroscopic polarization mechanism:

– *Electronic polarization*: Under an electric field, the electron cloud of an atom becomes distorted and does not remain centered on its nucleus (Figure 3.3). This effect produces an induced polarization, because it ceases when the field disappears (there is also atomic polarization due to the movement of the nuclei, but this is weak compared to electronic polarization).

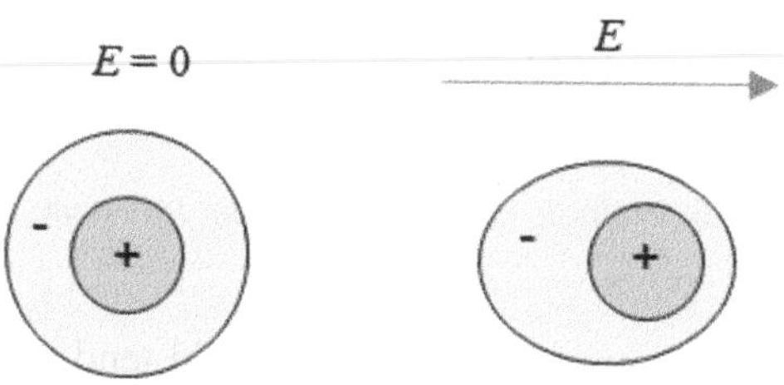


**Figure 3.3.** *Representation of the electronic polarization of an atom*

– *Ionic or molecular polarization*: This phenomenon is caused by the relative movements of oppositely signed ions within one molecule under the action of an external field (molecules are bonded to each other and cannot move). This also produces an induced polarization (Figure 3.4).

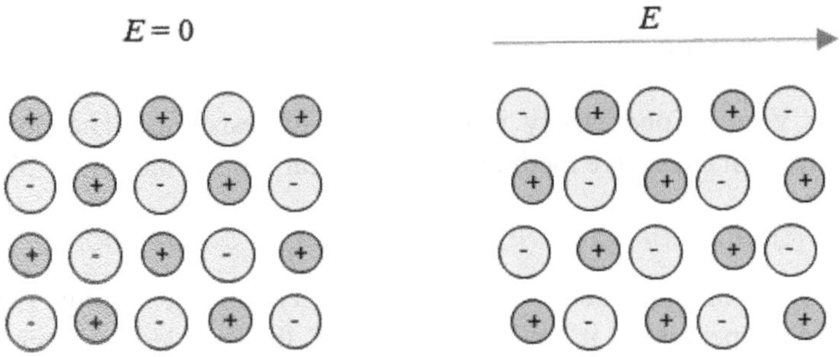


**Figure 3.4.** *Representation of ionic polarization*

– *Orientation polarization*: This type of polarization occurs in the presence of polar molecules. These are molecules that naturally have a positive charge center separate from their negative charge ones. This is the case for ionic molecules and asymmetrically structured molecules, which are formed of different and unaligned atoms. In such a set-up, the most charged nuclei attract more electrons, creating a permanent microscopic dipole, as shown by the water molecule in Figure 3.5.

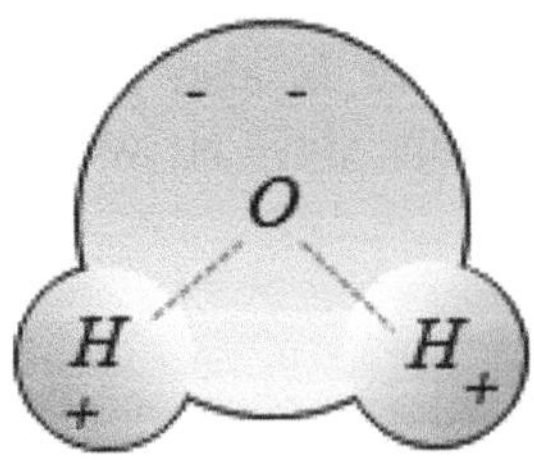


**Figure 3.5.** *Water is composed of polar molecule ($H_2O$)*

The forces exerted by an external electric field tend to orientate this type of molecule, as well as electronic and ionic polarization. In the absence of an electric field, these dipoles are orientated randomly, and the overall resulting field is zero (Figure 3.6). This polarization does not occur in solid media, where molecules remain strongly bonded.

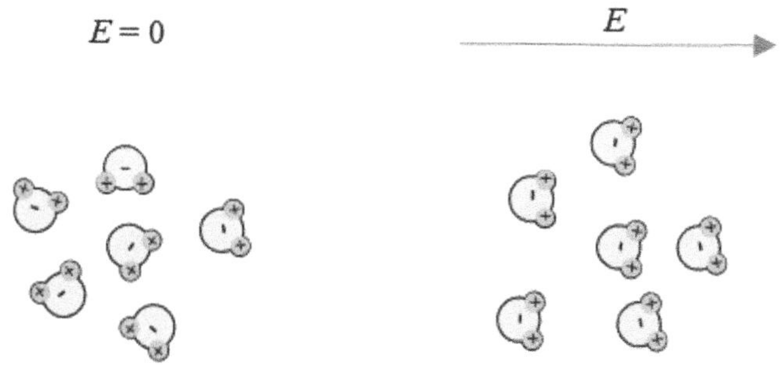


**Figure 3.6.** *Representation of the orientation polarization of polar molecules*

– *Macroscopic polarization or space charge polarization*: A polarization phenomenon occurs macroscopically in heterogeneously structured media with insulated conductive areas. Their free charges migrate distances restricted to an area when an electric field is applied. The imbalance of charges polarizes the conductor, inducing an internal electric field opposite to the external electric field. This polarization is established more slowly than those discussed previously, as the distances to travel are longer.

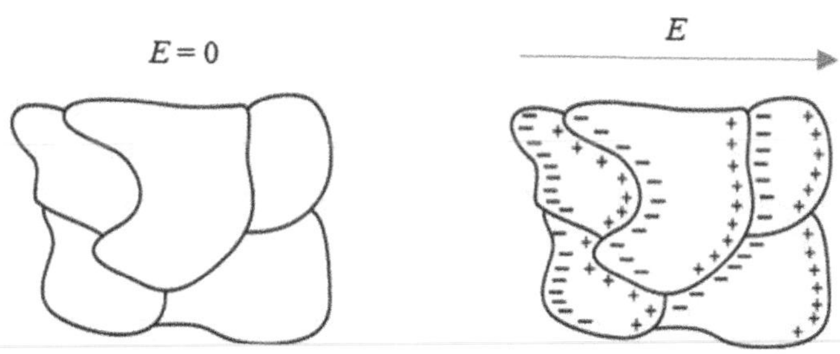


**Figure 3.7.** *Representation of macroscopic polarization*

EXAMPLE OF THE SPACE CHARGES POLARIZATION.– When a biological cell or tissue is subjected to an external electric field, its ions are distributed under its insulating membrane. The polarized structure induces an internal electric microfield in opposition to the external field. This cell does not move, because it is globally neutral. Charges outside the cell will be attracted to the interfaces by the trapped charges. Induced electric microfields form chains in this way through the whole medium (Figure 3.8).

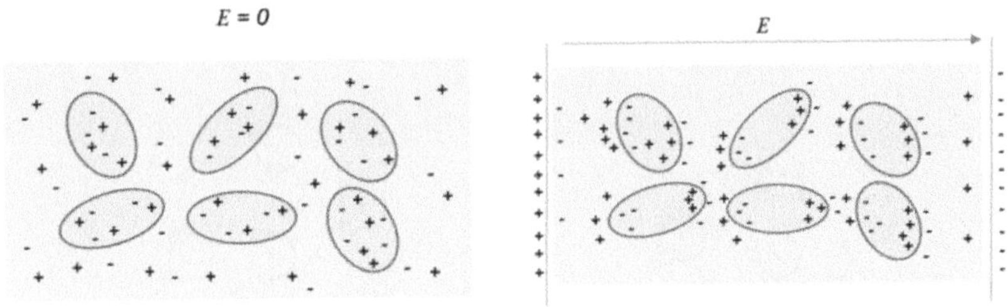


**Figure 3.8.** *Representation of the space charge polarization*

Polarization mechanisms depend on the medium composition. At the lowest frequencies, they combine to form an overall polarization. They disappear one after the other as the frequency increases, due to the inertia of the particles and the viscosity of the medium. Orientation and space charge polarizations are the most significant mechanisms up to around 20 GHz.

#### 3.1.1.2.1. Displacement current and electric displacement field

The movement of charges over infinitesimal distances along with an external alternating electric field produces a macroscopic current called a displacement current, although no particle actually migrates. This current exists only in the presence of a time-varying electric field. An electric displacement field is associated with this current. It is the resulting local field, weaker than the external field, which polarizes the charges. This polarization is proportional to the electric field. It is not present if there is no external field.

#### 3.1.1.2.2. Dielectric losses

Dielectric losses also exist. These are the result of particle friction effects caused by the viscosity of the medium. They are very weak at relatively low frequencies and high frequencies, domains where polarization phenomena disappear. Therefore, dielectric losses are centered on a resonance or relaxation frequency particular to each particle. They result in an attenuation of the electric field penetrating the medium. The lost energy is transformed into heat.

Dielectrics have defects, such as impurities. Therefore, a very weak conduction (or leakage) current independent of the frequency also flows through them. This current also causes losses, which are added to the dielectric losses. A medium is called "dielectric" if the displacement current is much greater than the conduction current.

Conduction is defined as "effective" if it is representative of these losses in their entirety.

These phenomena explain why even dielectrics have electrical properties. The physical quantity associated with dielectric properties is called dielectric permittivity. This indicates the ease with which electric charges contribute to the various polarizations. It is the relationship between the electric displacement field and the electric field applied to the medium.

Dielectric permittivity is measured in farads per meter ($F \cdot m^{-1}$) and has the symbol ε (*epsilon*). Relative permittivity $\varepsilon_r$, also called dielectric constant, a value normalized by that of the vacuum, is usually used. We have:

$$\varepsilon = \varepsilon_r . \varepsilon_0 \quad [3.2]$$

where $\varepsilon_0$ is the permittivity constant of the vacuum, with the value:

$$\varepsilon_0 = \frac{1}{c^2 . \mu_0} \ F \cdot m^{-1}$$

i.e. around $8.85 \times 10^{-12}\ F \cdot m^{-1}$ or $\frac{1}{36 \cdot \pi} . 10^{-9}\ F \cdot m^{-1}$

with $\mu_o = 4 \cdot \pi \cdot 10^{-7}\ H \cdot m^{-1}$ and $c$ = speed of light ($\approx 3 \times 10^8\ m \cdot s^{-1}$).

$\varepsilon_r$ is a dimensionless quantity. It is greater than or equal to 1 for ordinary materials.

Dielectric permittivity diminishes naturally and almost monotonically with the frequency, as the polarization phenomena fade. It may be extremely variable at certain frequencies (Figure 3.9).

Permittivity allows the displacement current density $J_D$ to be linked to the electric field. For a sinusoidal field $E$ with pulse $\omega = 2 \cdot \pi \cdot f$, we have:

$$J_D = \omega \cdot \varepsilon_r \cdot \varepsilon_0 \cdot E \quad [3.3]$$

This relationship indicates that the displacement current may increase with the frequency, although the permittivity ($\varepsilon_r$) may decrease.

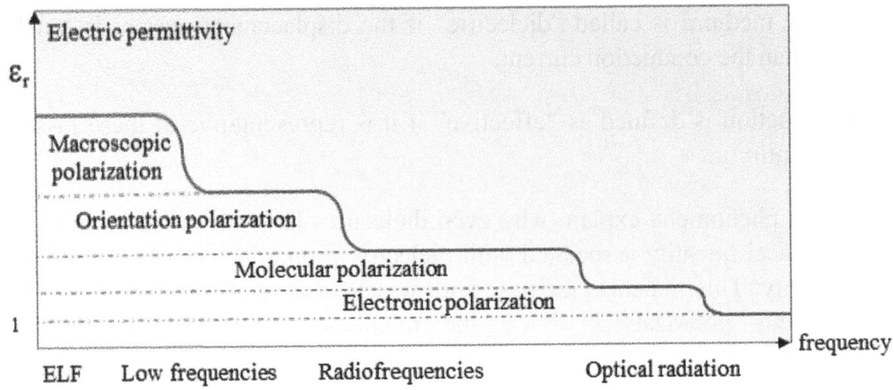


**Figure 3.9.** *Decrease in electric permittivity with frequency*

A usual factor which makes it possible to quantify dielectric losses is associated with permittivity. It is *tan(δ)* (*tangent delta*). This is the ratio between the effective conduction of the medium (representing ohmic losses plus losses due to friction) to the medium polarization ($\omega.\,\varepsilon_0.\,\varepsilon_r$). This ratio is valid only for a sinusoidal field. To simplify the expression notation, the polarization and conduction current phenomena are combined in a single parameter, complex permittivity.

In a material, the total current density, calculated using equations [3.1] and [3.3], becomes:

$$J_T = J_C + i.J_D = (\sigma + i.\ \omega.\varepsilon_0.\varepsilon_r).E \quad [3.4]$$

The letter $i$ indicates that $J_T$ is a complex number ($i = \sqrt{-1}$). This means the conduction current ($J_C$) and displacement current ($J_D$) are not in phase. Conductivity $\sigma$ combines all medium losses.

This phenomenon can be represented using a current applied to a resistor and capacitor connected in parallel (Figure 3.10). The conduction current circulates through the resistor, while the displacement current enters or leaves the capacitor. No charge passes through the latter and it exists only if the current is variable over time. This can be illustrated by a pair of parallel electrodes with surface area $S$, separated by a material sample of thickness $l$, conductivity $\sigma$ and relative permittivity $\varepsilon_r$.

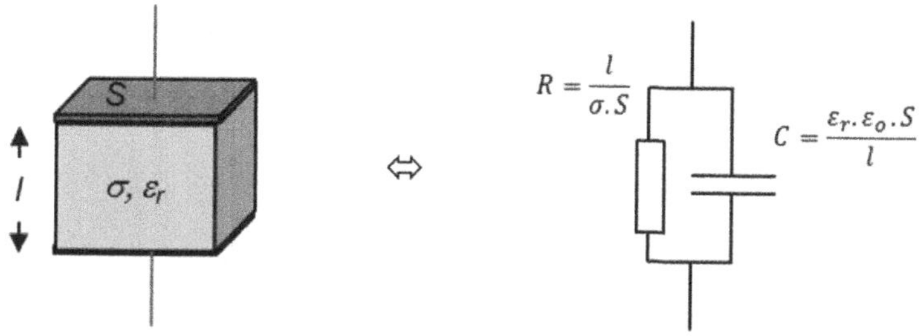


**Figure 3.10.** *Dielectric sample and equivalent electric circuit made up of a resistor and capacitor in parallel*

The conduction current is dominant when the frequency nears 0 and the displacement current may become greater at high frequencies.

Table 3.2 summarizes the relative dielectric permittivity of a few media.

| Medium | Relative permittivity at 20°C (68°F) |
|---|---|
| Vacuum | 1 |
| Mineral oil | 2.2 |
| Paper | 3–7 |
| Glass | 4–7 |
| Ethyl alcohol | 24 |
| Water ($f$ < 10 GHz) | 78.4 |

**Table 3.2.** *Relative permittivity of various media*

The vacuum is the perfect "dielectric", since it is not subject to polarization phenomena.

From a macroscopic point of view, dielectric permittivity is involved in the study of the propagation of electromagnetic waves. It allows the reflection and transmission coefficients between two media and characterizes the attenuation of waves by absorption to be determined.

The polarization of the medium by an electric or electromagnetic field results physically in energy storage. The result is a reduction in propagation speed ($v = c/\sqrt{\varepsilon_r}$) and therefore an increase in wavelength. In optics and wave propagation at radioelectric frequencies, permittivity allows the refractive index to be determined.

### 3.1.2. *Matter exposed to a magnetic field*

The ability of a material to modify a magnetic field, i.e. to divert or strengthen the magnetic flux lines, is characterized by its magnetic permeability denoted by the Greek letter $\mu$ (mu). This is expressed by the product of the vacuum permeability $\mu_0$ and the relative permeability $\mu_r$ in the same way as for electric permittivity:

$$\mu = \mu_r \cdot \mu_0 \quad [3.5]$$

$\mu_0$ is a universal physical constant ($\mu_0 = 4 \cdot \pi \cdot 10^{-7}$ H·m$^{-1}$). Vacuum permeability[1] is sometimes called the magnetic constant.

The permeability of ferromagnetic materials, such as iron, is high ($\mu_r \sim$ 10,000) and can be much greater for some alloys. It may depend on the frequency and magnetic field strength.

Magnetic induction and magnetic fields can be linked through permeability [1.6].

Matter can be divided into two categories: magnetic and non-magnetic. Magnetic matter consists of magnetic dipoles (sensitive to magnetic fields), unlike non-magnetic matter.

#### 3.1.2.1. *Non-magnetic matter exposed to a magnetic field*

A magnetic field interacts with a non-magnetic material only if it is conductive. The external magnetic field then induces an electric field, causing electric currents to circulate through a closed circuit (loop). These are eddy currents: they are higher on the edge of the exposed surface and zero at its center. They increase the temperature through the Joule effect. This is the principle of induction heating. These currents do not exist in insulating material ($\sigma \approx 0$).

1 The vacuum has electric and magnetic properties ($\varepsilon_0$ and $\mu_0$), just like materials. These properties make it possible to calculate propagation phenomena, in particular at the interface of another medium. It is treated as a specific case of matter in electromagnetism.

The induced current in turn generates a magnetic field opposite to the one that created it (Lenz's law), like all moving electric charges. The induced field is lower by several orders of magnitude than the inducing field in weakly conductive materials.

From a microscopic point of view, the strength of the induced electric field is proportional (Faraday's law [1.5]):

– to the surface area of material subjected to the magnetic field;

– to the magnetic flux density, also called magnetic induction, the product of the magnetic permeability of the material and of the magnetic field ($B = \mu.H$ [1.6]);

– to the time variation of the magnetic flux or to its frequency for a sinusoidal field.

The induced current is proportional to the electric conductivity of the material.

The electric field is at its maximum when the magnetic induction is perpendicular to the surface.

A magnetic field also exerts a force on electric charges, provided that they are moving. The strength of this force is proportional to the electric charge, to its movement speed and to the magnetic induction level $B$ (magnetic flux density). This force is perpendicular to both the direction of the speed and the direction of the field (Lorentz's law). Its mathematical formula is:

$$F = q.v \times B \qquad [3.6]$$

$F$ is expressed in Newton and $B$ in tesla, $q$ is the charge in coulomb and $v$ is the speed in $m{\cdot}s^{-1}$. Magnetic induction acts only on the perpendicular component of speed. The force is at its maximum when the charge moves perpendicular to the induction and zero when the charge follows the induction lines. The movement speed is not affected.

### 3.1.2.2. *Magnetic matter exposed to a magnetic field*

Some media are also sensitive to magnetic fields, because they have magnetic dipoles. The resulting mechanisms are not discussed in this book, because they do not exist in a significant way in biological tissues.

However, magnetic biological microstructures exist (formed from magnetosomes). They are present in many living organisms, such as humans, but our perception of them has been lost during our evolution.

### 3.1.3. *Summary of matter*

Table 3.3 summarizes the main electrical characteristics of materials.

| | Key points of materials | |
|---|---|---|
| Conductor | Material with lots of free electric charges | The charges are electrons and/or ions |
| Electrical conductivity (σ) | Characterizes the mobility and number of free charges in conductors | Links the electric field and the current density through Ohm's law |
| Dielectric | Material subject to microscopic polarization phenomena, causing displacement currents and losses | Often called insulators |
| Electrical permittivity (ε) | Main characteristic of dielectrics | Makes it possible to calculate propagation in media (speed, wavelength, reflection coefficient) and losses through absorption |
| Magnetic permeability (μ) | Relative permeability corresponds to a multiplying magnetic field factor | Magnetic permeability characterizes a material's ability to modify a magnetic field |

**Table 3.3.** *Electrical properties of common materials*

## 3.2. Interaction with biological tissues

### 3.2.1. *Electrical characteristics of biological tissues*

Biological tissues are both conductors and dielectrics. They are constituted of free electrons and ions that can migrate in response to the application of an electric field. Furthermore, they have the properties of dielectric materials with polarization mechanisms. Some ions can be trapped on both sides by the membrane of a tissue or organ, reducing the quantity of charges that are able to move. This phenomenon produces space charge polarizations that are much greater than that which occur at the cellular level.

The electrical properties of tissues explain their behavior toward external fields. They also make it possible to predict the strength of induced electric fields and the attenuation of electromagnetic fields propagating there.

Although the structure of tissues is complex, they can be considered homogeneous from a macroscopic point of view. Conductivity and permittivity have therefore been studied for each of them (or almost). This has facilitated the development of a parametric tissue model valid over a large frequency range [GAB 96d]. The parameters are highly diverse, because they depend first and foremost on the water content of the tissue, which is not uniformly distributed in the various extracellular media (an adult body is 70% water, which is the major constituent part of most tissues).

Blood, cerebrospinal fluid, the lungs and other bodily fluids are tissues with a very high water content (>90%). They are composed of cells suspended in a liquid medium. Cellular and extracellular media contain ions. Muscles and internal organs, such as the liver, brain and kidneys, are tissues with a high water content (>80%), while adipose tissues and bones belong to the category of tissues with a low water content (<50%). Bones consist of cells fixed in a solid medium. Skin is formed of adjacent cells containing relatively little water.

Figures 3.11(a) and 3.11(b) compare the electrical properties of blood (which has a very high water content), muscle (which has a high water content) and fat (which has low water content) in the 10 Hz to 100 GHz band [GAB 96a, GAB 96b, GAB 96c].

The electrical properties of biological tissues can be seen clearly on these curves. It is noted that there are:

– notable differences in properties between the tissues;

– a significant variation in the conductivity and permittivity around certain frequencies. These variations are called dispersions;

– permittivities diminishing as the frequency increases;

– conductivities increasing with the frequency;

– permittivities much greater than that of classic materials, in particular at the lowest frequencies;

– conductivities between conductors and insulators. They depend on the water content (conductivity includes ohmic conductivity and dielectric losses).

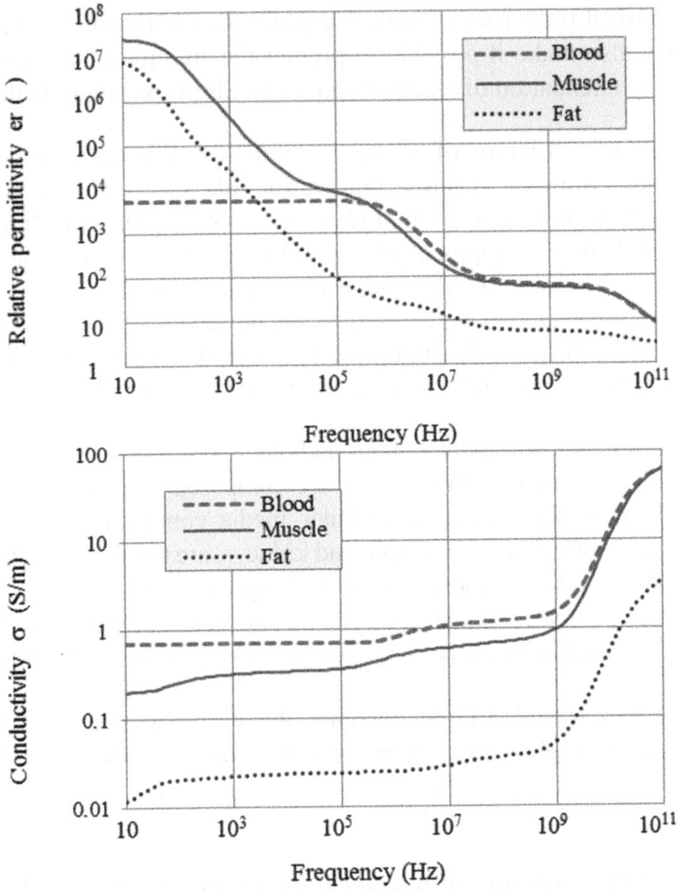


**Figure 3.11.** *a) Electrical conductivity of some biological tissues; b) electrical permittivity of some biological tissues*

Permittivity dispersions are explained by the disappearance of polarization mechanisms with the frequency. The three most significant dispersions can be clearly distinguished in muscles [WHO 93].

*At low frequencies (f < 100 kHz)*, understanding of the permittivity dispersion of low-frequency tissues remains incomplete. Nevertheless, the cycle is slow enough to allow the charges to accumulate on both sides of the cell membrane. This is related to an ionic diffusion phenomenon and the polarization of large membrane structures. The capacitive effect produced causes a high relative permittivity, exceeding around 10 million under

100 Hz. This phenomenon, which is dominant at low frequencies, does not exist in blood, which is a barely structured liquid tissue. Many dispersion frequencies may overlap, forming an overall dispersion whose extent depends on the composition of the tissues. *Dispersion in this area is called* $\alpha$ *dispersion.*

Furthermore, given that cells do not conduct a current at low frequencies (their membrane acts as an insulator), the conductivity of tissues depends only on the conductivity of the extracellular medium and the volume density of the cells. Thus, tissues with a high liquid electrolytic (water) content and a low cell density are the best conductors (e.g. blood).

Although permittivity is high, the displacement current proportional to the frequency may be negligible compared to the conduction current ($\sigma >> \omega\varepsilon_r\varepsilon_0$). The conduction current is due solely to the conductivity of the extracellular medium.

With an average conductivity of close to 0.2 $S{\cdot}m^{-1}$, tissues can be considered good conductors at these frequencies in comparison to the air.

*In the radiofrequency (RF) range from 100 kHz to 100 MHz*, conductivity improves with frequency, because the current circulates in both the cellular and extracellular media, as the capacitive effect introduced by the membranes diminishes. This also causes a decrease in permittivity. Orientation polarization mechanisms always persist, due to dipolar elements, including proteins and macromolecules. *Dispersion in this area is called* $\beta$ *dispersion.*

*Beyond a few gigahertz*, the relative permittivity of the tissues converges around 80, a value corresponding to water. *This is* $\gamma$ *dispersion.* Only the orientation polarization of water molecules in tissues with high water content is visible, because other molecules are too inert. In this frequency domain, the conduction current is approximately equal to the displacement current. Tissues are considered high-loss dielectrics.

*At frequencies over 20 GHz*, the orientation polarization mechanisms fade, allowing ionic and electronic polarizations to appear gradually. Conductivity increases greatly.

The structure and composition of biological tissues explains the disparity of their electrical properties and variations. In practice, physical and physiological factors also influence these properties. Examples include hydration rate, physiological state and, to a lesser degree, temperature. Strictly speaking, the orientation of fibrous tissues in relation to the direction of the field should be taken into account (the electrical properties of many tissues are anisotropic, i.e. they may vary depending on the direction of the fibers).

NOTE.– Cells consist of a conductive interior insulated by a membrane, which may, however, allow ions to pass through under specific electric fields conditions. Electrochemotherapy uses this property [PER 10].

### 3.2.2. *Summary of biological tissues*

Table 3.4 summarizes the main properties of biological tissues.

From this summary, we can deduce that interactions between fields and human bodies depend on frequency.

| Key points of biological tissues |
|---|
| Biological tissues have different properties to those of the materials used in the domain of electricity or electronics. |
| They do not have measurable magnetic properties. |
| Each tissue has its own electrical properties. |
| Significant differences in properties exist between tissues. |
| Tissues vary significantly in conductivity and permittivity at certain frequencies. |
| Tissues are considered good conductors at low frequencies, although their permittivity can be very significant. |
| Tissues are therefore considered a dielectric loss medium at high frequencies. |
| From a microscopic point of view, tissues are formed from cells with similar electrical properties. They are therefore distinguished mainly by their extracellular medium. |
| From a macroscopic point of view, tissues are considered homogeneous, although their composition and structure are complex. |
| The electrical properties of tissues make it possible to determine how a field or wave penetrates tissues and beyond in a human body. |
| The electrical properties of tissues are used to precisely model the human body, for research and medical diagnoses or to assess the electromagnetic fields exposure level. This is the domain of numerical dosimetry. |

**Table 3.4.** *Electrical properties of biological tissues*

### 3.2.3. *Penetration of fields in biological tissues*

Although conductivity and permittivity allow the behavior of a tissue subjected to a low-frequency electric or magnetic field to be characterized, they also make it possible to predict how tissues will react to a high-frequency electromagnetic field.

When an electromagnetic incident wave changes its propagation medium, a reflected wave and a refracted wave (transmitted with a deviation in direction) are created at the interface. We can experience this by directing a light ray toward a glass surface.

Figure 3.12 illustrates this phenomenon for an electromagnetic plane wave propagating through the air toward a planar medium with large dimensions in relation to wavelength.

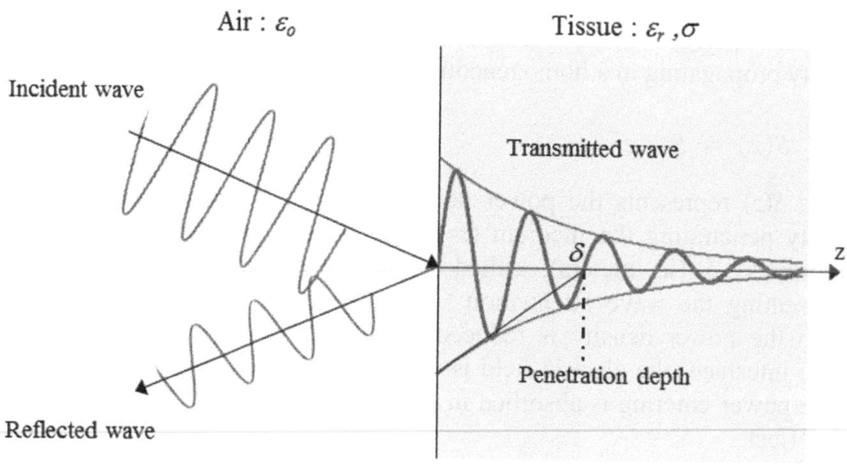


**Figure 3.12.** *Representation of a wave propagating between two media with different dielectric properties*

The field level transmitted in the medium is the (vectorial) difference between the incident field level and the reflected field level. The reflection coefficient is determined by the ratio between the reflected wave and the incident wave. This coefficient depends on the angle of incidence, the difference in the electrical properties and the polarization of the wave (orientation of the electric field plane, section 1.3.2).

Without going into mathematical demonstration, the more different the characteristics of the media, the more significant the reflection, leading to a weaker wave being transmitted. At low frequencies, where biological tissues are essentially conductors, while the air is only dielectric, the electric field is almost totally reflected.

At very high frequencies, tissues become dielectric (with losses) and reflect more than half of the wave (reflection coefficient >50%). This means that less than half of the incident wave penetrates the tissues (half of an electric field corresponds to a quarter of the power).

As in optics, the wave propagation speed is modified when it changes medium. The speed is reduced by a factor of $\sqrt{\varepsilon_r}$ in tissues in relation to propagation in the air.

Furthermore, a medium with a certain degree of conductivity will absorb part of the energy transmitted, causing the wave to be attenuated. The power density propagating in a homogeneous tissue decreases exponentially:

$$S(z) = S_0 \, . \, e^{-\frac{2z}{\delta}} \quad [3.7]$$

where $S(z)$ represents the power density at a depth of $z$ meters; $S_0$ is the density penetrating the medium ($z = 0$); $e \approx 2.718$ is the base of natural logarithms. $\delta$ (in meters), called "penetration depth", is a parameter representing the wave attenuation (Figure 3.12). It indicates the depth at which the power density is reduced to around 14% from its level at the inside interface (the electric field is reduced to 37%). This means that 86% of the power entering is absorbed in this thickness. This effect is also called skin effect.

The penetration depth is related to the conductivity and permittivity of the medium, as well as to the frequency, by a complex formula. It is zero for perfect conductors. For other materials, it decreases when the frequency increases. Figure 3.13 illustrates this thickness for the three previously mentioned tissues (blood, muscle and fat).

Penetration depth is negligible at frequencies lower than 10 MHz. The fields penetrating the body reach all the tissues, because they are only very slightly attenuated. Fat absorbs the fields less than other tissues do, since it is less conductive. The combined effect of the frequency and the increase in conductivity makes the penetration depth decrease rapidly. Its average value is 2 cm at 3 GHz, and absorption becomes superficial.

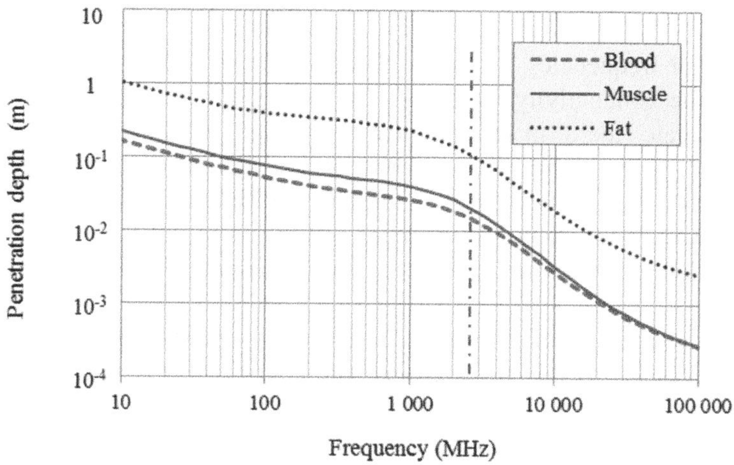


**Figure 3.13.** *Penetration depth between 10 MHz and 100 GHz for various tissues (very high, high and low water content). The vertical dashed line corresponds to the microwave ovens and Wi-Fi frequency*

At low frequencies, the wave is greatly reflected, but the penetrating wave is barely attenuated by the tissues. At high frequencies, it is relatively less reflected, but the penetrating wave is greatly attenuated.

Computer-aided engineering (digital simulation) can be used to precisely quantify the phenomena, taking into account complex forms with limited dimensions or non-homogeneous media (such as human bodies).

The behavior of an electromagnetic field on the boundary between two media is determined by the continuity equations derived from Maxwell's equations. It can notably be deduced from this that the electric field is distorted in the vicinity of a body. It is practically perpendicular to the external surfaces, regardless of its initial orientation (Figure 3.14).

In a static state, the ratio of the internal field to the external field is equal to the ratio of the conductivities (air/tissues). Also, the internal field is zero, because the conductivity of the air is considered to be zero and there are no permittivity-related phenomena.

More generally, the low-frequency internal electric field is always lower by several orders of magnitude than the external electric field. The conductivity of the tissues protects them in this range of frequencies.

At higher frequencies, dielectric phenomena can no longer be negligible, although the conductivity increases. The waves penetrate the human body and propagate there in a complex way. The waves are reflected or transmitted, to an extent that depends on the dielectric nature, forms and orientations of the tissues and body organs. Multiple reflections may lead to standing wave phenomena (resonance) or to the waves being focused on a limited volume, which may cause a higher than average local field level.

Only numerical simulation using an electrically representative anthropoid model makes it possible to determine the energy absorbed by the various parts of the body with sufficient precision. This method also allows non-uniform radiation sources to be taken into consideration.

At frequencies higher than a few gigahertz, wave penetration is reduced to the cutaneous and subcutaneous tissues. Assessment of the electromagnetic field absorption by internal tissues is interesting only for a few specific applications.

### 3.2.4. *Wave absorption and specific absorption rate*

The physical quantity characterizing the absorption of an electromagnetic field is the specific absorption rate (SAR). The notion of rate refers to a quantity per unit of time, absorption refers to energy absorption and the term "specific" is related to unit mass. SAR is expressed in watts per kilogram ($W \cdot kg^{-1}$).

To be precise, SAR represents the variation in energy absorbed per unit of mass or, more simply, the power absorbed per unit of mass [IEE 05]:

$$\text{SAR} = \frac{d}{dt}\left(\frac{dW}{dm}\right) = \frac{d}{dt}\left(\frac{dW}{\rho_m \cdot dv}\right) \quad (\text{W} \cdot \text{kg}^{-1}) \qquad [3.8]$$

where $\rho_m$ represents the volume mass in $kg \cdot m^{-3}$ and $W$ is the energy absorbed. $dW/dt$ is the quantity of energy per unit of time, which corresponds to an energy rate, i.e. a power.

The absorbed energy is transformed into thermal energy, resulting in temperature elevation in the exposed tissues. The expression of SAR depending on temperature is simpler to conceptualize:

$$\mathrm{SAR} = \mathrm{C}\,.\ \frac{\Delta T}{\Delta t} \quad (\mathrm{W{\cdot}kg^{-1}}) \qquad [3.9]$$

where:

– C is the specific heat capacity of tissues, expressed in $\mathrm{J{\cdot}kg^{-1}{\cdot}K^{-1}}$. This is the quantity of energy required to raise the temperature of a unit mass by one degree;

– $\Delta T/\Delta t$ is the temperature variation at the start of exposure (initial slope), as SAR must be determined without considering caloric dissipation phenomena.

SAR is also expressed as a function of the rms current density $J$ ($\mathrm{A{\cdot}m^{-2}}$) or of the rms electric field $E$ ($\mathrm{V{\cdot}m^{-1}}$) present in the tissues:

$$\mathrm{SAR} = \frac{J^2}{\sigma.\rho_m} = \frac{\sigma.E^2}{\rho_m} \quad (\mathrm{W{\cdot}kg^{-1}}) \qquad [3.10]$$

where:

– $\sigma$ is the effective electrical conductivity of the tissue in siemens per meter ($\mathrm{S{\cdot}m^{-1}}$), representing the ohmic and dielectric losses;

– $\rho_m$ is the volume mass ($\mathrm{kg{\cdot}m^{-3}}$).

These relationships indicate that SAR can be determined by various measuring methods (experimental dosimetry on a representative physical model).

SAR is a function of the electromagnetic field strength penetrating the human body, as well as of the dielectric properties of biological tissues. This field depends on exposure configuration (electromagnetic environment), frequency, polarization and the dielectric characteristics of the tissues.

The relationship between the decrease in absorbed power and the depth [3.7] can be formulated as follows with SAR:

$$\mathrm{SAR(z)} = \mathrm{SAR_0}\,.\ e^{-\frac{2z}{\delta}} \qquad [3.11]$$

where SAR(z) is SAR at depth z, $\mathrm{SAR_0}$ its value at the interface (z = 0) and $\delta$ is the "penetration depth" in the same unit as z.

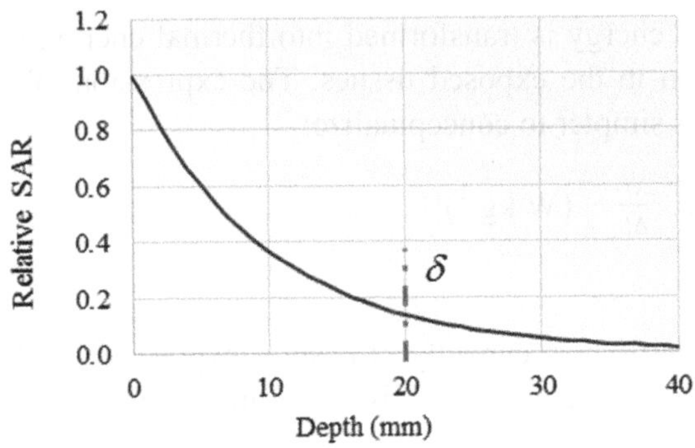


**Figure 3.14.** *Decrease in SAR as a function of the depth from the internal surface area of a homogeneous medium*

SAR assessment is the technique generally used to check the exposure level when the person is at close proximity to a high-frequency radiating source (such as a mobile phone).

One whole body SAR and local SARs, i.e. limited to one part of the body, are defined. This point is discussed in Chapter 4.

### 3.2.5. *Summary of field absorption*

| Key points of wave absorption |
|---|
| At low frequencies, the electric field does not penetrate the human body, or penetrates very little. |
| A large part of the electromagnetic incident field is reflected by the body due to the conductivity of the tissues. |
| SAR, expressed in $W \cdot kg^{-1}$, characterizes energy absorption per unit of mass of a tissue. |
| SAR can be estimated from temperature elevation, absorbed power, internal electric field and current density. |
| The penetration depth of the electromagnetic waves in the tissues is reduced when the frequency increases. |
| SAR is not an appropriate value for very high frequencies (> a few GHz), as absorption is considered superficial. Power density (power per surface unit) is the relevant physical quantity. |
| SAR is not an appropriate physical quantity at low frequencies ($f$ < 100 kHz), as the fields do not penetrate the body significantly, and the heating of tissues is therefore neither measurable nor quantifiable. |

**Table 3.5.** *Penetration and absorption of electromagnetic waves by tissues*

## 3.3. Coupling with the human body

Coupling is involved in interaction mechanisms between electromagnetic fields and the body of an exposed person. Three fundamental coupling mechanisms are well established:

- low-frequency electric field coupling;
- low-frequency magnetic fields coupling;
- high-frequency electromagnetic fields coupling.

These coupling mechanisms depend, on the one hand, on the characteristics of the field (frequency, spatial uniformity, propagation and polarization direction, etc.) and, on the other hand, on the bodily characteristics of the person, such as posture, size and morphology.

The biological effects and their consequences, which depend on the strength and on the waveform, are discussed in Chapters 4 and 5.

### 3.3.1. *Low-frequency electric field coupling*

At low frequencies, living organisms, including plants, distort the spatial distribution of an electric field. Figure 3.15 illustrates this phenomenon around a human body.

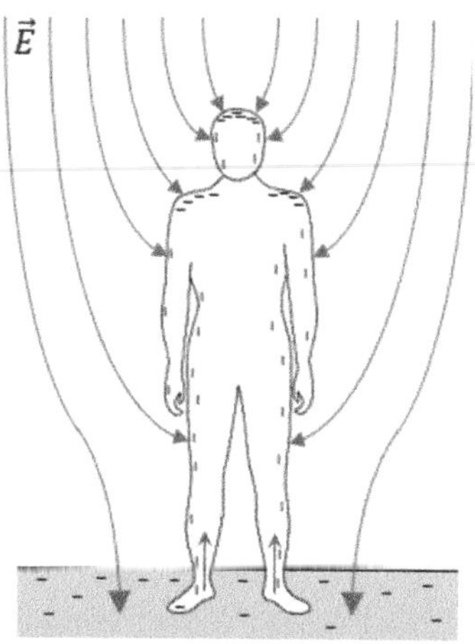


**Figure 3.15.** *Body exposed to a low-frequency electric field*

It will be noted that the electric field lines are almost perpendicular to the surface of the skin, due to the good electric conductivity of biological tissues in comparison to the air: only the normal component (perpendicular) at the surface of the field remains, since the tangential component is zero for the conductors.

The electric charges present in an organism will be attracted to its surface or pushed as the external field alternates. This phenomenon causes an internal electric current in a direction parallel to the field lines. It instantaneously follows variations in the field and ceases when the new distribution of charges balances the external field (electrostatic equilibrium). This new distribution of charges counterbalances the external field and the internal electric field becomes zero.

When a human body is connected to the Earth or to a mass with enough available free charges, it becomes electrically charged (Figure 3.15).

When the body is insulated, it becomes polarized through capacitive coupling (Figure 3.16). The induced currents are less significant.

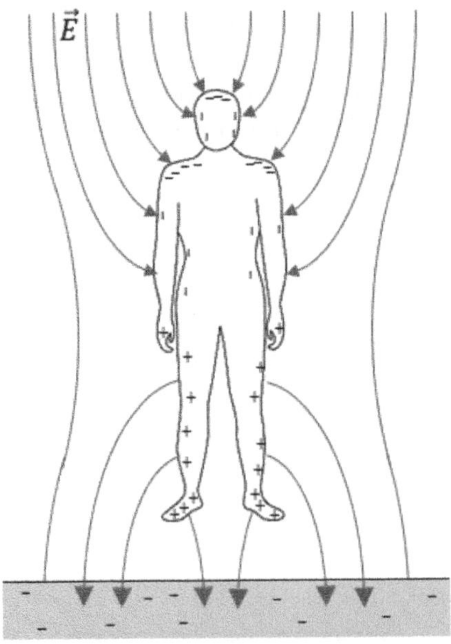


**Figure 3.16.** *Insulated body exposed to an electric field*

The size and posture of the body have a great influence on the fields and currents induced. The current densities present in a body touching the ground and located in a uniform vertical field of 10 kV·m$^{-1}$ at 60 Hz [WHO 84] are shown in Figure 3.17.

The accumulation of electric charges on the surface can locally accentuate the external electric field. It is at its maximum above the head. If the person moves his hand toward the source of the field, the field can reach very high values (fields must be measured in the absence of any person, including the measuring operator).

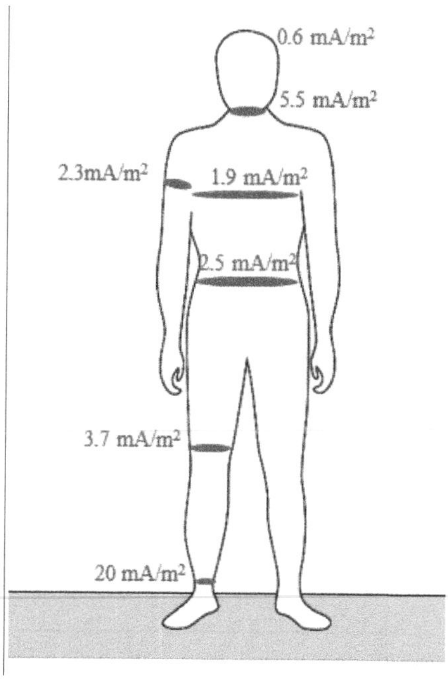


**Figure 3.17.** *Body exposed to a uniform vertical electric field of 10 kV/m at 60 Hz. The current density is more significant when the cross-section is reduced*

The alternation of electric charges and their distribution on the surface of the body determine the level of the currents. These are distributed depending on the conductivity of the biological tissues. Their densities depend on the surface being crossed. Hence, they are also more significant in the neck and ankles, especially when the body is touching the ground (Figure 3.15).

Internal electric fields can be approximated from the average conductivity of low-frequency tissues, $\sigma = 0.2\ \text{S}\cdot\text{m}^{-1}$ [3.1]. These quantities are no longer relevant at frequencies above 10 MHz, where wave absorption becomes dominant.

Some examples of coupling coefficients, the ratios between the induced electric field or current density and the external electric field, appear in Chapter 4.

In brief, when a human body is exposed to a low-frequency electric field, the laws of physics allow the conclusions [WHO 07] summarized in Table 3.6 to be reached.

| Key points of interactions between electric fields and human body |
|---|
| External electric fields attract electric charges to the surface of the body. Electric charges are concentrated more significantly on body parts that have small diameters (e.g. fingers) or that protrude. |
| The flow of these charges forms an induced current and electric field. |
| The induced current and electric field are mostly in the direction of the external field. |
| The current density is higher when the conductive cross-section is small (ankle, knee, neck). |
| The induced currents and electric fields are more significant when the body is in contact with an electrical ground. |
| The induced currents and electric fields are more significant when the person is standing in a vertical field. |
| The induced currents and electric fields depend mainly on the size, shape and position of the person. |
| The currents resulting from the movement of these charges are distributed in the body in a non-homogeneous way, depending on the conductivity of the tissues and their forms. |
| The distribution of currents in a body in electrical contact with the ground is determined by the size and shape of the body (including its posture), rather than by the conductivity of the tissues. The induced electric fields can be calculated from the current densities. |
| The internal electric field is $10^{-4}$ to $10^{-7}$ times less high than the undisturbed external electric field. It is zero for a static field (in a steady state). |

**Table 3.6.** *Human body exposed to low-frequency electric fields*

### 3.3.2. *Low-frequency magnetic field coupling*

The presence of a body in a magnetic field does not disturb it, unlike for electric fields. As the magnetic permeability ($\mu_r$) of the tissues is very close to 1, like the air, the internal magnetic field in the body is practically identical to the external magnetic field, and this field is not distorted.

Magnetic fields induce electric fields according to Faraday's law [1.5]. Electric currents occur in a loop perpendicular to the direction of the magnetic field, since the human body is considered a conductor at low frequencies. These are eddy currents.

In the presence of a uniform magnetic field, the internal electric field and current density are zero at the center of the exposed area. They take on an increasingly significant value as they near the edges, and the highest values are encountered in the cutaneous tissues. They are distributed non-homogeneously, depending on the electric conductivity of the tissues. These quantities are not relevant at frequencies above 10 MHz, because absorption phenomena take over.

Figure 3.18 is a simplified representation of eddy currents in the case of a time-varying magnetic field whose flux lines are directed frontally toward a body.

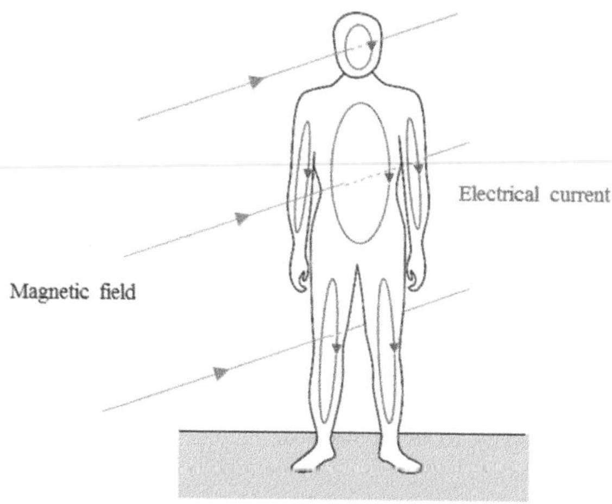


**Figure 3.18.** *Currents caused by a time-varying low-frequency magnetic field in a human body (the magnetic flux is perpendicular to the frontal plane)*

Some examples of coupling coefficients, the ratios between the induced electric field or current density and the external magnetic fields appear in Chapter 4.

In brief, when a human body is exposed to low-frequency magnetic fields, the laws of physics allow the conclusions [WHO 07] summarized in Table 3.7 to reached.

| Key points of magnetic fields and human body interactions |
|---|
| Induced electric fields and currents are perpendicular to the magnetic flux axis. |
| They increase with the size of the person (more significant exposed surface area). |
| They are more significant for a frontal horizontal flux when the person is standing (the cross-section is larger). |
| They are less significant when the flux is vertical for a standing person (smaller cross-section). |
| They are weaker in the limbs, and especially the hands, than in the trunk. |
| They are more significant on the edges than at the center of the exposed surface area. |
| They are proportional to the time variation of the induction (magnetic flux density). For a sinusoidal waveform, they therefore increase when the frequency increases. |

**Table 3.7.** *Human body exposed to low-frequency magnetic fields*

### 3.3.3. *Electromagnetic field coupling*

A high-frequency electromagnetic incident field is reflected, absorbed or even diffused by a biological organism, because it has the characteristics of dielectric loss materials (section 3.2.1). The electromagnetic field coupling mechanism causes energy to be absorbed.

This absorption is a function of the frequency, the power and the polarization of the field. The size, morphology and posture also count for a body, just as the size and orientation of a RF antenna do. The absorbed energy agitates the polarized particles (through polarization mechanisms) in the tissues enough to transform them into thermal energy. This absorption is quantified by the SAR [3.8].

The proximity of the body to the radiating source can alter the emitted power, due to the mutual coupling phenomenon: an object close to a source modifies its characteristics. This generally results in a decline of the radiated power (this proximity is sometimes necessary for some applications, and is therefore taken into account when the source is designed. Its functioning is *optimum* only in this situation).

### 3.3.3.1. *Influence of frequency*

Figure 3.19 illustrates absorption depending on frequency at the upper part of the spectrum. The person is standing up and is entirely exposed to a vertically polarized uniform field (plane wave obtained at a great distance from the radiating source, see Figure 1.11) [WHO 93].

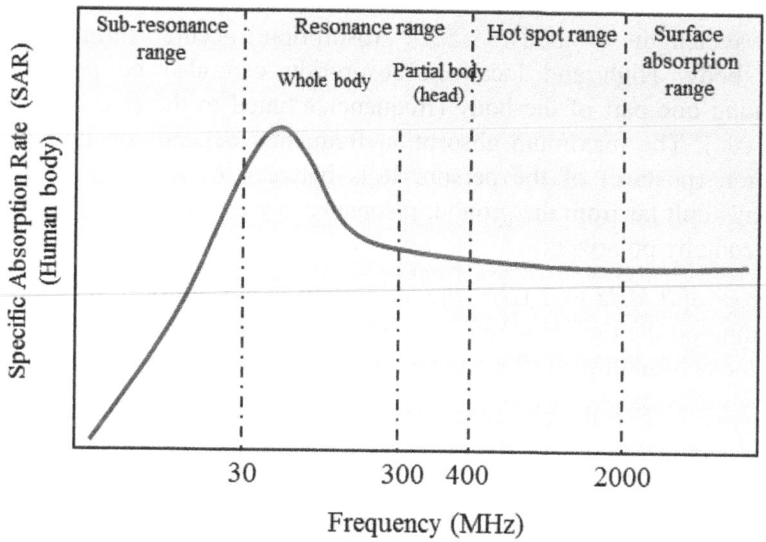


**Figure 3.19.** *Absorption of the electromagnetic field as a function of the frequency*

The absorption of energy by the human body is differentiated as a function of the frequency:

– *below 100 kHz*, the absorption is negligible and the temperature increase is not measurable. The induced fields and currents are dominant;

– *from 100 kHz to a few megahertz*, the body is more conductive than dielectric. The electric field and the induced current are mainly created through electrical coupling and/or magnetic coupling, according to the type of source (sources with a dominant electric field: HF heating, diathermy applicator; source where magnetic fields are dominant: induction heating; RF antennae emit proportionally as many electric fields as magnetic fields to maximize their propagation). Absorption greatly increases up to around 20 MHz. The inhomogeneity of the tissues and the small sections crossed by the induced currents create a denser absorption in the neck and ankles. Furthermore, the currents are concentrated, due to the reduced conductivity of bones and cartilage. The absorption of the field becomes homogenized in the whole body when the frequency increases;

– *between 20 MHz and 300 MHz*, absorption is the highest, since the frequencies are tuned to the dimensions of the human body (half-wavelength ≈ body size). Absorption occurs throughout the whole body. High and localized absorption can also be produced by resonating one part of the body (frequencies tuned to the size of the head, arms, etc.). The maximum absorption frequency depends on the size and extension (posture) of the person. It is between 60 and 100 MHz in a standing adult far from the ground; resonance is weak when the electric field is horizontally polarized;

– *from 300 MHz to 2 GHz*, the absorption is not uniform. It is localized depending on the dielectric and dimensional properties of the tissues, due to phenomena of multiple reflection and internal focalization in the body;

– *above 2 GHz*, the absorption level is maintained at a high level, but as the frequency increases, the penetration depth of the waves is greatly reduced. At frequencies above several gigahertz, absorption concerns only the skin and subcutaneous tissues. It is superficial and no longer volumic.

The curves in Figure 3.21 indicate the average SAR for the whole body of an adult and children exposed to a uniform and vertically polarized electric field of 1 $V{\cdot}m^{-1}$ (simulation results) [DIM 02].

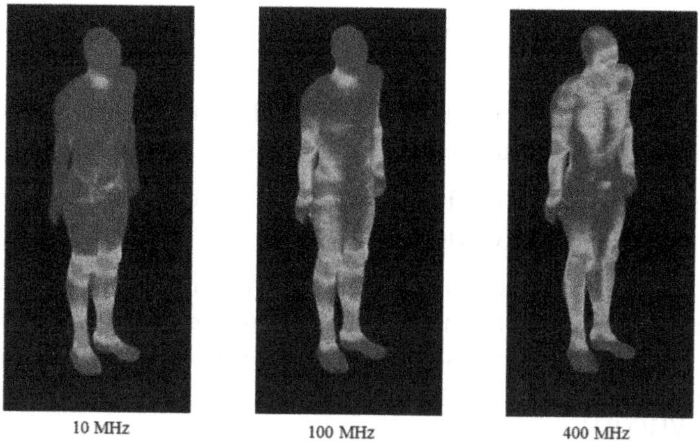


**Figure 3.20.** *Distribution of the SAR for a vertical polarized plane wave at various frequencies (Dr R. Findlay, private communication). For a color version of this figure, see www.iste.co.uk/staebler/electromagnetic.zip*

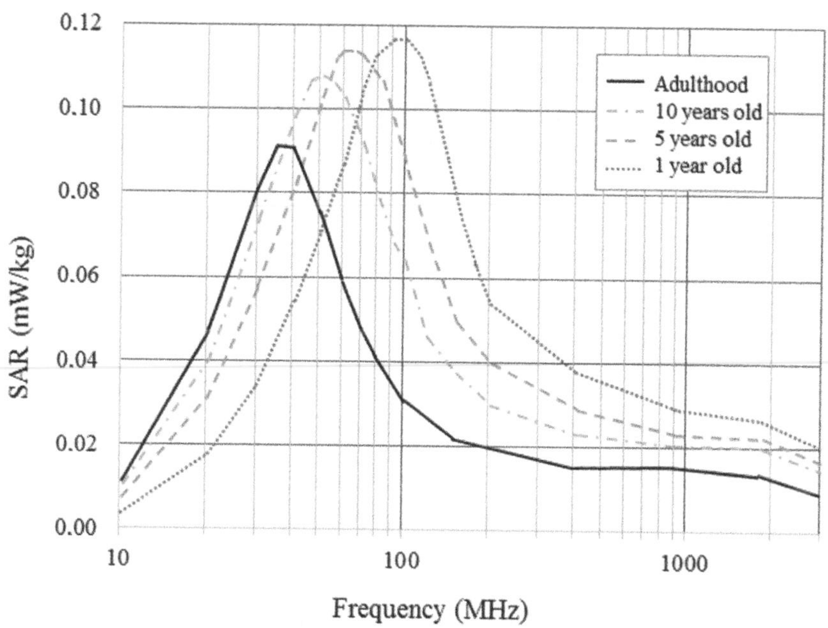


**Figure 3.21.** *Influence of the size of the body on the power absorbed during exposure to an electric field of 1 V·m$^{-1}$*

When a human body is exposed to electromagnetic fields, the laws of physics lead to the following general observations:

– the absorption of an electromagnetic field is more significant when the electric incident field is parallel to the body axis;

– the absorption of an electromagnetic field is weaker when it propagates from the head to the feet (or vice versa);

– the absorption is at its maximum at the resonance frequency of the organism, i.e. when the half-wavelength *in the body* is comparable to its size. This frequency is higher in a smaller person. It can be approximated by the relationship $fo = 114/\text{size}$ ($fo$ in MHz and size in meters) [WHO 93];

– the absorption increases when the object is narrow. A person standing up with their arms in the air will absorb more of an electromagnetic field than a person crouching down with their arms folded. There is a strong dependence between absorption at frequencies lower than 2 GHz and the size of the body, its posture and the polarization of the wave;

– absorption is not homogeneous, even in the presence of a uniform field. Tissues with high water content (cerebrospinal fluid, vitreous humor, etc.) absorb more electromagnetic fields than tissues with low water content (fat, bone, etc.);

– the penetration depth of electromagnetic waves in a tissue decreases when its conductivity increases and when the frequency increases (skin effect);

– normal clothes have very little influence on absorption (they are good dielectrics and therefore relatively imperceptible to electromagnetic fields).

It can be demonstrated that a body standing on the ground or on a conductive plane, not necessarily in electrical contact (insulated by a few cm), will be tuned to a frequency of around half of what it would be if he were far from the ground. The resonance frequency can reach a minimum value of 30 MHz.

### 3.3.4. *Summary of coupling mechanisms*

| Key points of coupling mechanisms |
|---|
| Biophysics distinguishes three fundamental coupling mechanisms dependent on the nature of the field: electric field, magnetic field and electromagnetic field. |
| Low-frequency electric fields induce superficial charges. The change in these charges leads to internal currents and electric fields in the body, which is considered a conductor. The distribution and intensity of these two quantities depend on the exposure conditions: intensity, direction and frequency of the field, as well as size, position and body posture. |
| Low-frequency magnetic fields induce electric fields. These cause electric currents proportional to the conductivity of the tissues to circulate. The strength of the induced electric fields is proportional to the exposed surface area and the magnetic induction frequency. |
| Electromagnetic fields at frequencies above 100 kHz may be significantly absorbed in the tissues. This energy loss is converted into heat. Exposure to a uniform field generally leads to non-uniform energy absorption in the body. |

**Table 3.8.** *Summary of the coupling mechanisms of electric, magnetic and electromagnetic fields*

# 4

# Biological Effects of Electromagnetic Fields

Electromagnetic fields can act directly on an exposed body, causing direct biological effects. These effects may be *thermal* at high frequencies, due to the absorption of the fields, or *non-thermal* at low frequencies, due to the presence of induced electric fields and currents inside the body.

A biological effect is a physical, biochemical or behavioral change in an organism in response to a stimulus. These changes are reversible if they occur within the limits of homeostasis, i.e. within the limits that allow the organism to retain its internal balance, despite the presence of constraints. These limits depend on the organism, its state and the environmental conditions. Biological effects are not necessarily caused by external stimulation (such as an electromagnetic field). They are also the natural response to an internal stimulus, e.g. sweating after warming up during a physical effort. A biological effect does not necessarily lead to a health effect[1]. Health will be affected if the effect is prolonged beyond the point at which the human body can normally compensate or endure it.

The direct biological effects of electromagnetic fields are divided into sensory effects and health effects. Sensory effects can cause transient sensory perceptions and minor changes in brain function in the exposed person. Health effects are adverse to the health.

---

1 The WHO defines health as a state of complete physical, mental and social well-being and not merely the absence of disease or infirmity.

There are also indirect effects. These are caused by the presence of an object in an electromagnetic field. Burns produced by a contact current, disturbance of active medical devices, spark discharges, and attraction of ferromagnetic elements are examples of this kind of effect. These effects may become the cause of a safety or health hazard.

The main effects are broken down as shown in Figure 4.1 and will be discussed in more detail throughout this chapter.

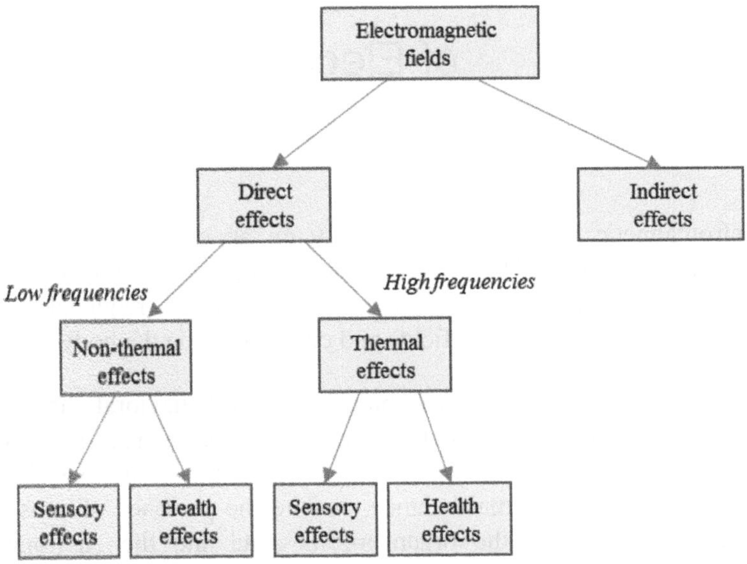


**Figure 4.1.** *Effects caused by electromagnetic fields*

## 4.1. Direct effects

Two main types of direct biological effects are observed:

– non-thermal effects: these effects are produced from 0 Hz to 10 MHz. Currents and electric fields induced by an electric or magnetic field may stimulate biological tissues;

– thermal effects: these effects occur from 100 kHz to 300 GHz. The absorption of electromagnetic fields causes tissues to heat up.

The two effects are unconnected, but can be produced simultaneously by frequencies between 100 kHz and 10 MHz. This frequency range is called "intermediate frequencies". For non-thermal effects, static or quasi-static fields (frequencies lower than 1 Hz) are distinguished from low-frequency fields (1 Hz to 10 MHz). Non-well-established direct effects are discussed in section 4.1.5.

### 4.1.1. *Direct effects of low-frequency fields*

The biological effects generated by low-frequency fields ($1 \text{ Hz} < f < 10 \text{ MHz}$) are well established, having been recognized for decades and agreed upon internationally by the scientific community. Electric and magnetic fields induce currents and electric fields inside an exposed body, as explained in Chapter 3. These signals are added to the natural physiological signals and are liable to stimulate the skin sensory receptors, retina, nervous tissues and muscle tissues. The effects observed range from perception to a fatal outcome. They can be classified as follows:

- perception;
- pins and needles;
- tingling sensation;
- discomfort, well-being disorder;
- embarrassment, uneasy feeling;
- muscle contraction;
- loss of muscle control;
- pain;
- change in cardiac function.

Effects occur when the fields or induced electric currents reach certain thresholds, which depend on the duration of stimulation [LAP 1907, LAP 1909] and on the sensitivity of the exposed person. Levels lower than the natural physiological signals are not strong enough to produce detectable effects. Although these signals are weak, they are strong enough to be measured outside the body, for example using an electrocardiograph (ECG) or electroencephalograph (EEG).

The effects that can be generated by low-frequency fields are divided into sensory and health effects. The former include direct stimulation of the central nervous system (CNS) of the head, stimulation of cerebral functions and cutaneous perceptions. The latter may arise from stimulation of the peripheral nervous system (PNS), which controls the skeletal muscles, or direct muscle stimulation. They can, therefore, affect the health and safety of the exposed person.

The field levels associated with the effects indicated below are by and large given for exposure to a uniform electric or magnetic field. Higher levels are required at close proximity to sources with major spatial variations (Figure 1.6).

#### 4.1.1.1. *Sensory effects*

Sensory effects include cutaneous perceptions, minor changes in cerebral function and direct stimulation of the CNS of the head.

##### 4.1.1.1.1. Cutaneous perception

When a human body is exposed to a variable and strong electric field, free electric charges are attracted or repelled toward the skin, including the hairs. The force exerted [1.2], then causes these hairs to vibrate, which can be perceived. Most people are able to discern this effect from a field strength of 20 $kV{\cdot}m^{-1}$. This threshold falls to 5 $kV{\cdot}m^{-1}$ for the most sensitive people. Beyond a field of 25 $kV{\cdot}m^{-1}$, the effect becomes unpleasant, or even painful, due to micro electric shocks caused by the corona effect [TEN 91]. On the ground, fields of 20 $kV{\cdot}m^{-1}$ can be found only under the busbar of an electrical interconnection substation with a very high voltage of 400 kV, an area prohibited to the public.

##### 4.1.1.1.2. Stimulation of the CNS

The CNS consists of the encephalon, commonly called the brain, the spinal cord, the optic nerves and the retinas. The role of the CNS is to interpret sensory information, including that transmitted by the PNS, and to control the organism and its movements. The CNS also records signals and is where mental functions and thought occur.

The most sensitive tissues to external fields in the human body belong to the CNS in the head, which excludes the spinal cord in the context of sensory effects.

During exposure to a magnetic field, the first effect, called the "critical effect", is the induction of magnetophosphenes. These correspond to the perception of small flickers on the edge of the visual field, which may become flashes of light.

These effects are harmless to the health and are present only during exposure of the head. Sensitivity to these perceptions peaks at 20 Hz. Therefore, all that is needed is an induced electric field between 50 and 100 $mV \cdot m^{-1}$ in the retina. These quantities lead to a minimum current density of around 10 $mA \cdot m^{-2}$. These values are reached under magnetic induction between 5 and 10 mT, according to the coupling conditions (induction uniformity and direction mainly). Sensitivity to sensory effects decreases at lower and higher frequencies [ICN 10]. At 50 Hz, the level required for this effect to be observed is around 15 mT [LEG 15].

An external electric field induces a very weak internal electric field with a ratio of around 2 $mV \cdot m^{-1}$ for 1 $kV \cdot m^{-1}$ at 50 Hz (the coupling coefficient is between $10^{-5}$ and $10^{-6}$). To reach CNS stimulation level, one must be exposed to electric fields of a level well above those that trigger painful cutaneous effects. Furthermore, from 10 $kV \cdot m^{-1}$, potentially dangerous spark discharges can appear between the body and a conductive environment. For these reasons, no study appears to have been carried out on the sensory effects associated with low-frequency electric fields.

#### 4.1.1.1.3. Stimulation of cerebral functions

For higher exposure levels, direct cerebral stimulation by a magnetic field can disturb the processing of visual information and motor coordination. Despite the funny side of loss of movement control, in an uncontrolled situation these disorders can be dangerous.

A high magnetic field inducing a high electric field (>100 $V_{peak} \cdot m^{-1}$) is found in transcranial magnetic stimulation. This is used in neurological therapy and cognitive neuroscience.

### 4.1.1.2. *Health effects*

#### 4.1.1.2.1. Stimulation of the PNS

The PNS is the part of the nervous system formed of all the motor and sensory nerves, connecting the CNS to the organism. In particular, it controls the skeletal muscles.

The PNS is less sensitive than the CNS at frequencies under 400 Hz. In the most sensitive people, an internal field of around 2 $V \cdot m^{-1}$ must be induced to obtain a sensation, and a field of around 5 $V \cdot m^{-1}$ to provoke electrostimulation of these nerves. These levels are induced by an external electric field of several hundred kilovolts per meter [BAM 11] or magnetic field levels of several tens of milliteslas (the uncertainty associated with the indicated values is significant).

A nerve integrates an electric field lengthwise, so the risk of stimulation is greater when the exposed surface is larger (induced electric fields are also more significant, see section 3.3.2: magnetic fields coupling).

The PNS stimulation threshold does not depend on the frequency up to 3 kHz, after which point it increases. As indicated previously, the CNS is less sensitive than the PNS below 400 Hz. The two systems have comparable sensitivities just above 400 Hz, and beyond 3 kHz the PNS is more sensitive [ICN 10].

#### 4.1.1.2.2. Direct stimulation of muscles

Muscles are less sensitive than nervous tissues [ICN 10]. Direct stimulation of a muscle can cause a contraction, just as nervous stimulation would do.

Nerve or muscle stimulation can produce uncontrolled movements, causing a health and safety risk (an unintentional movement can result in injury through loss of balance or from a blunt object). A situation like this cannot occur in everyday life and is highly unusual in a work environment (e.g. if safety instructions are not complied with).

#### 4.1.1.2.3. Cardiac stimulation

Cardiac function deserves particular attention, as changes in it are dangerous or even fatal for the person (risk of ventricular fibrillation, i.e. cardiac dysrhythmia, corresponding to the rapid and disrupted contraction of the ventricles). The threshold at which this effect is triggered is higher than the stimulation threshold of other muscles. Exposure to a magnetic field of 200 mT at 50 Hz is required [IEE 02].

Levels well above those that cause cardiac fibrillation are required to heat tissues through the Joule effect. These levels must be superior to 1 T for a time-varying magnetic field. A current density of more than 10 $A{\cdot}m^{-2}$ is therefore induced, corresponding to an internal electric field of around 50 $V{\cdot}m^{-1}$.

The effects caused are temporary (unless the outcome is fatal!), ceasing once exposure to the ambient fields stops. Nerve stimulation is immediate and the threshold does not, therefore, depend significantly on the duration of the field. The instantaneous field value must be taken into consideration below 10 MHz [ICN 10]. Table 4.1 summarizes the effects, current densities and electric fields induced by magnetic fields of 10–300 Hz [BER 88]. The values are indicative.

| Effects | Current density (1) | Electric field induced (1) (2) | External magnetic field |
|---|---|---|---|
| Physiological levels | <1 $mA{\cdot}m^{-2}$ | <5 $mV{\cdot}m^{-1}$ | |
| Established effects, magnetophosphenes<br>Possible effects on the CNS | 10–100 $mA{\cdot}m^{-2}$ | 50–500 $mV{\cdot}m^{-1}$ | ~10 mT |
| Established effects on the CNS<br>Possible health effects | 100–1,000 $mA{\cdot}m^{-2}$ | 500–5000 $mV{\cdot}m^{-1}$ | ~ 50 mT |
| Proven health effects<br>Ventricular fibrillation and extrasystoles | >1,000 $mA{\cdot}m^{-2}$ | >5 $V{\cdot}m^{-1}$ | ~200 mT |
| (1) Order of magnitude.<br>(2) Calculated with $\sigma$ average = 0.2 $S{\cdot}m^{-1}$. | | | |

**Table 4.1.** *Summary of effects caused by magnetic fields*

In short for an exposure to low-frequency fields, a distinction is drawn between sensory effects, caused by stimulation of CNS without consequences on the health, and health effects, caused by stimulation of PNS. They cease when stimulation stops. These well-established effects result in two types of exposure limit values being used to protect people.

### 4.1.2. *Direct effects of static magnetic fields (<1 Hz)*

Biological effects may be produced during exposure to a static or a very low-frequency magnetic field ($0 \leq f < 1$ Hz).

When the person is moving, the static field may appear variable if it is not uniform or if its direction changes. In this case, the effects are identical to those of the low frequencies presented in the previous section, but with strength above 2 teslas (value disproportionate to the Earth's 50 μT magnetic field).

The main direct effects caused by static magnetic fields may be [HPA 08]:

– magnetophosphenes;

– vertigo and nausea;

– stimulation of the peripheral nerves;

– metallic taste sensation;

– neurocognitive effects (subtle changes in attention and concentration, and disorientation).

Magnetophosphenes are produced first and foremost through rapid movements of the head or eye in its socket. The excitable tissues of the CNS are not very sensitive to quasi-static frequencies due to various biological mechanisms. The strength of these fields must, therefore, be much greater than that which would be required at 20 Hz (minimum sensitivity frequency) to produce effects.

Vertigo and nausea are caused by excitation of the vestibular apparatus, a balance organ located in the inner ear and part of the CNS of the head. Its stimulation can be interpreted as a movement of the body inconsistent with its visual perception and its gravity perception.

A metallic taste sensation is likely due to an imbalance of the chemoreceptor cells located on the tongue. The induced currents present when the head moves in a magnetic field free metal ions through electrolysis of saliva.

Most of these effects are not dangerous in themselves, but may become annoying and disrupt activity if they persist. As well as effects created by movement, a strong static magnetic field creates currents and electric fields through the blood vessels, due to the circulation of ions (Lorentz force [3.6]). At these strengths, it also attracts infinitesimal quantities of the magnetic particles in tissues and modifies some chemical reactions. These effects are not considered dangerous to the health when magnetic induction remains lower than 7 teslas. Beyond this value, there are too few research results to draw a conclusion.

Meanwhile, a static magnetic field produces magnetohydrodynamic phenomena (behavior of a fluid in the presence of a magnetic field), which slow blood flow. Arterial tension increases and the cardiac rhythm is changed. These effects are visible on an electrocardiogram. They disappear instantaneously when we leave this kind of magnetic field. The metallic taste sensation is no longer perceptible after a few minutes.

Calculating the electric field induced by movements is difficult, since many parameters are involved (dimensions, sizes, shapes, field gradient, etc.). To this must be added the lack of dielectric characterizations of tissues below 10 Hz.

### 4.1.3. *Direct effects at frequencies between 100 kHz and 300 GHz*

#### 4.1.3.1. *Effects on the health*

A human body exposed to an electromagnetic field absorbs part of the power transported by this field. It is transformed into heat. This is the only proven biological effect at frequencies between 10 MHz and 300 GHz. Between 100 kHz and 10 MHz, heating and tissue stimulation can coexist with no connection between them.

The energy absorbed by a body is proportional to the strength of the radiation. This depends on the coupling and duration of exposure. The heat induced by electromagnetic fields is different to other thermal phenomena, because it is produced inside the body and does not naturally penetrate the skin. Energy distribution is also specific and its absorption rapid. This unusual heating can cause a disorder in the person in response to the change in internal metabolism, even without a detectable temperature increase. This disorder is known as "thermal stress". The heat deposited by the radiation is added to that produced by the organism's metabolism.

The absorbed energy is transformed into heat. This phenomenon is quantified by the specific absorption rate (SAR), measured in watts per kilogram [3.8].

The temperature rises by less than 1 °C when an entirely exposed human body absorbs a power of 4 $W \cdot kg^{-1}$. A new thermal balance is reached in around 30 min.

Body temperature is regulated mainly through the diffusion of heat transported through blood flow and, to a lesser extent, conduction through muscles and adipose tissues. The physiological response to hyperthermia beyond a certain threshold is an increase in cardiac rhythm and vasodilation. The resulting increase in blood flow evacuates the heat more rapidly to the skin, where it is dispersed into the surrounding air through thermal radiation, convection or evaporation (sweating).

When body temperature is excessive, the dispersion of calories through sweating becomes dominant. This phenomenon is related to temperature, humidity and air speed. If heating does not stop when the homeostatic limits are reached, there is a risk of hyperthermal collapse. In the worst case, the outcome is fatal. Tissue necrosis can also occur during exposure to intense localized fields.

The electromagnetic field absorption by biological tissues depends on the frequency and on their dielectric properties. It will be more significant in tissues with a high water content (see section 3.2.1). Furthermore, some tissues, such as the eyes and testicles, disperse less heat because they are not vascularized.

Figure 4.2 shows the temperature evolution of two tissue categories. Vascularized tissues are distinguished by three phases of hyperthermia. During the first phase, the temperature increases by a few degrees over several minutes. Thermal balance follows for several dozen minutes in the vascularized tissues. Finally, the thermoregulation system can no longer disperse enough heat. When it collapses, the internal temperature increases rapidly. The decline in temperature elevation slope for non-vascularized tissues is due to thermal conduction toward adjacent tissues.

Referring to relationship [3.9], the SAR is calculated from the initial slope ($t = 0$) of the curves in Figure 4.2.

There is no thermal health risk for people entirely exposed to radiation leading to absorption of less than 4 $W \cdot kg^{-1}$. This value is comparable to that of a relaxing person, who naturally disperses around 1 $W \cdot kg^{-1}$, and a person in good health doing physical exercise, who disperses from 3 to 5 $W \cdot kg^{-1}$, a situation where his internal temperature may rise by more than 2 °C.

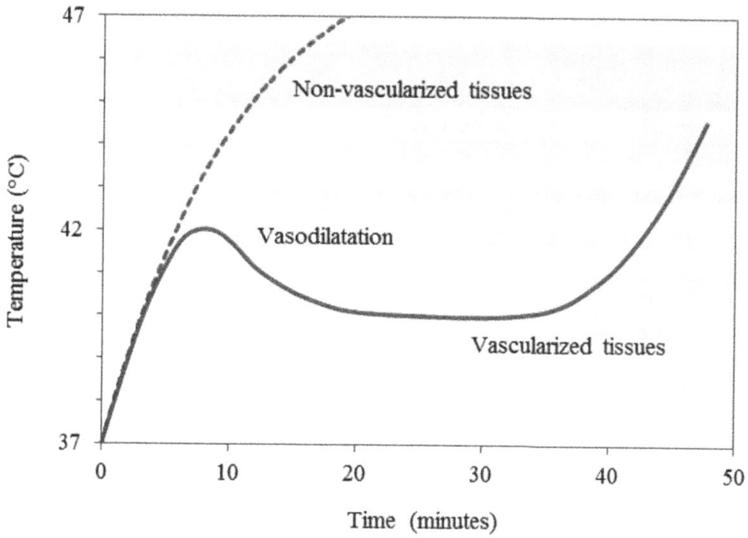


**Figure 4.2.** *Temperature variation of two categories of biological tissue exposed to an electromagnetic field [AND 01]*

A higher absorption level is required to generate an irreversible effect when only one part of the body is exposed. Local heat can disperse in a larger volume through blood flow and thermal conduction. We therefore differentiate between exposure of the whole body, exposure of part of it and exposure of the limbs. A level of around 100 $W.kg^{-1}$ is required to trigger a proven effect during localized exposure.

The thermal effects on an adult person in a state of relaxation, exposed to an electromagnetic field, are summarized in Table 4.2.

The electromagnetic field penetrates the body less deeply the higher the frequency and the more conductive the tissues (penetration depth, see section 3.2.3).

The average penetration depth is 2 cm at 3 GHz. Beyond 10 GHz, this phenomenon is superficial (<0.3 mm) and the electromagnetic field behaves rather like visible optical radiation, apart from the fact that it passes through clothes. The level of absorbed power per mass unit (SAR) no longer has any meaning. It is the power per surface unit, i.e. the power density in W·m$^{-2}$, which becomes relevant for quantifying exposure. Furthermore, the energy is absorbed by a smaller volume, so the temperature rises faster.

| Effects | Whole body SAR (W·kg$^{-1}$) |
|---|---|
| No appreciable temperature elevation | 0.1 |
| Slight temperature elevation (<0.5 °C) | 0.4 |
| Basal metabolism in a relaxed state (without exposure) | 1 |
| Non-harmful temperature elevation (<1 °C) | 1–4 |
| Health risk (>1 °C) | >4 |
| Thermal damage to certain organs | 100[a] |
| [a]Localized exposure. | |

**Table 4.2.** *Biological effects depending on absorption level (SAR) for an exposed adult in a relaxed state*

Heating is observed when the power density exceeds 1 250 W·m$^{-2}$ or 25 mW·cm$^{-2}$ (as a point of comparison, the Sun provides a maximum power of 1,000 W·m$^{-2}$ to the Earth's surface).

The average powers (or rms values) must be limited, as the energy absorbed is cumulative. The calculation must be performed by integrating the power over the exposure duration.

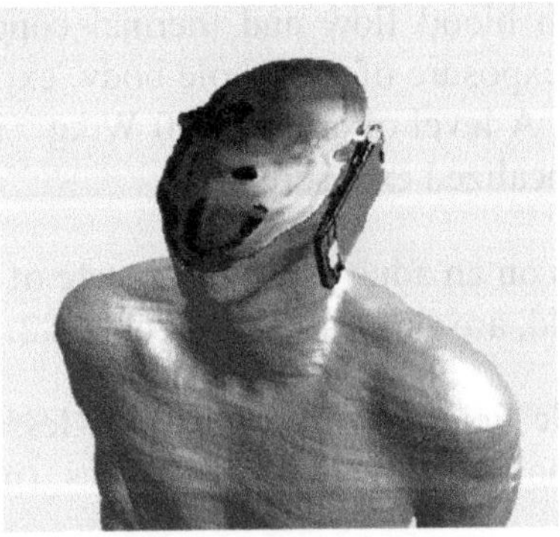

**Figure 4.3.** *Head exposure during a mobile telephone call. Highest levels occur in the outer layers of the head around the ear (source: IT'IS Foundation, ETH Zürich [ITI 10]). For a color version of this figure, see www.iste.co.uk/staebler/electromagnetic.zip*

Physical measurements are possible on uniform humanoid models (called phantoms), i.e. models filled with a type of physiological liquid. Digital anthropomorphic models make it possible to obtain absorption levels through simulation with great precision. These levels should be compared with the SAR. These assessment techniques are standardized. They are introduced in Chapter 8.

#### 4.1.3.2. *Sensory effect: auditory perception*

Although it is not possible to directly discern the signals of a frequency beyond a few tens of a kilohertz, there is an exception. A pulsed radiofrequency radiation, like that emitted by radar, can produce a perceptible acoustic sensation. This phenomenon appears during exposure to electromagnetic fields with a very high peak power level, around 1 $kW \cdot m^{-2}$, even under a relatively low average power (<1 $W \cdot m^{-2}$).

The sensitivity threshold depends on the carrier frequency of the signal and the modulation. It is minimal in the UHF band (0.3–6 GHz), then increases rapidly beyond this range. Pulses lasting a few microseconds are perceived the most.

This phenomenon may be due to very small thermoelastic cells that dilate in response to the instantaneous absorbed power. This mechanical dilation is propagated through bone conduction and then picked up directly by the cochlea, located in the inner ear. This explains why even people who have lost their auditory conduction, i.e. lost the functions of the middle ear (eardrum and ossicles), perceive this phenomenon.

Therefore, some exposed people may have the impression that they can hear a radiofrequency emission. This is a sensory effect that has no known adverse health-related consequences.

### **4.1.4. *Summary of direct biological effects***

Each year, almost a thousand results of studies into exposure to electromagnetic fields are published, bringing the total number of publications to more than 25,000. No other physical or chemical risk has probably been so extensively studied. The acute effects are well known and allow exposure limits to be defined.

The glossary at the end of this book provides definitions of the various effects and exposure categories:

– thermal, non-thermal, sensory, health, acute, chronic effects, etc.;

– short-term, long-term exposure, etc.;

– weak, medium, high intensity exposure, etc.

Table 4.3 summarizes the effects that may be caused in the short term by electromagnetic fields.

| | Key points of direct effects |
|---|---|
| Biological effects | Biological response to environmental factors, possibly with no health effect. |
| Low-frequency electric and magnetic fields | The main effect at frequencies lower than 10 MHz is the stimulation of sensory organs and nervous and muscle tissues. These effects are produced above certain thresholds. |
| Limits for low frequencies | Electric fields and induced currents must be limited to avoid harmful irreversible health effects. |
| High-frequency electromagnetic fields | The main effect at frequencies above 100 kHz is the heating of exposed tissues through absorption of the field. The temperature elevation is discernable only beyond a certain threshold. |
| Limits for high frequencies | The absorption of waves by tissues must be limited to avoid excessive heating, which is harmful to the health. |
| Intermediate frequencies | Between 100 kHz and 10 MHz, stimulation effects and thermal effects can be produced independently from one another. |
| Response time | The proven effects are acute, i.e. immediate. |
| Weak fields | No adverse short-term physiological effect has been observed in people exposed to weak fields. |

**Table 4.3.** *Summary of short-term effects caused by electromagnetic fields*

## 4.1.5. *Long-term effects of electromagnetic fields*

### 4.1.5.1. *Introduction*

Electromagnetic fields produce thermal effects that cause biological tissues to heat up. This can be harmful to the health when the energy absorption greatly exceeds the basal metabolism level of the human body. Furthermore, low-frequency electric and magnetic fields are likely to trigger non-thermal effects, such as the stimulation of sensory and nervous tissues.

These effects are acute. They occur in the short term (from a few milliseconds to a few hours) and disappear spontaneously when exposure ceases, provided that they have not caused irreversible disorder (cardiac fibrillation, tissue necrosis, etc.). Such situations are encountered only by accident in the workplace and should never occur in everyday life.

The effects have been thoroughly studied and scientifically proven. The exposure limit values have been refined over many decades to ensure a high level of protection of people (general public and workers).

However, the following questions might be asked:

1) Is prolonged exposure to these levels, lower by a factor of 1,000 or more than the thresholds that trigger the first acute effects, likely to produce biological effects that may damage people's health or well-being? These levels are around a few tenths of a volt per meter or a few microteslas ($1\ \mu T = 10^{-6}$ T).

2) Are there effects due to radiofrequencies which cause non-thermal effects? Are there effects due to ELFs other than those created by stimulation?

3) Are there pathogenic effects due to such exposure?

The evolution of species and populations, particularly humans, is the result of permanent adaptation to a generally favorable environment, including natural electromagnetic waves (solar radiation, cosmic rays) and static magnetic and electric fields (see section 2.1). Lightning, an electromagnetic wave with considerable power, may have created life on Earth, if it did not originate from interstellar space.

4) Is human physiology adapted to weak artificial radiation? Can it cause harmful health effects?

Although there is international consensus on the short-term effects of electromagnetic fields, long-term exposure to a very weak level raises questions that preoccupy both scientists and the public. The suggested responses are the subject of endless controversies, which often spark passionate, not to say heated, debates.

For some, the answers available today do not yet seem sufficient, although few physical or chemical risks have been the subject of so many years of research, publications and conferences. They are regularly summarized by internationally renowned scientific groups [AFS 10, ANS 13, ICN 09b, IAR 13, WHO 93, SCE 15, GAU 16].

It is recognized that ionizing rays, such as far UVs and X-rays, have enough energy to ionize an atom. Cell degradation can follow, possibly causing cancer, whether as a result of high levels over a short duration or low levels over a long period (dose concept)[2]. The boundary between ionizing and non-ionizing fields is very clear.

The debate focuses on non-ionizing radiation, notably on electromagnetic fields with frequencies lower than 300 GHz, which are not strong enough to cause an acute effect. Is this radiation more dangerous than light or infrared rays, given that there are continuous physical phenomena in this part of the electromagnetic spectrum? It will be recalled that the boundary between the electromagnetic domain and the optical domain has been arbitrarily determined at a wavelength of 1 mm (300 GHz) (see section 1.4.2).

#### 4.1.5.2. *Studies into long-term effects*

There are three main types of study into the long-term effects of electromagnetic fields: epidemiological studies, *in vivo* biological studies and *in vitro* biological studies [ZWE 09].

##### 4.1.5.2.1. Epidemiological studies

On a large scale, studies into the direct effects of electromagnetic fields on humans are essentially epidemiological studies for ethical reasons. Epidemiology is based on the observation of a population in real living conditions. The epidemiologist compares the frequency of an illness within a group of people exposed to a suspect agent with that of a group of

---

2 The term *dose* is used improperly in electromagnetic radiation to define the product of the field strength multiplied by the exposure duration. In toxicology and radioprotection, the dose is the product of the active biological parameter multiplied by the time. The active parameter is the SAR or induced electric field for electromagnetic fields [ICN 09].

unexposed people. These are comparative statistical studies, which look for a correlation between the emergence of illnesses and environmental factors. The difficulty lies in isolating the required factor, especially when it is combined with other factors, such as lifestyle, and when the emergence of an illness affects very few people.

There are two main types of epidemiological studies on humans:

– *case–control study*: the background of people with an illness and the background of people without this illness are compared. This is a retrospective study;

– *cohort study (or follow-up study)*: the study focuses on two groups of people. Only one of the groups is exposed to the suspect agent. This group is monitored in the long term, i.e. over several years or even decades, and is compared to the non-exposed group.

The uncertainties in these studies remain significant, because many factors are involved. Examples include the level and duration of exposure, the number of samples (size of the groups of people), lifestyles, eating habits, etc.

Such studies have been conducted for more than half a century in the field of electromagnetism.

#### 4.1.5.2.2. *In vitro* biological studies

*In vitro* biological studies are experimental studies that focus on cellular cultures of varying complexity. They are conducted under perfectly controlled exposure conditions in order to remove factors that could skew the results. Such studies can be carried out in the long term, i.e. over many cellular multiplications.

These studies aim to detect all cellular alterations from the early stages. The mechanisms of action are expected. However, they do not make it possible to monitor the biological evolutions that could result from this, with consequences for health, such as the emergence of cancer.

#### 4.1.5.2.3. *In vivo* biological studies

*In vivo* biological studies are also experimental studies. They involve the operating mechanisms of a complete (and complex) biological system. They are appropriate for the observation of the overall effects of exposure on living animals or even humans. Analysis can cover continuous exposure of many generations of animals (provided that their lifetime is compatible with the search duration). The results are extrapolated to humans as far as possible.

Laboratory studies on humans include double-blind experiments. The exposed person and the operator have no knowledge of the exposure conditions. The operator therefore cannot unconsciously influence the responses that the examiner will analyze. They may be single-blind. In this case, the subject alone does not know his exposure level. They may also be open (exposure conditions are known to everyone), for example, the impacts of electromagnetic fields on cardiac pacemakers in real conditions.

A few tens of thousands of studies of this type have been carried out since the 1930s, covering the extent of the domain of electromagnetic fields. Many regularly updated monographs exhaustively and rigorously summarize the results produced, take stock of knowledge and suggest avenues for research. The conclusions of these monographs, briefly introduced below, coincide.

#### 4.1.5.3. *Long-term effects caused by low-frequency electromagnetic fields*

Low-frequency environmental radiation mainly comes from the power grid, as well as from devices powered directly by the 50 Hz mains. The latter radiate very locally, unlike power lines. The general tendency of these devices is the reduction of their radiation level. By far the best examples are the flat screens of televisions and handheld electric battery-operated devices, which have replaced cathode ray tube screens, and devices connected to the mains (see section 2.2.2.5).

At low frequencies, studies into people exposed to levels much lower than those that trigger acute effects have mainly focused on cancer. Additional research has focused on the neuroendocrine system (deregulation of hormone levels), neurodegenerative disorders (Alzheimer's disease, some forms of sclerosis), cardiovascular disorders and reproductive functions, as well as on development. A summary of the conclusions [SCE 15, ICN 10, IAC 13] is presented hereafter.

#### 4.1.5.3.1. Carcinogenic effects: analysis of the carcinogenicity mechanism

The publication of Wertheimer and Leeper's report in 1979 [WER 79] sparked controversy into the increase in cancer risk due to long-term exposure to low-frequency magnetic fields with low strength, and particularly exposure at 50 or 60 Hz. Their conclusion was based on the distance between homes and power lines. An average daily threshold of 0.4 μT was suggested in 2000. Many additional studies have been carried out over the three following decades. They have taken into account the magnetic field level or the electric current.

Causal links between exposure and cancers are not confirmed in adults, but given the gaps in this type of study, it is impossible to disregard an increase in the risk of leukemia in children.

The possible role of ELF magnetic fields as a cancer promoter through reduction in immune capacities has not been established [IAR 13b]. No credible biophysical or biological mechanism has been validated below 5 mT. More thorough research still needs to be carried out. The International Agency for Research on Cancer (IARC) has therefore classified magnetic fields as possibly carcinogenic to humans, 2B[3].

Studies in a work environment are more difficult to conduct, since people may be exposed to other physical agents or to chemical products.

---

3 Classification of the International Agency for Research on Cancer (IARC):

- agents that are carcinogenic to humans: group 1;
- agents that are probably carcinogenic to humans: group 2A;
- agents that are possibly carcinogenic to humans: group 2B;
- agents that are not classifiable as to their carcinogenicity to humans: group 3;
- agents that are probably not carcinogenic to humans: group 4;

2B classification means that there is limited evidence of carcinogenicity in humans and less than sufficient evidence of carcinogenicity in experimental animals; or inadequate evidence of carcinogenicity in humans but sufficient evidence of carcinogenicity in experimental animals; or inadequate evidence of carcinogenicity in humans and less than sufficient evidence of carcinogenicity in experimental animals together with supporting evidence from mechanistic and other relevant data. Biases, confusion factors and the fact that the results are obtained randomly are not excluded and do not allow a conclusion to be drawn.

#### 4.1.5.3.2. Effects on reproductive function and development

Recent results have confirmed previous reports, i.e. the absence of an effect of ELF fields on human reproductive functions [SCE 15].

Generally speaking, the results of studies systematically show the absence of an effect of electric and magnetic fields on development, including during pregnancy.

#### 4.1.5.3.3. Effects on the neuroendocrine system

Studies conducted on volunteers and epidemiological studies do not show that low-frequency electric and magnetic fields affect the neuroendocrine system in a way that would be likely to have an impact on human health.

#### 4.1.5.3.4. Cardiovascular disorders

Convincing data does not suggest that there is an association between exposure to low-frequency fields of levels often encountered in daily life or the professional environment and cardiovascular disorders.

#### 4.1.5.3.5. Neurodegenerative disorders

Epidemiological studies do not provide convincing proof and therefore do not make it possible to conclude that there is an increased risk of neurodegenerative disorders (Parkinson's disease, Alzheimer's disease, multiple sclerosis) due to exposure to low-frequency magnetic fields.

#### 4.1.5.3.6. Others

The results of studies on professionals exposed to 50 Hz electric or magnetic fields over long periods of time are not statistically significant enough to conclude that there is a risk of brain cancer or leukemia. Large-scale additional studies, carried out on many hundreds of thousands of electricity company employees, have not shown cardiovascular effects or negative psychological effects.

#### 4.1.5.3.7. Summary of long-term effects caused by low frequencies

The results suggest that, overall, there are no links between exposure to low-frequency fields and health disorders, or that they are fairly unconvincing, uncertain or, in the extreme, inconclusive, because they are incoherent and contradictory. Other symptoms, such as headaches and fatigue, which are said to be caused by electric or magnetic fields, are not proven.

#### 4.1.5.4. *Long-term effects due to RF (100 kHz–300 GHz)*

The main sources of RF radiation are radio and television broadcasting transmitters operating between 100 kHz and 2 GHz, mobile telephony base stations and mobile telephones. Distance from these sources and their transmission power are the two main factors determining exposure.

The conclusions of epidemiological studies and of laboratory studies in the radiofrequency domain can be distributed as follows [SCE 15, ANS 13, GAU 16, IARC 13].

##### 4.1.5.4.1. Cellular and animal studies

Recent studies into cells and animals reveal no negative effect at exposure levels below those that produce thermal effects.

Healthcare organizations have found no convincing proof demonstrating the presence of harmful health effects.

##### 4.1.5.4.2. Possible carcinogenic effects: the risk of cancer in humans

Many epidemiological studies have been conducted on increasingly large populations to reduce the uncertainties. They are mainly focused on the risk of brain cancer and cancer of the head (glioma, meningioma, neuroma).

The emergence of leukemia, melanoma (cutaneous and ocular) and other malignant illnesses due to radiofrequency fields has been studied.

Recognized healthcare organizations believe that proof of a link between exposure to high-frequency electromagnetic fields and cancer is either negative, insufficient, or far from conclusive.

##### 4.1.5.4.3. Non-carcinogenic effects on the CNS

Cognitive functions, sleep disorders, biological rhythms (circadian) and neurological disorders (Alzheimer's disease, epilepsy, etc.) have been studied.

Although physiological modifications have been observed in sleep, studies suggest that they have no consequences on the health.

#### 4.1.5.4.4. Non-carcinogenic effects outside the CNS

The results available on reproductive functions and development (fertility, behavior, teratogenicity, etc.) do not suggest that pathological effects may exist.

There is not enough evidence to conclude that exposure to high-frequency weak electromagnetic fields might have an effect on the cardiovascular system, the immune and endocrine systems and the ocular system in humans.

#### 4.1.5.4.5. Others

Some studies into well-being (in the general public) show a link between the perceived symptoms and the estimated distance to the mobile telephony base station antenna, rather than the real exposure level.

#### 4.1.5.4.6. Summary of long-term effects caused by high frequencies

There is no coherent evidence, across all the studies carried out, to support the conclusion that long-term exposure to weak radiofrequencies (lower than the limits established to protect humans against thermal effects) has a long-term harmful effect on humans. These studies do not show an increased risk of physiological or pathological effects, or of specific disorders. Temporary reactions and adaptation responses have been observed, although they cannot be shown to cause health effects, as they are located within normal physiological variations. The risk, if it exists, must be low.

The absence of a conclusion does not, however, mean that any possibility of risk can be completely excluded. On this basis, the IARC has also classified radiofrequency transmissions as possibly carcinogenic (2B).

The question of an increase in the risk of acoustic neuroma in people who use their mobile telephones intensively remains without a clear answer but a risk might exist.

Today, research is focusing on the possibility that metabolic changes following exposure to high-frequency electromagnetic fields might foster the development of tumor cells when they are initially present.

### 4.1.5.5. *Electromagnetic hypersensitivity*

Electromagnetic hypersensitivity (EHS) is the common name for idiopathic environmental intolerance attributed to electromagnetic fields (IEI-EMF)[4].

People who claim to have EHS suffer from various symptoms, which they attribute to exposure to electromagnetic fields, even very weak ones. No clinical sign is observed and the symptoms are medically unexplained. Some of the most commonly cited symptoms are fatigue, burning sensations, headaches, insomnia and mood disorders. They may cause distress or disorders that seriously affect quality of life.

The medical community, including the World Health Organization (WHO), recognizes these symptoms, but is unable to establish a causal link to exposure to electromagnetic fields [WHO 04].

The consistent results of multiple clinical trials carried out in double blind have shown that people who claimed to have EHS were incapable of distinguishing real exposure to electromagnetic fields from simulated exposure, in the same way as a person who does not claim to have EHS. Recent meta-analyses of observation and challenge results support this conclusion.

The lack of convincing proof of a causal link between exposure to electromagnetic fields and self-declared symptoms, along with the absence of a biological mechanism, suggests that exposure to radiofrequencies does not cause these effects, but that the problems are due instead to a *nocebo* effect, the opposite of the *placebo* effect (increase of reported symptoms after exposure to an inert treatment).

Although EHS syndrome is not recognized by the medical community, responsibility must be taken for these people, whose suffering cannot be denied.

Research in this domain is difficult, because, on the one hand, there are no clear diagnostic criteria, nor a scientific basis for linking the various reported symptoms, and, on the other hand, subjective reactions may be

---

4 Idiopathic environmental intolerance (IEI) is a generic term that covers a certain number of disorders with similar non-specific symptoms, which remain medically unexplained and produce effects detrimental to human health [WHO 05].

involved, independent of the direct effects of the fields. Work is continuing nonetheless [LAG 15].

#### 4.1.5.6. *Summary of long-term effects*

The risk of an adverse health effect following exposure to weak but continuous electromagnetic fields can be only slight, if it exists at all. The increase in the risk of developing a pathology cited in the studies is generally lower than the uncertainty, which sometimes produces contradictory results and makes conclusions impossible.

The number of studies has been decreasing for several years as a result of this observation, and countries that do not have sustainable funding are slowing down their research activities. They are continuing, but the avenues to be explored are almost infinite, which can discourage researchers. Are there particularly sensitive frequencies? Do particular exposure profiles promote a long-term effect? A modulation? A strength? Do fields promote another carcinogenic agent or a latent problem?

It is incorrect to think that electromagnetic fields are adverse just because they have not been proven not to be. It is important to remember that it is impossible to prove that something does not exist.

As scientific studies have not produced certainty as to a possible increase in certain pathologies, studies have recently been interested in the perception of risk associated with sociological and cultural factors, etc. [FRE 15].

| | Key points of long-term effects |
|---|---|
| Controversy | There is controversy surrounding long-term effects after long-term exposure to levels much lower than the threshold that produces acute reactions. |
| Long-term effect | Despite a great deal of research, today nothing indicates that exposure to weak electromagnetic fields is harmful to human health.<br>Today, no dose–response relationship has been determined for a mechanism that is possibly carcinogenic due to electromagnetic fields. |
| Problems relating to quality of life | Idiopathic environmental intolerance attributed to electromagnetic fields syndrome, commonly called electromagnetic hypersensitivity (EHS), is recognized by the WHO, although there is no cause–effect link with exposure to electromagnetic fields. |

**Table 4.4.** *Summary of long-term effects caused by electromagnetic fields*

Should real scientific proof emerge, the regulatory texts would doubtless adapt quickly. In the meantime, it is socially difficult to suggest restrictions limiting exposure without quantitative element(s) and without service degradation(s).

Nevertheless, electromagnetic fields save more lives today that they might be destroying.

## 4.2. Indirect effects

### 4.2.1. *General*

Indirect effects are the result of the presence of objects in the electromagnetic field, which may become the cause of a health or safety hazard. Indirect effects are in addition to direct effects.

Indirect effects can be produced over a wide range of frequencies. Some may be barely perceived or cause unpleasantness, while others may be fatal. Some examples are as follows:

– malfunctioning of active medical devices, either implanted or body-worn (pacemaker, medication pump, etc.);

– coupling with passive implants (orthopedic prostheses, etc.) or objects worn by choice (jewelry, body art, etc.) or otherwise (metal fragments, etc.);

– attraction and projection of ferromagnetic objects in a static magnetic field;

– ignition of electronic explosive devices (detonators);

– spark discharge, causing fires or explosions;

– combustion of radiofrequency absorber material near very powerful antennae;

– passage of an electric current causing an electric shock or burn during contact with an object polarized by a field, although it is galvanically isolated from an electric source.

Disturbance of electronic devices by external fields lies within the application domain of electromagnetic compatibility (EMC), which is not covered by texts relating to human exposure to electromagnetic fields.

Sections 4.2.2 and 4.2.3 deal more specifically with contact currents and the attraction of objects by static magnetic fields.

### 4.2.2. *Indirect effects caused by contact currents*

A contact current is not directly produced by the radiated field, but by the charges induced on the surface of a conductive object located in an electric, magnetic or electromagnetic field. Variable fields induce voltages and currents in these objects through inductive or capacitive coupling. A powered-down electric circuit (e.g. in maintenance), a fence, or a construction vehicle beneath a power line can become dangerous if they are not earthed, sometimes at several points (Figure 4.4). In some cases, it is the person who is electrically charged, while the object is connected to the ground.

When the object is touched, the accumulated charges will flow through the person. The physiological effects depend on the contact surface, for example touching with a fingertip or grasping with the whole hand, on the quantity of accumulated charges and on the permanent current frequency. The humidity and conductivity of the skin are also parameters that determine the strength of this current. These effects may be disagreeable, annoying or painful, or they may have lethal consequences [BAM 11, BER 88].

**Figure 4.4.** *Current flowing through a body on contact with an insulated object exposed to an electric field*

In general, two phases following contact, which are not necessarily exclusive, can be distinguished:

– current discharge over a fraction of a millisecond;

– permanent contact current.

A concentration of high charges flowing may cause nerve stimulation or muscle contraction.

Furthermore, burning through the Joule effect is possible, in particular when the current density is excessive due to a smaller contact surface (fingertip). A burn is less likely when contact is made with the whole hand.

Contact currents are more specifically perceptible on large objects, which are capable of accumulating enough electric charges, or on objects tuned in frequency, i.e. of a size equal to half a wavelength of the electromagnetic radiation.

These effects also have thresholds that depend greatly on the people, the properties of the objects and the frequency. These effects exist at least up to 110 MHz (there is insufficient data to draw a conclusion beyond this point). The lowest sensitivity thresholds are found at lower frequencies. Beyond 100 kHz, they do not depend much on the frequency and the effects are less perceptible to the touch, but there is still a real risk of burning or electric shock.

Furthermore, an initial electric discharge may be accompanied by a spark discharge (electric arc). These discharges occur when a person approaches a conductive object that is strongly electrically charged, without touching it or just before touching it. Sensitive people may perceive them from an electric field of 1 $kV \cdot m^{-1}$. The discomfort level is around double or triple.

Contact current limits (Ic) have been established by considering all these effects and taking into account a wide range of human sensitivity.

Sparks may also cause an explosion in a sensitive atmosphere.

Indirect effects are common in everyday life, but they are barely perceived and generally inconsequential. The simplest solutions for avoiding this discomfort are to earth objects that may charge, or move them away. However, in some situations, these effects cannot be avoided. One solution involves equalizing the potentials (equipotential link) and wearing insulating gloves, for example.

Contact currents circulating through the body (indirect effects) are separate from currents induced directly in the exposed body (direct effects). These effects may be present simultaneously between 100 kHz and 10 MHz. The ambient electric field levels required to produce indirect effects may be weaker than those that may cause a direct effect.

### 4.2.3. *Indirect effects caused by static magnetic fields*

The main indirect effect caused by very strong static magnetic fields is the attraction and projection of ferromagnetic objects. This requires the presence of a field source greater than 100 mT. This level is equal to 2,000 times the Earth's magnetic field level. In MRI rooms, a field of a few teslas (section 2.2.1.1) can turn a pair of scissors located several meters away into a projectile hurtling toward the tunnel of the device. This may result in serious injury or even a fatal outcome for the patient or medical personnel. The healthcare devices and equipment entering the room, including emergency equipment, must be appropriate (made up of ceramic, aluminum, titanium, etc.).

**Figure 4.5.** *Ferromagnetic objects are banned*

The presence of foreign ferromagnetic elements in the patient's body (ocular metal parts, cardiac valve, vascular clips, etc.) is an absolute contraindication for some MRI examinations. This also applies to tattoos and permanent make-up made from magnetite or iron oxide [BAM 11]. In addition to attraction, they may be constrained by a rotational torque.

Another effect of static fields is the disturbance of active medical implants such as cardiac pacemakers, neurostimulators and cochlear implants.

5

# Exposure Limits for Electromagnetic Fields

## 5.1. General considerations

Exposure limits for electromagnetic fields arise from established biological effects, such as those described in the previous chapter. The International Commission on Non-Ionizing Radiation Protection (ICNIRP) draws up guidelines, defining these limits based on studies into the potential risks associated with electromagnetic fields. This commission consists of scientists who do not have conflicts of interest, i.e. do not belong to commercial or industrial businesses, in particular those whose activity generates electromagnetic fields.

The World Health Organization, the International Labour Office, the International Radiation Protection Association and the European Union recognize this commission. It maintains relations with the most important international research and protection institutes in this domain.

ICNIRP regularly analyzes the results of scientific studies, and identifies new sources of exposure and new conditions of use for them. It also takes into account experiment feedback. It works meticulously and exhaustively to keep its knowledge of health effects up-to-date. Exposure limits are established on the basis of this work. ICNIRP takes into account biological effects that are considered established.

All the work is summarized in guidelines and recommendations, which are used internationally. Most standards and legislative documents are based on these publications.

The main biological effects used to establish the limits are central nervous system (CNS) stimulation, peripheral nervous system (PNS) stimulation and temperature increase in tissues. The first biological effect that may be adverse to health or safety is called the "critical effect". Exposure limits are deduced from them through consideration of the most unfavorable coupling conditions (full-body exposure in a uniform field, etc.). Furthermore, these limits have been established with a safety margin to take into account serious environmental conditions, uncertainty of results and knowledge, possible interpretations and methodological variations. This approach can be considered reliable. The physical quantities used to quantify these limits are called basic restrictions.

These physical quantities within the exposed body are difficult, or almost impossible, to measure. Reference levels have therefore been derived from them. These are external quantities, which are easier to assess.

Electric shocks or burns following contact with an object with a different potential (indirect effects) from the exposed person are also taken into account and have their own limit values.

Long-term effects, implied for levels much lower than the limit values, have not been adopted due to lack of scientific evidence.

Defining the population to be protected is one of the fundamental aspects of establishing the limits. The *general public* is distinguished from *workers (occupational exposure)*. The latter consists of adults who are generally exposed under known conditions, performing the activities required by their job during a working day. They are trained to be aware of the potential risk involved and are likely *a priori* to be medically supervised.

Wider considerations have been used for the general public, since this is composed of individuals in varying states of health and physical condition. The possible effects on the evolution of various illnesses must also be considered. In addition, the general public includes young children and elderly people, who have reduced thermoregulatory capacities. Furthermore, the general public may be exposed for 24 h a day, without being aware of it.

This distribution explains the need for a dual system of basic restrictions. Some are specific to workers, while others, which are more restrictive, are aimed at the general public. The latter have been established by applying an additional safety margin, and also taking into consideration a wider range of sensitivity to electromagnetic fields.

NOTE.– The exposure limits presented in this chapter are those applicable in legislation to the general public and to workers. They do not necessarily reflect the latest developments on the subject.

### 5.1.1. *Recommendation 1999/519/EC*

In 1999, the Council of the European Union published a recommendation [EC 99] to limit the exposure of the general public to electromagnetic fields. The purpose of this recommendation is to protect the general population from the established adverse effects which can occur following exposure to electromagnetic fields. This recommendation fully incorporates the limit values and terminology of the ICNIRP Guidelines published in 1998 [ICN 98].

This document is the current reference for the general public, who do not, as a rule, dispute the values protecting them from short-term effects. A minimum framework representing the state of scientific knowledge at this period was therefore provided.

Member states have partially or fully applied this recommendation by decree (e.g. *Décret français no. 2002-775 du 3 mai 2002 pour les radio-fréquences* [JOU 02]).

### 5.1.2. *Directive 2013/35/EU*

Directive 2013/35/EU (EMF Directive), on the minimum health and safety prescriptions regarding the exposure of workers to the risks arising from physical agents (electromagnetic fields), was published on June 29, 2013 [EC 01].

It was transposed into national legislation by the different member states of the European Union from July 1, 2016. This applies particularly to industrial, electrical energy and telecommunication companies, which use frequencies between 0 Hz and 300 GHz, intentionally or otherwise.

The EMF Directive, along with details of its specific features and its implementation, is introduced in Chapter 9. The exposure limits given in the EC Recommendation (general public) and the EMF Directive are presented in this chapter.

## 5.2. Low-frequency limits for direct effects

The primary key physical quantities used to define the exposure limits for preventing direct effects are called basic restrictions [ICN 98, ICN 10], exposure limit values[1] (ELVs) [EC 01] or maximum permissible exposure [IEE 05, IEE 02]. These values aim to prevent critical effects (first adverse effects) relating to exposure to electric, magnetic and electromagnetic fields in the 0 Hz–300 GHz frequency domain.

The basic restrictions preventing the stimulation of the CNS of the head are used in the EMF Directive, under the name "sensory effects exposure limit values", also called ELVSs in this book. The term "CNS of the head" is often shortened to "head", probably for reasons of writing simplification.

The basic restrictions preventing the stimulation of the PNS are called "health effects exposure limit values", also called ELVHs in this book.

"ELV" will be used to refer only to workers.

### 5.2.1. *Basic restrictions, ELVs at low frequencies*

For electric or magnetic fields between 1 Hz and 10 MHz, the main effect is the electrostimulation of excitable tissues by the induced fields and currents. The values determining the basic restriction in this domain are therefore physical quantities inside the exposed organism:

– the induced current density $J$ ($A \cdot m^{-2}$);

1 The EMF Directive and European standards relating to the exposure of workers prefer the term "exposure limit value (ELV)" to "basic restrictions", because it has traditionally been used for the management of risks due to physical agents, such as noise or vibrations. International standards use both terms, without distinguishing between them.

or:

– the internal electric field $E$ ($V \cdot m^{-1}$).

In its 1998 Guidelines, ICNIRP [ICN 98] used current density, while the 2010 Guidelines refer to the electric field [ICN 10], since it is the voltage (electric field over a certain distance in a conductive medium), which stimulates the nerve cells and other cells sensitive to electricity.

These basic restrictions were established based on the levels that cause adverse effects, taking into consideration a significant safety margin. They depend on the frequency to follow tissue sensitivity.

The basic restrictions for workers are different from those aimed at the general public. The latter are lower, because the general public may be more sensitive to electromagnetic fields.

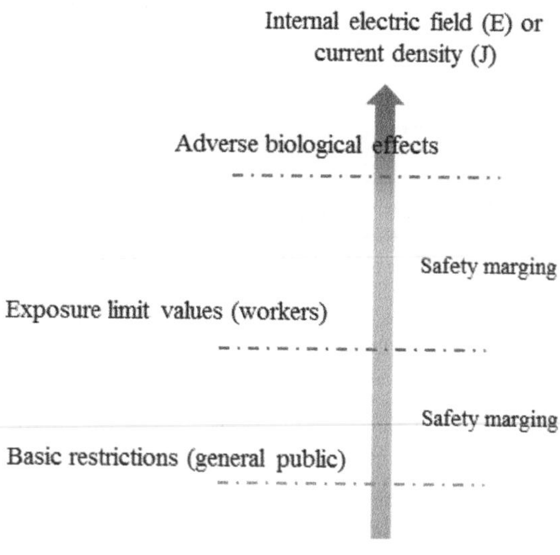


**Figure 5.1.** *Hierarchy of internal exposure limits*

As presented in section 4.1.1, the first risks of exposure to low-frequency fields are direct stimulation of the CNS of the head and stimulation of the PNS.

#### 5.2.1.1. *Limits for workers*

Magnetophosphenes are the first effects perceived due to the great sensitivity of retinal cells to induced electric fields. These cells belong to the CNS of the head. At a higher exposure level, some cerebral functions, such as motor coordination and the processing of visual information, may disturb the exposed person.

To protect people from these effects, ICNIRP defined a basic restriction for occupationals in 2010, at 50 mV·m$^{-1}$ in the 20 Hz range. This value rises toward the lowest and highest frequencies to follow the sensitivity curve of the CNS. At 50 Hz, the restriction reaches 100 mV·m$^{-1}$. There is no need to specify them beyond 400 Hz, because the field levels required to trigger a sensory effect are greater than those able to stimulate the PNS, with the possible consequence of a health effect.

Stimulation of the CNS is not dangerous to the health. This may disturb an exposed person if he has not been informed of it. Discomfort may be produced if the stimulation is sustained.

| Characteristics of exposure | | Frequency domain | Internal electric fields (V·m$^{-1}$ rms) | Internal electric fields (V·m$^{-1}$ peak) |
|---|---|---|---|---|
| Workers | Head tissues of the CNS (sensory effects) (4) | 1Hz–10Hz | 0.5/*f* | 0.7/*f* |
| | | 10Hz–25Hz | 0.05 | 0.07 |
| | | 25Hz–400Hz | 0.002.*f* | 0.0028.*f* |
| | All head and body tissues (health effects) (5) | 1Hz–3kHz | 0.8 | 1.1 |
| | | 3kHz–10MHz | 0.00027.*f* | 0.00038.*f* |

1) *f* is the frequency expressed in hertz (Hz).
2) The ICNIRP Guidelines define the rms values, while the EMF Directive gives peak values. The values are consistent for a sinusoidal field, where $E_{peak} = \sqrt{2} \cdot E_{rms}$.
3) In the range of frequencies over 100 kHz, the limits relating to thermal effects (due to high frequencies) presented in section 5.4 should also be taken into account.
4) These are spatial values in the central nervous system of the exposed person's head.
5) These are spatial values in the whole body of the exposed subject (head, trunk and limbs).

**Table 5.1.** *Basic restrictions or exposure limit values for workers according to Directive 2013/35/EU*

Stimulation of a peripheral nerve is likely to trigger a muscle contraction, which may indirectly damage the health or safety of the person (injury due to uncontrolled movement, etc.). The basic restriction preventing such stimulation was fixed at 800 mV·m$^{-1}$ for frequencies lower than 3 kHz. This value incorporates a safety margin, in order to prevent the first established effects, which are produced at around 6 V·m$^{-1}$. This limit increases linearly at higher frequencies, because the sensitivity of the nerve cells decreases. These values apply to the whole body, including the head.

Table 5.1 summarizes the ELVs that prevent the effects induced by electric and magnetic fields in the 1 Hz–10 MHz range. The limits in the 2010 ICNIRP Guidelines are rms values, given for a sinusoidal signal. The EMF Directive and its amendment [EC 10] refer to peak values, which should be compared with the instantaneous value of the signal. The ratio of peak value to rms value is √2 (this distinction, explained in Chapter 6, is fundamental).

Figure 5.2 illustrates the table (peak values).

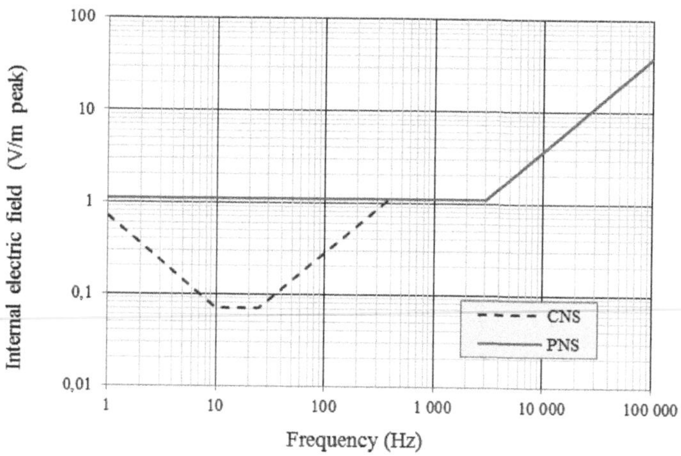


**Figure 5.2.** *Exposure limit values (basic restrictions), expressed in internal peak electric field strengths, to protect workers from sensory effects (CNS) and health effects (PNS)*

### 5.2.1.2. *Limits for the general public*

The limits for the general public relating to the CNS were reduced by a factor of 5 compared to those of workers, thus lowering the minimum threshold to 10 mV·m$^{-1}$. This factor is reduced to 2 for levels associated with the PNS, for which a significant safety coefficient has already been taken into account for workers. The values are summarized in Table 5.2.

| Characteristics of exposure | | Frequency domain | Internal electric fields (V.m$^{-1}$ rms) | Internal electric fields (V.m$^{-1}$ peak) |
|---|---|---|---|---|
| General public | Head tissues of the CNS (sensory effects) (4) | 1Hz–10 Hz | 0.1/$f$ | 0.14/$f$ |
| | | 10Hz–25 Hz | 0.01 | 0.014 |
| | | 25Hz–1,000 Hz | 0.0004.$f$ | 0.00057.$f$ |
| | All other tissues of the head and body (health effects) (5) | 1 Hz–3 kHz | 0.4 | 0.57 |
| | | 3 kHz–10 MHz | 0.000135.$f$ | 0.00019.$f$ |
| The notes from Table 5.1 also apply to this table. | | | | |

**Table 5.2.** *Basic restrictions for the general population (public) according to the 2010 ICNIRP Guidelines*

Table 5.2 can be summarized by Figure 5.3 (rms values).

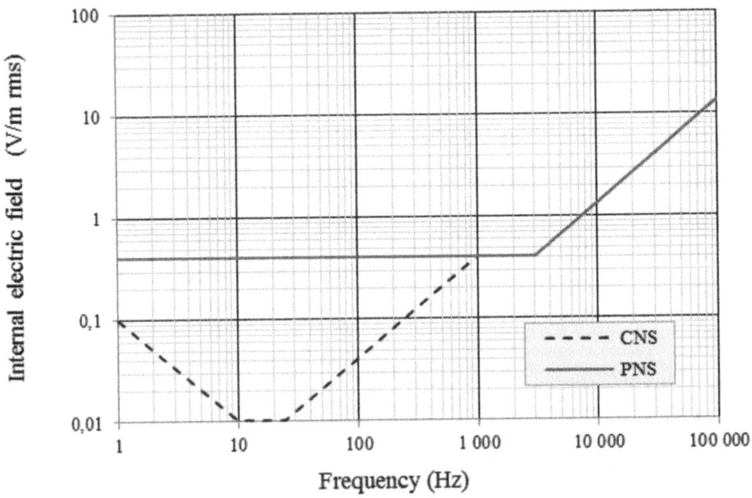


**Figure 5.3.** *Basic restrictions expressed in internal rms electric field strengths for protecting the general public from sensory effects (CNS) and health effects (PNS)*

The limits applicable to the general public determined by the 1999/519/EC Recommendation are those of the 1998 ICNIRP Guidelines. In these documents, sensory effects and health effects are not distinguished. Therefore, the lowest excitation values have been used. They are very conservative regarding the PNS.

The physical quantity taken into consideration in these two documents is the current density. The quantities and units corresponding to the exposure limit for the general public are summarized in Table 5.3 (the levels in the 1998 ICNIRP Guidelines concerning workers are no longer relevant since the publication of the EMF Directive).

| Characteristics of exposure | Frequency domain | Current density ($A.m^{-2}$ rms) |
|---|---|---|
| General public | <1 Hz | 8 |
| | 1Hz–4 Hz | 8/*f* |
| | 4 Hz–1 kHz | 2 |
| | 1 kHz–10 MHz | *f*/500 |

1) *f* is the frequency expressed in hertz (Hz).
2) Given the electrical inhomogeneity of the body, the current densities should be averaged over a section of 1 $cm^2$ perpendicular to the direction of the current.
3) For frequencies of up to 100 kHz, the peak current density is obtained by multiplying the rms value by $\sqrt{2}$ ($\approx$1.41). For pulses of tp duration, calculating the equivalent applicable frequency is recommended, for the calculation of basic restrictions according to the formula f = 1/2 tp.
4) For frequencies of up to 100 kHz and for pulsed magnetic fields, the maximum current density associated with the pulses can be calculated from the rise/fall time and the maximum variation speed of the magnetic flux density. The induced current density can then be compared to the applicable basic restriction (which means applying the weighted peak method introduced in Chapter 6).

**Table 5.3.** *Basic restrictions for the general public according to the 1998 ICNIRP Guidelines*

NOTE.– The limit value of 2 $mA{\cdot}m^{-2}$ from the 1998 ICNIRP Guidelines (general public) is consistent with the 10 $mV{\cdot}m^{-1}$ of the 2010 ICNIRP Guidelines, assuming an average tissue conductivity of 0.2 $S{\cdot}m^{-1}$ [3.1].

#### 5.2.1.3. *Peak values or rms values?*

A brief stimulus, lasting a few milliseconds, is enough to excite the CNS or PNS. For this, the EMF Directive follows the 2010 ICNIRP Guidelines in recommending taking into account the instantaneous values of stimuli. These are, therefore, peak values, which must be compared with the peak limit values. Assessment of exposure in the time domain is based on the instructions of the Directive (other methods are possible if they produce approximately equivalent and comparable results).

Recommendation 1999/519/EC (general public), based on work carried out before 1998, proposes using the weighted sum of the spectral components as a conformity criterion. This method can be used for signals stable in time. Practically, it is used only for a few specific applications, since it is difficult to implement for signals with complex waveforms. This technical point is detailed in Chapter 6.

### 5.2.2. *Reference levels or action levels at low frequencies*

The basic restrictions are corporeal physical quantities (induced electric field or current density). These quantities should be measured in a living organism with all its physical and biological faculties, which is difficult, even if ethical aspects are disregarded. However, it is possible to obtain them through calculations on digital models (heterogeneous anthropomorphic models) in which all the organs and their electrical properties are incorporated. This is the domain of dosimetry. This process requires specialized skills and specific simulation tools. To avoid this, reference levels have been derived from the basic restrictions in order to assess exposure in a practical way. The quantities corresponding to these limits can be measured or calculated at the position of the body (living area or workplace, but in the absence of the person).

To establish these levels, the strictest coupling conditions between the human body and the external field were used (field direction, position of the person, least suitable size, greatest exposed surface area, etc.) (see Tables 3.6 and 3.7). The uncertainties of the assessment techniques were considered.

These proposed limits are conservative and represent a warning level. This caution means a maximum protection level can be guaranteed.

NOTE.– Compliance with the reference levels guarantees compliance with the corresponding basic restrictions, in all cases. Exceeding them does not automatically mean exceeding the basic restrictions.

The term “reference levels” is used for the exposure of people (general public and workers), while the term “action levels (ALs)” is used exclusively by the EMF Directive for workers (in Europe), according to the "basic restrictions" and "ELVs" principle which is followed by this book. The reference levels applicable to low-frequency fields are:

– electric field $E$ (V·m$^{-1}$);

– magnetic induction B (μT) or sometimes magnetic field H (A·m$^{-1}$);

– current $I_c$ (A) due to contact with conductive objects (indirect effects).

These values are assessed in the working or living place, in the absence of the person, taking into account his position. This condition is required, because the body might unintentionally alter the radiation, rendering measurements non-reproducible and difficult to interpret (the limits were defined with these considerations in mind).

Measurements are not mandatory. Assessment through calculation is possible, in particular when the radiating sources are simple (e.g. single hanging cable). It is also possible to combine measurements and calculations. In very complex configurations, simulation and measurement are the only options.

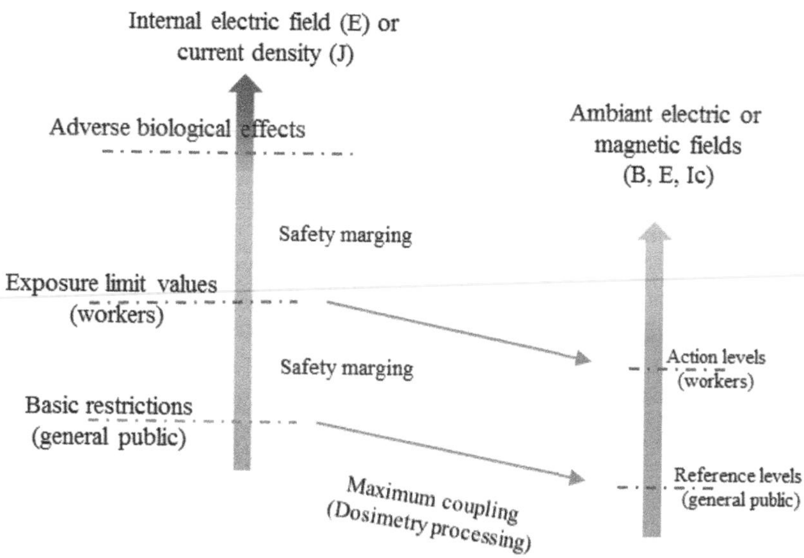


**Figure 5.4.** *General principles of exposure limits*

In Figure 5.4, it is not possible to place the induced electric field (or current density) on the same scale as external fields, because there is no direct link between these quantities. This link depends on the coupling coefficient (relationship between the internal field and the external field), which takes into consideration the direction and homogeneity of the field, the exposed body part, etc. This coefficient can vary by several orders of magnitude over the extent of the spectrum.

#### 5.2.2.1. *Exposure to low-frequency magnetic fields*

The reference levels (or ALs) associated with low-frequency magnetic fields were established from the basic restrictions (or ELVs) in the harshest exposure conditions. To simplify the assessment of exposure level, the Directive proposes a system with three types of AL for magnetic fields:

– *low ALs* aiming to prevent sensory effects. Compliance with these levels guarantees compliance with the ELVSs (sensory);

– *high ALs* aiming to prevent health effects in the head and trunk. Compliance with these levels guarantees compliance with the ELVHs (health);

– *limb ALs* aiming to prevent health effects in the arms and legs. Compliance with these levels guarantees compliance with the ELVHs (health).

Such an AL system does not exist in the 2010 ICNIRP Guidelines, although there are basic restrictions relating to sensory effects and basic restrictions relating to health effects.

The ELVs for the CNS (workers) of 100 $mV{\cdot}m^{-1}$ at 50 Hz (Table 5.1) associated with the most negative coupling conditions (induced electric field of 30 $mV{\cdot}m^{-1}$ for magnetic induction of 1 mT) led to a low AL of 1 mT to be defined in the 25 Hz–300 Hz band (Table 5.4 and Figure 5.5), taking a safety margin into consideration.

The same reasoning leads to high ALs in order to avoid stimulation of the PNS. The basic restrictions were set at 800 $mV{\cdot}m^{-1}$ rms (Table 5.1). A conservative coupling coefficient of 60 $mV{\cdot}m^{-1}$ per mT was used for the head and trunk at frequencies lower than 300 Hz. The level is therefore established at 6 mT at 50 Hz, also applying a safety margin.

From 300 Hz, the low and high action values meet (the critical effect becomes stimulation of the PNS). From 3 kHz, the joint effect of increasing coupling and loss of nervous tissue sensitivity with frequency leads to the ALs remaining constant.

Identical reasoning with an additional safety margin made it possible to define the reference level aimed at protecting the general public.

The EMF Directive distinguishes the head and trunk from the limbs for the PNS. The ALs (reference levels) of the latter are higher, because coupling is weaker. There must be a more significant magnetic field to induce the same level of electric field (the induced field is proportional to the exposed surface area). A conservative coefficient of 3 in relation to the body was added, raising the AL to 18 mT at 50 Hz.

Table 5.4 and Figure 5.5 summarize the ALs (applicable to workers).

Tables 5.5 and 5.6, as well as Figure 5.6, summarize the reference levels applicable to the general population (1998 and 2010 ICNIRP Guidelines).

| Frequency range | Low AL (B) (μT rms) | High AL (B) (μT rms) | Limbs AL (B) (μT rms) |
|---|---|---|---|
| 1 Hz–8 Hz | $200{,}000/f^2$ | $300{,}000/f$ | $900{,}000/f$ |
| 8 Hz–25 Hz | $25{,}000/f$ | $300{,}000/f$ | $900{,}000/f$ |
| 25 Hz–300 Hz | 1,000 | $300{,}000/f$ | $900{,}000/f$ |
| 300 Hz–3 kHz | $300{,}000/f$ | $300{,}000/f$ | $900{,}000/f$ |
| 3 kHz–10 MHz | 100 | 100 | 300 |

1) $f$ is the frequency expressed in hertz (Hz).

2) The ALs are rms values equal to the peak values divided by √2 for sinusoidal fields. The weighted peak values for magnetic induction must be compared with the peak action values for a field of any waveform. This point is discussed in detail in Chapter 6.

3) The ALs are the maximum values calculated or measured in the workplace in the absence of the worker. This leads to a conservative exposure assessment and automatic compliance with the ELVs in all non-uniform exposure conditions.

**Table 5.4.** *Action levels for exposure of workers to magnetic fields between 1 Hz and 10 MHz according to the 2013/35/EU Directive*

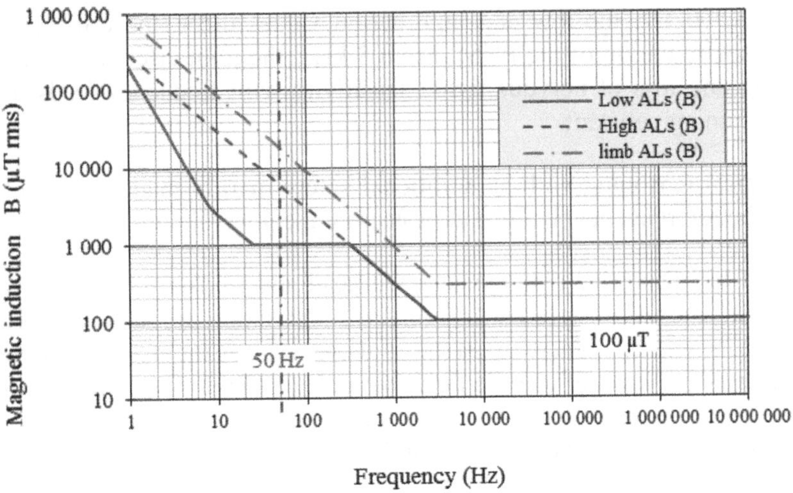


**Figure 5.5.** *Action levels for magnetic fields (rms values)*

| Frequency range | Magnetic induction (µT rms) |
|---|---|
| 1 Hz–8 Hz | $40{,}000/f^2$ |
| 8 Hz–25 Hz | $5{,}000/f$ |
| 25 Hz–400 Hz | 200 |
| 400 Hz–3 kHz | $80{,}000/f$ |
| 3 kHz–10 MHz | 27 |

1) $f$ is the frequency expressed in hertz (Hz).

2) The reference levels are rms values equal to the peak values divided by √2 for sinusoidal fields. The weighted peak values of magnetic induction must be compared with the peak action values for a field of any form. This point is discussed in detail in Chapter 7.

3) In the frequency range exceeding 100 kHz, the limits specific to radiofrequencies (presented in section 5.4) must be taken into account.

**Table 5.5.** *Reference levels for exposure of the general population to magnetic fields between 1 Hz and 10 MHz according to the 2010 ICNIRP Guidelines*

| Frequency range | Magnetic induction (μT rms) |
|---|---|
| <1 Hz | 40,000 |
| 1 Hz–8 Hz | $40,000/f^2$ |
| 25 Hz–800 Hz | $5,000/f$ |
| 800 Hz–150 kHz | 6.25 |
| 150 kHz–10 MHz | $920,000/f$ |

1) *f* is the frequency expressed in hertz (Hz).

2) The magnetic induction values may be exceeded provided that the basic restrictions are met and indirect adverse effects can be excluded.

3) For frequencies of up to 100 kHz, the peak current density is obtained by multiplying the rms value by √2 (≈1.414). For pulses of tp duration, calculating the equivalent applicable frequency is recommended, for the calculation of basic restrictions according to the formula f = 1/2 tp.

**Table 5.6.** *Reference levels for exposure of the general population to magnetic fields between 0 Hz and 10 MHz according to the 1998 ICNIRP Guidelines*

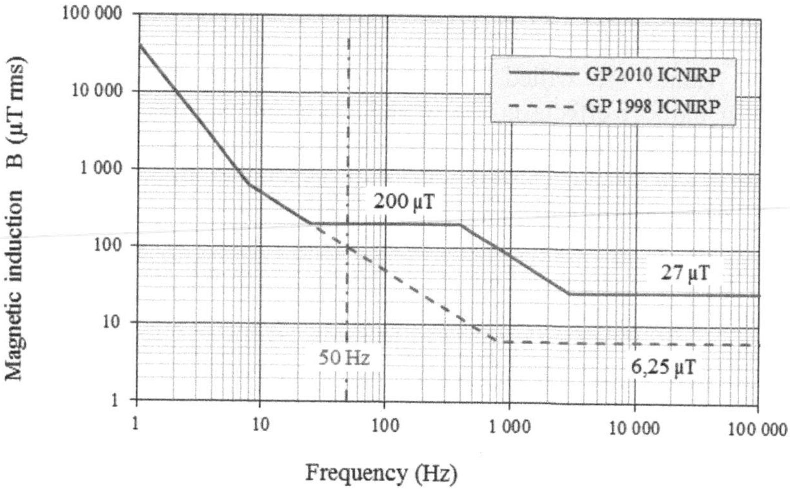


**Figure 5.6.** *Reference levels in rms value for magnetic (induction) fields*

The 1998 Guidelines (or EMF Recommendation [EC 99]) draw a subtle distinction between the reference levels for thermal effects and those for non-thermal effects. For the latter, they must remain constant beyond 150 kHz, as shown in Figure 5.6 [ICN 98].

#### 5.2.2.2. *Exposure to low-frequency electric fields*

The reference levels for electric fields were established in a different way from those for magnetic fields.

The coupling coefficient between the external electric fields and electric fields induced in the brain is around 2 $mV{\cdot}m^{-1}$ per $kV{\cdot}m^{-1}$ at 50 Hz. This coefficient becomes approximately 20 $mV{\cdot}m^{-1}$ per $kV{\cdot}m^{-1}$ in the skin [ICN 10].

Defining the AL at 20 $kV{\cdot}m^{-1}$ at this frequency means the margin is sufficient to guarantee both compliance with the sensory effect ELV (100 $mV{\cdot}m^{-1}$) and compliance with the health effect ELV (800 $mV.m^{-1}$) for workers. The levels decrease with the frequency, because coupling becomes more significant (the induced current increases with the frequency). Beyond a few kilohertz, the levels stabilize, because the increase in coupling is offset by the loss in tissue sensitivity. These levels also prevent painful spark discharges (section 4.1.1.1).

These values are called "high ALs". As these are inadequate for preventing indirect effects in the workplace, such as spark discharges and contact currents, "low ALs" were established (10 $kV{\cdot}m^{-1}$ at 50 Hz). However, beyond these values, cutaneous perceptions (hair vibrations) are possible, without the ELVs being exceeded [EC 01].

NOTE.– The definitions of low ALs and high ALs for electric fields are different from those for magnetic fields.

The reference levels for the general public are reduced by a factor of 2–4 in relation to the level for workers, to take into account a wider range of people and to prevent the effects of epidermal perception.

The ALs or reference levels represent the maximum calculated or measured values in the position of the person's body. They produce a conservative exposure assessment in particular when the electric fields are not uniform (situations encountered beyond a few kilohertz).

Table 5.7 summarizes the ALs aimed at protecting workers. They are illustrated by Figure 5.7.

| Frequency range | Low (E) ALs ($kV.m^{-1}$ rms) | High (E) ALs ($kV.m^{-1}$ rms) |
|---|---|---|
| 1 Hz–25 Hz | 20 | 20 |
| 25 Hz–50 Hz | 500/*f* | 20 |
| 50 Hz–1.64 kHz | 500/*f* | 1,000/*f* |
| 1.64–3 kHz | 500/*f* | 0.61 |
| 3 kHz–10 MHz | 0.17 | 0.61 |

1) *f* is the frequency expressed in hertz (Hz).

2) The low (E) AL and the high (E) AL are average quadratic values (rms) of the electric field strength, equal to the peak values divided by √2 for sinusoidal fields. These are values obtained in an unperturbed condition. Exposure to non-sinusoidal fields or multiple-frequency fields is discussed in Chapter 6.

3) For frequencies over 100 kHz, the limits specific to high frequencies (section 5.4) must also be taken into account.

**Table 5.7.** *Action levels for exposure of workers to electric fields between 1 Hz and 10 MHz according to the 2013/35/EU Directive*

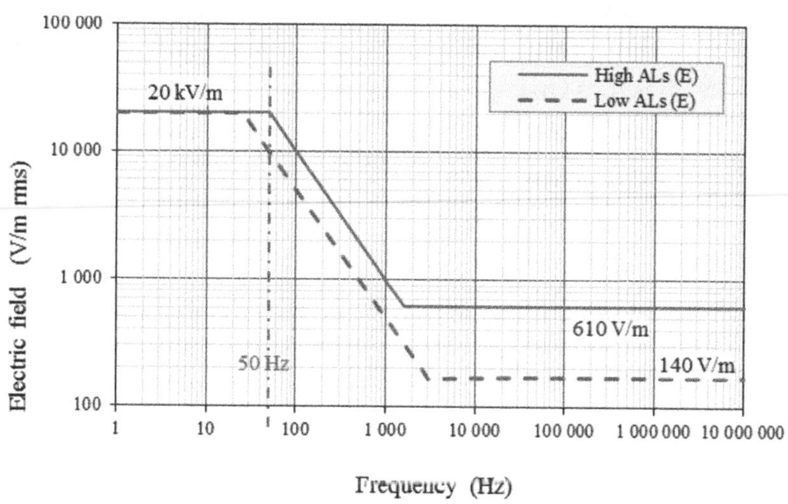


**Figure 5.7.** *Action values for electric fields applicable to workers (rms values)*

Tables 5.8 and 5.9 summarize the reference levels applicable to the general population according to the 2010 and 1998 ICNIRP Guidelines.

| Frequency range | Electric fields ($kV.m^{-1}$ rms) |
|---|---|
| 1 Hz–50 Hz | 5 |
| 50 Hz–3 kHz | 250/*f* |
| 3 kHz–10 MHz | 0.083 |

1) *f* is the frequency expressed in hertz (Hz).
2) The reference values are average quadratic values (rms) of the electric field strength, equal to the peak values divided by √2 for sinusoidal fields. These are values obtained in an unperturbed condition.
3) For exposure to non-sinusoidal fields or multiple-frequency fields, refer to Chapter 6.
4) For frequencies over 100 kHz, the limits specific to high frequencies (section 5.4) must also be taken into account.

**Table 5.8.** *Reference levels for exposure of the general public to electric fields between 1 Hz and 10 MHz according to the 2010 ICNIRP Guidelines*

| Frequency range | Electric fields ($kV.m^{-1}$ rms) |
|---|---|
| 1 Hz–25 Hz | 10 |
| 25 Hz–3 kHz | 250/*f* |
| 3 kHz–1 MHz | 0.087 |
| 1 MHz–10 MHz | $0.087/\sqrt{f}$ |

1) *f* is the frequency expressed in hertz (Hz).
2) The field strength values may be exceeded, providing that the basic restrictions are met and adverse indirect effects can be excluded.
3) For frequencies of up to 100 kHz, the peak value is obtained by multiplying the rms value by √2 (≈1.414). For pulses of tp duration, calculating the equivalent applicable frequency is recommended, for the calculation of basic restrictions according to the formula f = 1/2 tp.
4) No field value *E* is given for frequencies lower than 1 Hz, which correspond to static electric fields. Electric surface charges are not perceived at strengths lower than 25 $kV \cdot m^{-1}$. Spark discharges, a source of stress and discomfort, should be avoided.

**Table 5.9.** *Reference levels for exposure of the general public to electric fields between 1 Hz and 10 MHz according to the 1998 ICNIRP Guidelines*

Figure 5.8 illustrates Tables 5.8 and 5.9 (rms values).

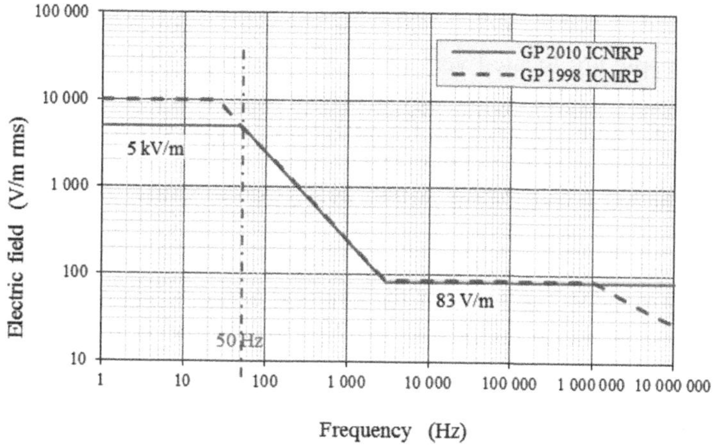


**Figure 5.8.** *Reference levels for electric fields applicable to the general public*

These curves show that in 2010, ICNIRP reduced the electric field reference level to 5 kV·m$^{-1}$ below 25 Hz, a domain that the general public will probably never meet.

As for magnetic fields, the 1998 Guidelines (or EC Recommendation) draw a subtle distinction between the reference levels applicable to electric fields, according to whether the effects are thermal or non-thermal. For the latter, they must remain constant beyond 1 MHz [ICN 98].

#### 5.2.2.2.1. Exposure to non-sinusoidal fields

The limits presented above are given for sinusoidal and continuous (permanent) fields. In many situations, voltages are switched at a few kilohertz and fields are distorted by harmonics. The instantaneous values should therefore be considered. Exposure assessment must be based on the method using the weighted peak technique, as specified by the 2010 ICNIRP Guidelines and the EMF Directive. This method is described in Chapter 6.

#### 5.2.2.2.2. Non-uniform electric or magnetic fields

The reference levels were defined in conditions of exposure to electric or magnetic fields with quasi-uniform spatial variation. This generally occurs when people are quite far away from the source, where the fields are not strong.

When people are located very close to a source, a few centimeters away (as is typical for workers), the field distribution is non-uniform and localized to one part of the body, making the reference levels particularly protective. When the localized field "moderately" exceeds these levels, it is not very likely that the basic restrictions will be exceeded [ICN 10].

The ICNIRP Guidelines indicate that measured field values may be spatially averaged in the place of the exposed body or one of its parts when the distance to the source exceeds 20 cm. In the case of a strongly inhomogeneous field encountered a few centimeters away from the source, verification in relation to the basic restrictions is almost the only option; this point is discussed in Chapter 8 (one possible alternative is to reduce the exposure by other means, see Chapter 9). The ICNIRP Guidelines leave scope for establishing standards or simplified recommendations to avoid using this process (dosimetry). The practical guide associated with the EMF Directive [CE 04] uses this principle. Some standards, such as [STD 07, STD 08], have already taken this option.

#### 5.2.2.3. *Simultaneous exposure to electric and magnetic fields*

Although, strictly speaking, the exposure levels of the electric field and the magnetic field should be added together, the 2010 ICNIRP Guidelines suggest that they should not be combined. The situations where it would be legitimate to do this are exceptional, because, on the one hand, the distribution and direction of electrically and magnetically induced fields are very different and, on the other hand, their maximum values would have to meet at the same point and be in phase. These conditions can be produced only exceptionally with one or several sources [ICN 10]. [ZOP 09] illustrate this situation perfectly.

#### 5.2.2.4. *Summary of the limits for low-frequency fields*

The principal reference documents for the protection of people against low-frequency fields are the 1998 ICNIRP Guidelines (general public) and the 2010 ICNIRP Guidelines (workers) in Europe. These are the main cornerstones of standards and legislative documents. Limits have been established to protect people from the emergence of adverse direct effects:

– *limits relating to sensory effects*: these aim to protect people from the effects of induced electric fields on the CNS of the head, i.e. perception of retinal phosphenes or minor changes in brain functions;

– *limits relating to health effects*: these aim to protect people from electrical stimulation of tissues of the PNS, whatever the exposed body part (head, trunk and limbs).

The general public and workers are distinguished. The general public, which presents a wider diversity of physical and health conditions, is protected by lower levels than those for workers by an additional margin of between 2 and 5.

The following summary (Table 5.10) can be made up of the exposure limits for low-frequency electric or magnetic fields (1 Hz–10 MHz).

| KEY POINTS OF LOW-FREQUENCY EXPOSURE LIMITS |
|---|
| Basic restrictions, also called exposure limit values, have been established to protect people from sensory and health effects. These effects are due to the stimulation of nervous tissues by low-frequency electric or magnetic fields. |
| The basic restrictions have been established with a safety margin in relation to the critical effects (first established effects). |
| The basic restrictions are defined in terms of the current density and electric field within the body. |
| Reference levels or action levels, measurable quantities, have been derived from the basic restrictions, in order to facilitate exposure assessment. Levels are defined for electric and magnetic fields. |
| The reference levels depend on the frequency, according to the sensitivity of the body to the critical effects and variations of the coupling coefficient. |
| The reference levels correspond to unperturbed field values, i.e. calculated or measured in the living place or workplace in the absence of people. |
| Exposure assessment in relation to reference levels is conservative under the conditions of non-uniform fields. |
| Sensory and health effects may appear instantaneously. Basic restrictions and reference levels should therefore not be compared with the average values in time, although they are given as rms values. |
| Compliance with the reference levels guarantees compliance with the basic restrictions. |
| Exceeding the reference levels does not automatically mean exceeding the basic restrictions. |
| The general public and worker exposure limits are distinguished. As the general public is more sensitive to EMF than workers, it is protected by lower levels. |

**Table 5.10.** *Summary of low-frequency exposure limits*

The protection system for workers, in particular regarding exposure to magnetic fields, is complex. Between the ELVs and ALs, there are five different thresholds. The hierarchy of these thresholds is explained in Chapter 9.

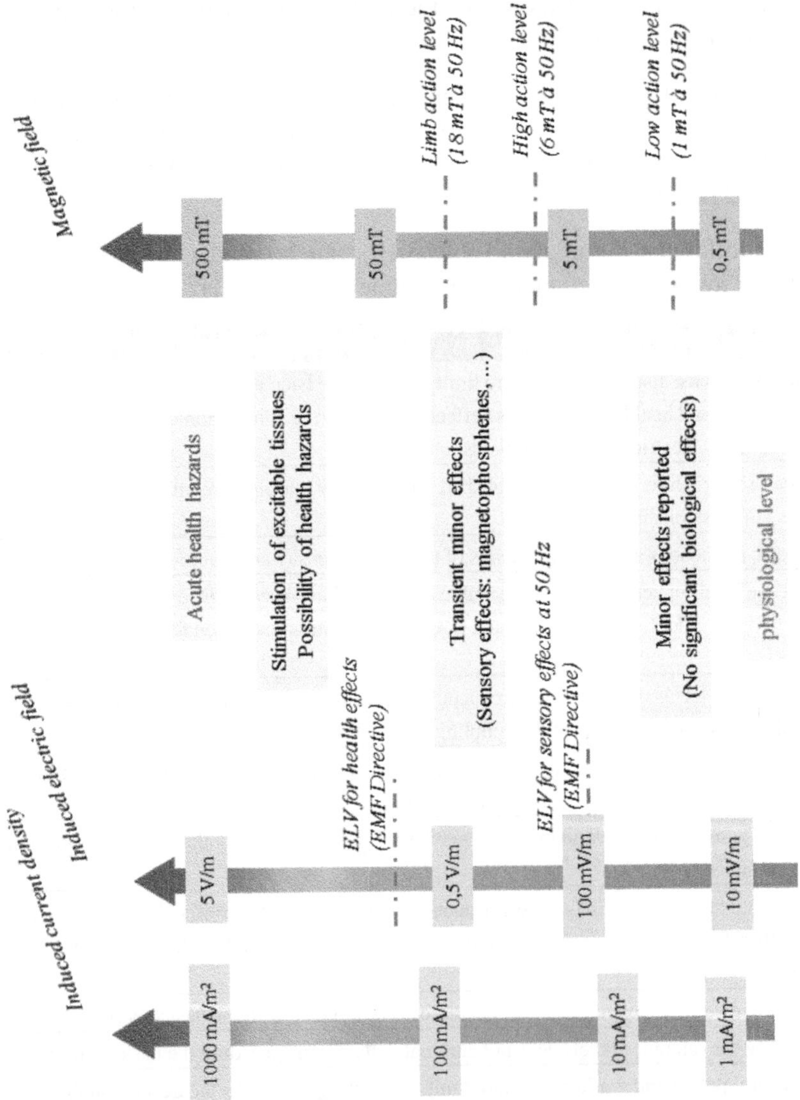


**Figure 5.9.** *Biological effects, induced fields, magnetic fields and limits*

Figure 5.9 summarizes the main biological effects occurring at low frequencies, induced electric fields, current density and the limits applicable to magnetic fields, as defined by the EMF Directive. The values mentioned are indicative, as they are orders of magnitude. There is no direct link between them. These links depend on the conductivity of the tissues and coupling mechanisms (Chapter 3).

It has been noted that the reference levels from the 1998 ICNIRP Guidelines were increased in the 2010 ICNIRP Guidelines to conform to the most recent scientific studies. Meanwhile, the method for comparing the measured or calculated fields and the limit has been clarified. Overall, this results in a better degree of protection for exposed people.

## 5.3. Limits for magnetic fields <1 Hz for direct effects

Limits have been defined to prevent sensory effects (vertigo, nausea, etc.) and the stimulation of the PNS due to exposure to a static magnetic field and a slowly varying magnetic field (frequency lower than 1 Hz).

### 5.3.1. *Basic restrictions for static magnetic fields*

The basic restrictions were established to limit exposure to static magnetic fields. The EMF Directive adopted the guidelines proposed by ICNIRP in 2009 [ICN 09a]. Table 5.11, for occupational exposure, is the result. The values are limits for exposure outside the body, and therefore measurable in the workplace.

| | Sensory effects ELVs |
|---|---|
| Normal working conditions | 2 T |
| Localized exposure of limbs | 8 T |

| | Health effects ELV |
|---|---|
| Controlled working conditions | 8 T |

**Table 5.11.** *Exposure limit values for external magnetic induction between 0 and 1 Hz*

Controlled working conditions include the provision of preventive measures, including movement control. The EMF Directive (published in 2013) could not clarify this aspect, because the relevant information was published by ICNIRP in 2014 [ICN 14].

It should be noted that the use of magnetic resonance imaging (MRI) equipment for medical purposes is exempt when certain conditions are fulfilled. These exemptions are detailed in legislative texts from the transposition of the EMF Directive.

### 5.3.2. *Basic restrictions for quasi-static magnetic fields*

In addition to exposure limits for static fields, ICNIRP has established a basic restriction for protecting people from effects due to movement in a static magnetic field thus perceived as a variable field [ICN 14]. It was set at 2 teslas ($\Delta B$) over every 3-s period. This limitation aims to prevent sensory effects.

| | Basic restrictions | | | |
|---|---|---|---|---|
| Critical effect | Vertigo due to movement in static field | Vertigo due to time-varying field | Stimulation of the PNS due to movement in static field and due to time-varying field | Phosphenes due to movement in static field and due to time-varying field |
| People | Workers | General public | Workers | General public |
| Physical quantities | $\Delta B$ (T) | $B_{peak\ to\ peak}$ (T) | Induced electric field ($V \cdot m^{-1}$ peak) | Induced electric field ($V \cdot m^{-1}$ peak) |
| | | | | |
| 0 Hz | 2 | | | |
| 0–1 Hz | | 2 | | |
| 0–0.66 Hz | | | 1.1 | 1.1 |
| 0.66–1 Hz | | | 1.1 | $0.7/f$ |

1) The maximum variation in magnetic induction $\Delta B$ is determined over any 3-s period.
2) For controlled exposure conditions, $\Delta B$ of 2 T may be exceeded. Assessment in relation to the basic restrictions for induced electric fields is established using the weighted peak method. The instantaneous values should be considered. This point is discussed in more detail in Chapter 6.

**Table 5.12.** *Basic restrictions for a static or quasi-static field (frequency lower than 1 Hz) [ICN 14]*

For the same reasons, a basic restriction ($B_{peak\ to\ peak}$) has been defined for exposure to a magnetic field with frequency lower than 1 Hz. This limitation is also confined to 2 teslas (peak-to-peak amplitude).

The basic restriction for preventing stimulation of the PNS is identical to that of low frequencies, i.e. 1.1 $V{\cdot}m^{-1}$ peak or 800 $mV{\cdot}m^{-1}$ rms, whether following movement in a static field or following exposure to a slowly varying field ($f < 0.66$ Hz). The safety margin used for low-frequency fields is thus retained. The basic restrictions for frequencies between 0.66 and 1 Hz are extrapolated from the values established for low frequencies. These values are summarized in Table 5.12.

### 5.3.3. *Reference levels for static magnetic fields*

In 2014, ICNIRP also published reference levels to guarantee that the basic restrictions expressed in induced electric fields are never exceeded, even in the harshest coupling conditions [ICN 14]. The reference level is a maximum field variation limited to 2.7 $T{\cdot}s^{-1}$ for workers. A frequency weighting was added for the general public, as per Table 5.13.

| | Reference levels | |
|---|---|---|
| Critical effect | Stimulation of the peripheral nervous system due to movement in static field and due to time-varying field | Phosphenes due to movement in static field and due to time-varying field |
| People | Workers | General public |
| | *dB/dt* ($T{\cdot}s^{-1}$ peak) | |
| 0–0.66 Hz | 2.7 | 2.7 |
| 0.66–1 Hz | 2.7 | 1.8/*f* |
| Reference level for a static or quasi-static field ($f < 1$ Hz), ICNIRP 2014. | | |

**Table 5.13.** *Reference levels for external magnetic induction between 0 and 1 Hz [ICN 14]*

NOTE.– Action levels for direct effects were not introduced in the EMF Directive, because the ELVs relating to the external values were deemed sufficient.

## 5.4. High frequency limits for direct effects

### 5.4.1. *High-frequency basic restrictions or ELVs*

Basic restrictions or ELVs have been established for high frequencies to protect people from thermal biological effects. These are direct effects. The quantities used are in relation to the power:

- SAR, specific absorption rate ($W{\cdot}kg^{-1}$);
- SA, specific absorption ($J{\cdot}kg^{-1}$);
- *S*, power density ($W{\cdot}m^{-2}$).

The SAR is used for quantifying the power absorbed per unit of mass [3.8]. The absorbed power must be averaged over a period of 6 min at frequencies lower than 6 GHz, taking into consideration the thermal time constant of the tissues.

The SA represents the energy absorption over a short period (energy = power × time). It makes it possible to limit exposure to electromagnetic fields with high peak amplitude, but low average value (pulsed fields).

The power density (*S*) represents the power received per unit surface area.

In order to establish the exposure limits at high frequencies, a safety coefficient of 10 was applied to the human homeostatic capacities, estimated conservatively at 4 $W{\cdot}kg^{-1}$ for the whole body (see section 4.1.3). The average whole-body SAR is thus fixed at 0.4$W{\cdot}kg^{-1}$ for workers. At this limit, body temperature could increase by no more than 0.1 °C in the most negative conditions.

This restriction was set at 10 and 20 $W{\cdot}kg^{-1}$ for localized exposure in the head or trunk and in the limbs, respectively, since heat is distributed rapidly throughout the rest of the body.

By way of comparison, the basal metabolism produces heat of around 1 $W{\cdot}kg^{-1}$ in a person. It may exceed 20 $W{\cdot}kg^{-1}$ peak during an athletic endeavor [DUR 86].

An additional safety margin of 5 is applied to general public, raising the safety coefficient to 50, thus encompassing a wider range of people and environments. The SAR for the general public is thus established at 0.08 $W{\cdot}kg^{-1}$ for the whole body, at 2 $W{\cdot}kg^{-1}$ for local exposure of the head and trunk and at 4 $W{\cdot}kg^{-1}$ for localized exposure of the limbs. The value for the head is the limit applicable (in Europe) to mobile telephones and other handheld devices.

The basic restrictions for workers and the general public are summarized in Table 5.14 for frequencies between 100 kHz and 10 GHz (volumic thermal effects). These values are drawn from the 1998 ICNIRP Guidelines. They were repeated in the 1999/519/EC Recommendation to protect the general public and in the EMF Directive to protect workers.

| Characteristics of exposure | Frequency domain | Average SAR whole body ($W \cdot kg^{-1}$) | Local SAR (head and trunk) ($W \cdot kg^{-1}$) | Local SAR (limbs) ($W \cdot kg^{-1}$) |
|---|---|---|---|---|
| Workers | 100 kHz–6 GHz | 0.4 | 10 | 20 |
| General public | 100 kHz–10 GHz | 0.08 | 2 | 4 |
| 1) The SAR should be averaged over 6 min.<br>2) The SAR must be assessed on 10 g of adjacent tissue. | | | | |

**Table 5.14.** *Exposure limit values (SAR) for exposure to a field between 100 kHz and 10 GHz*

The local SAR must be verified in any sufficiently small volume to offer a high level of protection while guaranteeing that the average SAR is not exceeded.

Beyond a few gigahertz, the incident power density (power radiated per surface unit) is relevant for characterizing exposure and establishing basic restrictions, since the waves induce only superficial heating. The limit for workers was established using a margin of 5 in relation to the critical effect, which is at 250 $W \cdot m^{-2}$ (section 4.1.3). An additional margin of 5 was applied to define general public reference levels. These are summarized in Table 5.15.

| Characteristics of exposure | Frequency domain | Power density ($W \cdot m^{-2}$) |
|---|---|---|
| Workers | 6 GHz–300 GHz | 50 |
| General public | 10 GHz–300 GHz | 10 |

**Table 5.15.** *Limit (S) for exposure to an electromagnetic field of over 6 GHz*

NOTE.– This limit (power density) is set at 6 GHz by the EMF Directive, but at 10 GHz in Recommendation 1999/519/EC. Other documents use 3 GHz as a boundary [IEE 05, ARP 02].

These basic restrictions are values external to the body and therefore measurable, which is why they are used for the reference levels. These values can be compared to the basal metabolism expressed in terms of power dissipated per unit of corporeal surface area, i.e. around 40 $W \cdot m^{-2}$; 10 and 50 $W \cdot m^{-2}$ correspond to a far electric field of 61 and 140 $V \cdot m^{-1}$, respectively, which ensures the continuity of the values with frequencies lower than 6 GHz.

Power density should be averaged over a surface area of 20 $cm^2$, without for any reason exceeding 20 times the power density authorized over a surface area of 1 $cm^2$.

Exposure must be assessed over a period that becomes shorter as the heated volume becomes more reduced, i.e. as the absorption becomes shallower (the skin effect intensifies with frequency). Below 10 GHz, exposure is determined over a period of 6 min. Beyond this point, power densities are averaged over a time interval of $68/f^{\,1.05}$ minutes ($f$ is the frequency in GHz). Thus, the duration is between 6 min for a frequency of 10 GHz and 10 s for the extreme frequency (300 GHz).

#### 5.4.1.1. *Sensory effects due to high frequencies*

A basic restriction and an ELV relating to sensory effects were established to prevent auditory effects caused by exposure of the head to pulsed microwave signals (section 4.1.3.2). The specific absorption or energy absorbed per unit of mass (SA in $J \cdot kg^{-1}$) is adequate for quantifying this value. It must be assessed over 10 g of adjacent tissue. These values are proposed by the 1998 ICNIRP Guidelines. The EMF Directive uses those applicable to workers.

| Characteristics of exposure | Frequency domain | Specific absorption ($mJ \cdot kg^{-1}$) |
|---|---|---|
| Workers | 0.3 GHz–6 GHz | 10 |
| General public | 0.3 GHz–10 GHz | 2 |

**Table 5.16.** *Exposure limit values (SA) for exposure to a field between 300 kHz and 6 GHz (1 joule = 1 W over 1 s)*

EXAMPLE 5.1.– For occupational exposure to a pulsed signal, the SA is limited to 10 mJ·kg$^{-1}$ in the head. This value is reached, for example, during exposure to an electromagnetic pulse with an emission duration of 10 μs associated with a peak SAR of 1,000 W·kg$^{-1}$ (1,000 W·kg$^{-1}$ × 10 μs = 10 mJ·kg$^{-1}$). The average corresponding SAR is 1 W·kg$^{-1}$ for a repetition period of 10 ms.

EXAMPLE 5.2.– A 2G GSM mobile telephone emits a modulated pulse every 4.6 ms (217 Hz). The emission duration is 1/8 of this period, at 0.577 ms. Taking an average SAR of 1 W·kg$^{-1}$ as a hypothesis, the peak SAR is 8 W·kg$^{-1}$. The SA takes on the value of 4.6 mJ·kg$^{-1}$ (8 W·kg$^{-1}$ × 0.577 ms).

This example shows that the ICNIRP Guidelines are conservative, since no auditory effect seems to have ever been reported following use of a mobile phone. The SA should cover only pulses lower than 50 μs, as specified in certain regulations (e.g. [ARP 02]). In practice, this value is in relation to exposure to long-range radar signals (military applications, aeronautical applications, etc.).

In medicine, tissues are heated locally through application of high-frequency signals, with the aim of producing a therapeutic effect (Chapter 2). The SAR values applied are around 10–50 W·kg$^{-1}$.

The SAR, which can be obtained by variations in tissue temperature, internal electric fields, etc., is not measurable in living people and is not representative in an inert body. The ethical aspects of this cannot be disregarded. Numerical and experimental dosimetry techniques applied to anthropomorphic models avoid this difficulty (Chapter 8).

### 5.4.2. *High-frequency reference levels and action values*

The SAR can only be determined in a laboratory, outside a real exposure configuration. To overcome all the assessment difficulties, and in particular to be able to carry out measuring on site, reference levels (or ALs) were introduced, as for low-frequency fields. The proposed quantities are:

– electric field $E$ (V·m$^{-1}$);

– magnetic field $H$ (A·m$^{-1}$) or magnetic induction $B$ (μT);

– power density $S$ (W·m$^{-2}$);

– induced current $Ii$ (A).

The electric field is generally the quantity chosen to measure radiation in the radiofrequency domain for practical reasons.

The reference levels are obtained from the basic restrictions, supposing that the absorption conditions of the harshest ambient field (uniform exposure of the body, normal wave incidence, least favorable polarization, etc.) are combined. The reference levels encapsulate all the absorption curves, in particular those that depend on the size and posture of the people (see section 3.3.3).

The base limit for protecting people from body temperature elevation is the SAR. By containing this defined constant SAR in a large frequency range (100 kHz–6 GHz) and taking into consideration the great variability in frequency of the incident fields coupling, limits which depend on frequency can be established. The more significant the absorption, the lower the reference level. Logically, these levels are the inverted image of the absorption curve.

Table 5.16 summarizes the ALs in the frequency range between 100 kHz and 300 GHz. These are the rms values of unperturbed fields. They are to be averaged over the volume occupied by the exposed body and over a 6-min period (or less beyond 10 GHz).

| Frequency domain | Electric field ($V{\cdot}m^{-1}$) | Magnetic field ($A{\cdot}m^{-1}$) | Magnetic induction (μT) | Power density ($W{\cdot}m^{-2}$) |
|---|---|---|---|---|
| 100 kHz–1 MHz | 610 | $1.6/f$ | $2/f$ | – |
| 1 MHz–10 MHz | $610/f$ | $1.6/f$ | $2/f$ | – |
| 10 MHz–400 MHz | 61 | 0.16 | 0.2 | – |
| 400 MHz–2 GHz | $3f^{1/2}$ | $0.008f^{1/2}$ | $0.01f^{1/2}$ | – |
| 2 GHz–6 GHz | 140 | 0.36 | 0.45 | – |
| 6 GHz–300 GHz | 140 | 0.36 | 0.45 | 50 |

1) $f$ is the frequency expressed in MHz.
2) For high-frequency pulses, the power density calculated over the duration of the signal must not exceed 1,000 times the average power density, or the rms electric field calculated over the duration of the pulse must not exceed 32 times the average corresponding value cited in this table.

**Table 5.17.** *Exposure of workers to high frequencies (action levels)*

The values of electric fields, magnetic fields and power density are connected by the far field relationships given in [1.15] and [1.16].

The reference level beyond 6 GHz is identical to the basic restriction (power density is determined by directly measuring the ambient field).

The following table brings together the reference levels aimed at the public in the 100 kHz–300 GHz frequency range [ICN 98].

| Frequency domain | Electric field ($V \cdot m^{-1}$) | Magnetic field ($A \cdot m^{-1}$) | Magnetic induction ($\mu T$) | Power density ($W \cdot m^{-2}$) |
|---|---|---|---|---|
| 100 kHz–150 kHz | 87 | 5 | 6.25 | – |
| 150 kHz–1 MHz | 87 | $0.73/f$ | $2/f$ | – |
| 1 MHz–10 MHz | $87/f$ | $0.73/f$ | $2/f$ | – |
| 10 MHz–400 MHz | 28 | 0.073 | 0.092 | 2 |
| 400 MHz–2 GHz | $1.375 f^{1/2}$ | $0.008 f^{1/2}$ | $0.0046 f^{1/2}$ | $f/200$ |
| 2 GHz–300 GHz | 61 | 0.16 | 0.20 | 10 |

1) $f$ is the frequency expressed in MHz.
2) For pulses of frequencies higher than 10 MHz, the power density averaged over the duration of the signal must not exceed 1,000 times the corresponding rms value (or 32 times if the assessment is characterized by electric or magnetic fields).

**Table 5.18.** *Reference levels (general public) at high frequencies*

Curve 5.10 shows the reference levels of high-frequency electric fields for workers (ALs) and the general public. It is the symmetry along the frequency axis of the absorption curve envelope illustrated by Figure 3.20.

In situations of exposure to multiple-frequency fields, the absorbed powers have to be added up. This point is discussed in detail in Chapter 6.

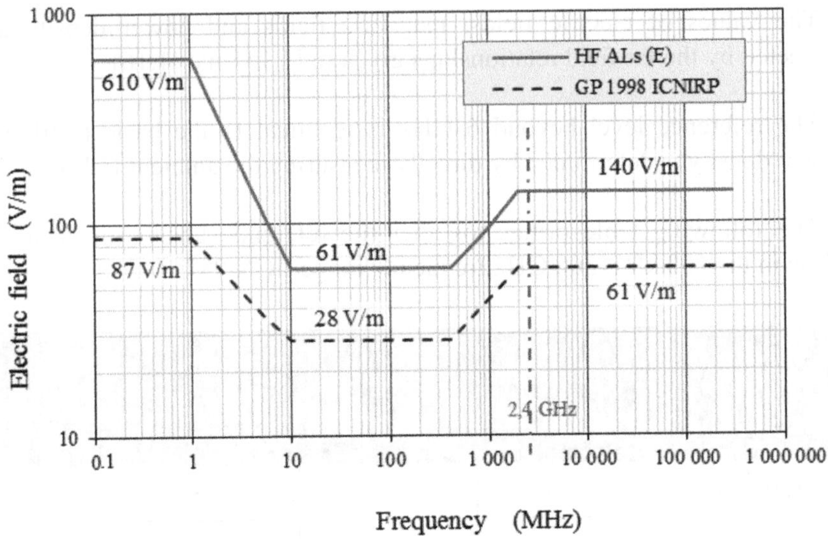


**Figure 5.10.** *Limits for HF incident electric fields*

### 5.4.3. *Limits for induced currents*

When the frequency is lower than a few tens of megahertz, the SAR averaged over the body is low and the heating is concentrated in the limbs, particularly in the ankles and wrists (Figure 3.19). This heating is due to the currents induced by the electromagnetic fields, although there is not necessarily any physical contact with the source or the environment.

The reference levels for electric fields are generally sufficient to ensure compliance with the limits of local SARs. However, in very disadvantageous conditions (person very close to the ground, field orientated unfavorably and only one body part exposed), the local SAR may be excessive in the limbs, even with fields below the reference levels (when measurable).

Reducing the reference levels to take into account these very specific situations, encountered only in an occupational environment (use of a high-frequency welding press or work near radio broadcasting antennae, see Chapter 2), may prove detrimental to all other situations.

To protect people working almost in contact with intense radiating sources, reference levels defined in currents induced (*Ii*) in the limbs have been proposed, as have limits for contact currents (indirect effects). An induced current may flow through one or several limbs of the person. These currents, expressed in amperes, have been established from current densities, themselves derived from SAR values [3.10].

The domain covered extends from 10 to 110 MHz, including the FM radio band. The maximum frequency is imposed by lack of data at greater frequencies, rather than by an absence of effects, and heating is not significant below 10 MHz.

| Frequency domain | Exposed people | Current induced in any limb (mA rms) |
|---|---|---|
| 10 MHz–110 MHz | Workers | 100 |
| | General public | 45 |
| These values should be averaged (rms) over any period of 6 minutes. | | |

**Table 5.19.** *Limits for currents induced (Ii) in the limbs (ICNIRP 1998)*

These limits are applied alongside the basic restrictions (SAR) or reference levels for the whole body. Compliance with the 100 mA limit ensures compliance with the limbs SAR in the event of occupational exposure (20 $W \cdot kg^{-1}$), even if the electric field strength exceeds the AL (it is 61 $V \cdot m^{-1}$ between 10 and 400 MHz).

The interest of these limits lies in being able to verify the exposure level directly by measuring currents that are non-invasive to the exposed person in a very near field. These are areas where measuring the electric field is inappropriate, mainly due to mutual coupling between the measuring antenna and the source of the radiation or surrounding metal structures (Chapter 8).

An exposed person may also suffer an electric shock or burn following contact with an object with a significant electric surface charge. This is why limits for contact currents also exist (section 5.5.1).

### 5.4.4. *Summary of limits for high frequencies*

The main biophysical mechanism at high frequencies is the absorption of electromagnetic energy, leading to temperature elevation in biological tissues. The limits published by ICNIRP in 1998 [ICN 98] aim to protect exposed people from adverse thermal effects. These limits were also used in the EMF Directive [CE 01]. The following summary (Table 5.20) has been undertaken in the frequency domain between 100 kHz and 300 GHz.

| Key points of high-frequency exposure limits |
|---|
| Limits of SAR and induced currents have been established for many years to protect people from thermal effects that could damage their health. |
| There is a whole body SAR (average value) and local SARs (head, trunk and limbs). |
| Reference levels, measurable external quantities, have been derived from the SAR, since this quantity, denoted by $W \cdot kg^{-1}$, cannot be directly measured or simply calculated. Compliance with these levels ensures compliance with the SAR. |
| The reference levels were established for a body entirely exposed to uniform fields. The values of fields measured in place of the body should be spatially averaged, provided that local SARs are complied with. |
| The reference levels and SARs are values averaged over a period of 6 min ($f < 6$ GHz). A peak reference level exists for pulsed fields. |
| Electromagnetic fields remain practically on the surface of the skin at frequencies higher than a few gigahertz. For this reason, the power density of the field ($W \cdot m^{-2}$) is substituted for the SAR, which is adapted to the energy absorbed per volume. The average value of this density is calculated over a time that reduces when the frequency increases. |
| Measuring induced currents (10 MHz $\leq f \leq$ 110 MHz) avoids calculating the SAR during specific exposures. |

**Table 5.20.** *Summary of high-frequency exposure limits*

## 5.5. Limits for indirect effects

The risks incurred by the indirect effects should be assessed separately from the potential risks relating to the direct effects. The previous reference levels do not cover these risks, which are produced only when many specific conditions are met simultaneously. Indirect effects may become the cause of a safety or health hazard.

There are three types of limits which aim to prevent some of these effects:

– limits for contact currents;

– limits for low-frequency electric fields;

– limits for static magnetic fields.

### 5.5.1. *Limits for contact currents*

Limits aim to prevent shocks and burns from contact currents between a body and a conductive object located in a field when they are not at the same potential (e.g. only one of them is connected to the earth). These limits do not apply to live conductors, i.e. directly powered by an energy source.

In order to prevent these indirect effects, reference levels in electric current strength (A) have been established, taking into account the fact that for children and women, the critical thresholds are, respectively, half and two-thirds of the levels corresponding to an adult man. For the general public (GP), the thresholds have therefore been divided by two. These limits do not prevent perception, but painful electric shocks [ICN 98, ICN 10, EC 01]. Perception of a contact current is not harmful in itself, but it can be considered a discomfort. A distinction is made between the general public and workers (Table 5.21).

| | Maximum contact current (mA rms) | |
|---|---|---|
| Frequency domain | General public | Workers |
| Up to 2.5 kHz | 0.5 | 1.0 |
| 2.5 kHz–100 kHz | 0.2.$f$(kHz) | 0.4.$f$(kHz) |
| 100 kHz–110 MHz | 20 | 40 |
| 1) $f$ is the frequency expressed in kHz. | | |

**Table 5.21.** *Reference levels for currents due to contact with conductive objects located in an electromagnetic field*

The frequencies involved in contact currents extend up to 110 MHz. These limits are given for a steady-state current, i.e. with the person in continuous contact with a conductor. A spark discharge caused by the flow of a current is likely to form just before physical contact.

The value remains constant beyond 100 kHz, since the threshold which causes effects (from tactile perception and severe pain through to serious burns) is constant [ICN 98].

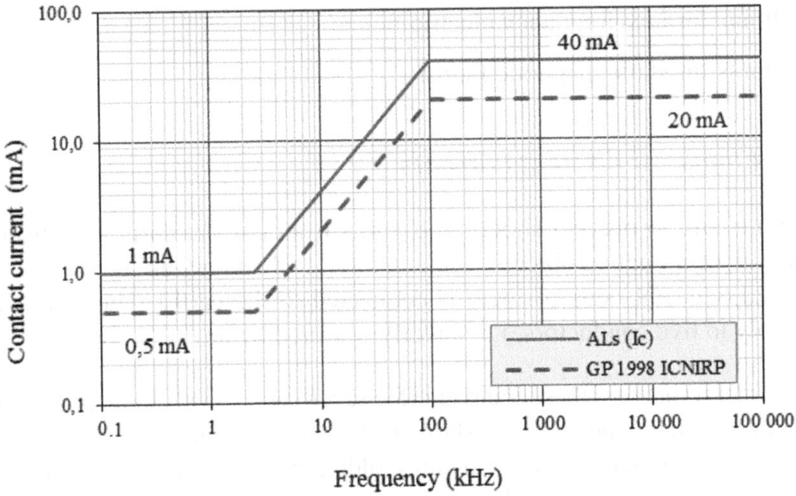


**Figure 5.11.** *Graphic representation of the limits for contact currents (ALs for workers; general public stands for general public)*

## 5.5.2. *Limits for electric fields, indirect effects*

As indicated in section 5.2.2.2, the EMF Directive specifies low ALs (10 kV·m$^{-1}$ at 50 Hz) in order to limit spark discharges in the working environment. Below the high ALs (20 kV·m$^{-1}$ at 50 Hz), these sparks are avoided, provided that prevention measures are taken, such as earthing, a connection between people and structures (equipotential link) or the use of personal protective equipment (insulating shoes, gloves or clothing).

## 5.5.3. *Limits for static magnetic fields, indirect effects*

Static magnetic fields can also produce indirect effects. These are risks of attraction or projection of ferromagnetic objects and interference with active implanted medical devices, such as pacemakers. The EMF Directive [CE 01] defines ALs to limit these effects.

| Risks | Action values ($Bo$) |
|---|---|
| Interference with active implanted medical devices | 0.5 mT |
| Attraction and projection of objects | 3 mT |

**Table 5.22.** *Action values for static magnetic induction*

Attraction and projection of objects may occur in the area close to the sources of strong fields (>100 mT). This situation is encountered in examination rooms equipped with an MRI machine or very close to electromagnets. There are therefore strict safety rules.

## 5.6. Summary of exposure limits

ICNIRP has developed guidelines on the basis of established effects that may affect the health of people exposed to electromagnetic fields. These guidelines were used in the 1999/519/EC Recommendation [REC 99] and the EMF Directive [CE 01] published by the European Commission with the aim of protecting the general public and workers, respectively.

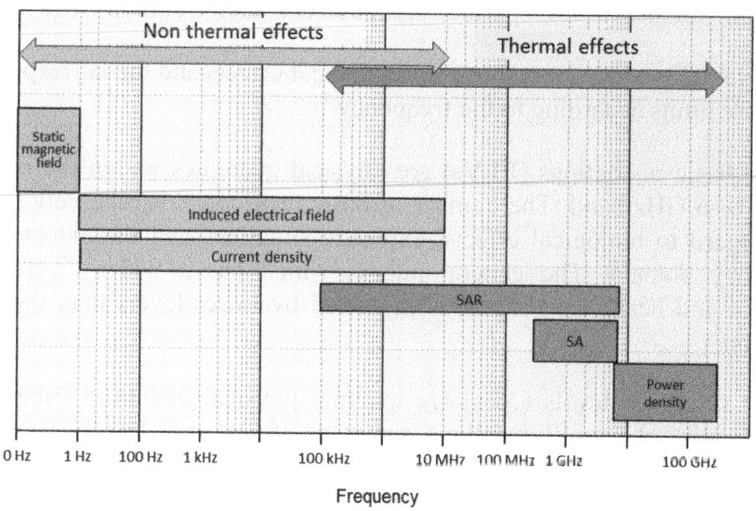


**Figure 5.12.** *Distribution of basic restrictions (exposure limit values)*

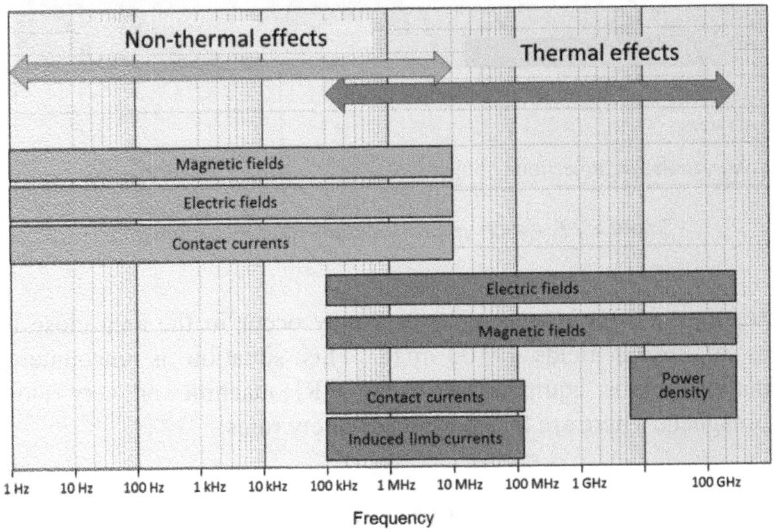


**Figure 5.13.** *Distribution of reference levels (action levels)*

The limits are different for thermal and non-thermal effects, but it is also important to consider how they are calculated. The instantaneous field value must be analyzed at low frequencies, while the relevant rms value, averaged over a six-minute (or less) period, should be considered at high frequencies.

Table 5.23 summarizes the main biological effects and the corresponding exposure limits according to the frequency.

The basic restrictions (ELVs) are physical quantities within the body in the 1 Hz–6 GHz band. The number of basic restrictions is relatively limited with regard to biological effects, exposure situations and the extent of the frequency domain. The general public, which has a wider diversity of physical and health conditions, is protected by lower levels than those for workers.

The reference levels (ALs) are external physical quantities established with the aim of simplifying the assessment of exposure levels. They vary depending on the frequency to follow tissue sensitivity and coupling variations. Compliance with these levels guarantees compliance with the basic restrictions. Indeed, they have been derived from those in the most negative exposure conditions (full body exposure to a uniform field, etc.). Thus, when a body part is exposed, these levels are conservative.

| Frequency domains | Main biological effects | Basic restrictions (exposure limit values) | Reference levels (action levels) |
|---|---|---|---|
| <1 Hz | Vertigo, nausea | Static magnetic field $B$ (T) | – |
| 1 Hz–10 MHz | Sensory stimulation<br>Nerve stimulation<br>Muscle stimulation | Electric field $E$ internal ($V \cdot m^{-1}$)<br>current density $J$ ($A.m^{-2}$) | electric field $E$ ($V \cdot m^{-1}$)<br>magnetic field $B$ (μT)<br>contact current $Ic$ (A) |
| 100 kHz–6 GHz | Heating | SAR ($W \cdot kg^{-1}$) | electric field $E$ ($V \cdot m^{-1}$)<br>magnetic induction $B$ (μT)<br>magnetic field $H$ ($A \cdot m^{-1}$)<br>power density $S$ ($W \cdot m^{-2}$) |
| 300 MHz–6 GHz | Auditory perception | SA ($J \cdot kg^{-1}$) | – |
| 100 kHz–110 MHz | Heating | SAR ($W \cdot kg^{-1}$) | contact current $Ic$ (A)<br>induced current $Ii$ (A) |
| 6 GHz–300 GHz | Superficial heating | power density $S$ ($W \cdot m^{-2}$) | power density $S$ ($W \cdot m^{-2}$) |

**Table 5.23.** *Summary of biological effects and exposure limits depending on frequency*

Exceeding the reference levels does not necessarily imply excessive exposure. It should be verified by referring to the basic restrictions.

Exposure limits may evolve in the future, in the wake of new knowledge in biophysics, electrophysiology and modeling.

There are no limits for unestablished effects, for which it would be difficult to propose values and an assessment method.

## 5.7. People at particular risk

People at particular risk include, non-exhaustively, those wearing active implantable medical devices (AIMD) or passive implantable medical devices (PIMD), those wearing external active medical devices and pregnant workers.

The limits aimed at protecting the general public, and especially those aimed at protecting workers from electromagnetic fields, are not always sufficient for protecting these people. If necessary, additional precautions must be taken to ensure their safety. They should declare themselves, however. It should be noted that active medical devices are increasingly prescribed, in particular in increasingly young populations.

### 5.7.1. *People with medical devices*

Electromagnetic fields can interfere with the nominal functioning of AIMD and medical devices worn on the body. They can also interfere with PIMD. These effects can produce a health risk for the holder.

| People at particular risk | Examples of medical devices |
|---|---|
| Wearers of active implantable medical devices (AIMD) | Pacemakers, cardiac defibrillators, cochlear implants, cerebral implants, neurostimulators, retinal encoders, implanted medication injection pumps |
| Wearers of passive implantable medical devices (PIMD) containing metal | Artificial joints, pins, plates, surgical staples, vascular clips, stents, cardiac valve prostheses, annuloplasty rings, metal contraceptive implants and AIMD casings |
| People wearing medical devices on the body | External medication pumps, hearing aids |

**Table 5.24.** *Example of people at particular risk wearing a medical device*

NOTE.– Recent implantable devices do not usually cause a particular problem with regard to magnetic fields due to their composition (titanium, plastic, etc.).

### 5.7.2. *Active medical devices*

Strong electromagnetic fields can disturb the nominal functioning of active medical devices. These devices contain electronic functions, which may incorporate software [EC 07].

The manufacturers of this equipment must take into account the risks relating to reasonably predictable environmental conditions, especially those relating to magnetic fields, external electric influences, electrostatic discharges, etc.

A system's ability to work in its electromagnetic environment, satisfactorily and without itself producing intolerable electromagnetic disturbances for anything within this environment, is due to the discipline of electromagnetic compatibility [STD 36]. A distinction is made between two additional aspects when the performances of electromagnetic compatibility *radiated* from a piece of equipment are characterized:

– it must not disturb its environment (radiated emission);

– it must work properly near a device radiating disturbances (radiated immunity or susceptibility).

The emissions of a medical device are not analyzed, since the voltages and currents used are extremely weak. Only immunity to ambient radiation is considered.

These devices have contradictory requirements. They must detect very weak physiological signals and resist strong induced fields and external fields. Standard EN 60601-1-2 [STD 34] defines minimum immunity prescriptions for active medical implants:

– electric fields:

- 10 $V{\cdot}m^{-1}$ from 80 MHz to 2.5 GHz (FM radio, mobile telephony, Wi-Fi, etc.) for devices that maintain life (e.g. pacemakers);

- 3 $V{\cdot}m^{-1}$ from 80 MHz to 2.5 GHz for other devices (e.g. cochlear implants);

– magnetic field strength at 50/60 Hz:

- 3 $A{\cdot}m^{-1}$ corresponding to magnetic induction of 3.75 μT.

#### 5.7.2.1. *Pacemakers*

These devices monitor cardiac activity by analyzing the electrical signals captured by one or several leads (probes). When the heartbeat is too slow, and if deemed necessary, the pacemaker sends electric impulses to one or more chambers (ventricle or auricle) to accelerate the heartbeat. The device is generally implanted under the skin below one of the two clavicles. The leads are connected to the heart through the veins. They are either monopolar (composed of a single electrode) or bipolar (with two electrodes).

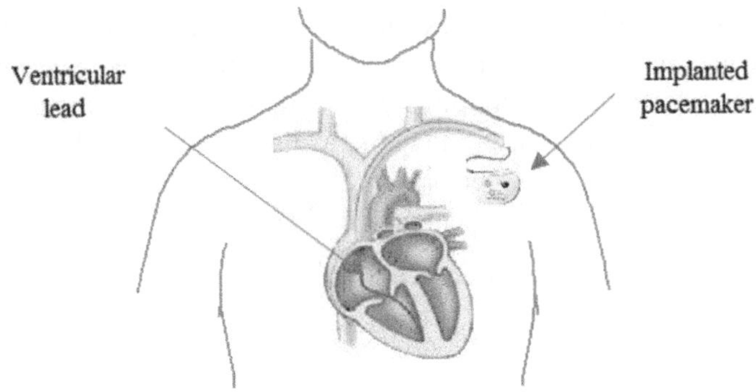


**Figure 5.14.** *Active implant: pacemaker with bipolar lead*

Wearers must follow the medical team's recommendations for taking care of the implant to prevent electromagnetic interference from impeding or disrupting its nominal functioning. Here are a few examples of advice, depending on the device:

– avoid putting a mobile phone into a pocket near the pacemaker;

– do not stand near antitheft detectors;

– contraindication for using electrotherapy or electrostimulation devices, etc.;

– do not practice arc welding or use devices that are marked as radiating magnetic fields.

Short-term inhibition of the pacemaker does not generally pose a problem.

New implants are compatible with some MRI examinations, provided that a few precautions are taken.

### 5.7.2.2. *Automatic implantable defibrillators*

An automatic implantable defibrillator (AID) is similar in appearance to a pacemaker, but with a slightly higher volume, due to a larger battery and an additional capacitor for loading the energy required for defibrillation. The AID treats tachycardia and ventricular fibrillation (disorders resulting in a very rapid heart rate), which can lead to a risk of death. If a heart rate irregularity is detected, an electric pulse is sent to the ventricle. The resulting shock enables the heart to regain its normal functioning.

The precautions for use are identical to those of a pacemaker.

#### 5.7.2.3. *Cochlear implants*

The cochlear implant enables a person to regain part of their hearing. It is usually composed of an outer casing, which picks up and processes sound, and a casing implanted under the skin behind the ear. This transmits electric signals to the auditory nerves through the cochlea. The external casing is connected to the implant by means of inductive coupling. This technique allows both the implant to be powered and the sound information to be transmitted.

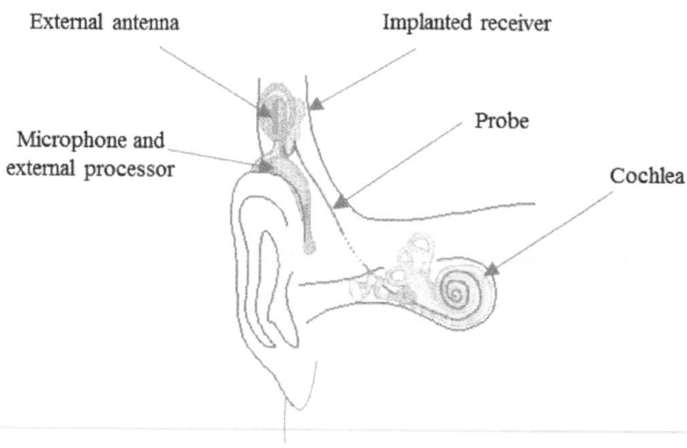


**Figure 5.15.** *Hearing aids: cochlear implant*

The wearer may receive the following recommendations:

– avoid electric shocks and neurostimulation;

– practice electrosurgery in compliance with the manufacturer's rules of caution;

– avoid any MRI examination.

Furthermore, a standard for testing and assessing electromagnetic compatibility between a mobile telephone and an auditory device (implantable or worn) has been established by the American National Standard Institute [STD 35]. Telephones are rated for hearing aid compatibility (HAC). A rate does not guarantee that an auditory device is working correctly with a telephone in particular, and some devices may work well with telephones that do not meet the HAC requirements. Before a mobile phone is bought, several should be tested for compatibility with hearing devices.

#### 5.7.2.4. *Malfunctions of active medical devices*

The malfunctions observed in a few active implantable or body-worn medical devices are summarized in Table 5.25 [MIL 15].

| Type of damage | Pacemakers | Implanted defibrillators | Infusion pump | Cochlear implants |
|---|---|---|---|---|
| Asynchronous pacing (back-up mode) | Possible | Possible | – | – |
| Inhibition of pacing | Possible | Possible | – | – |
| Pacing at maximum frequency | Possible | Possible | – | – |
| Inadequate defibrillation shock | – | Possible | – | – |
| Inhibition of defibrillation function | – | Possible | – | – |
| Pump stop | – | – | Possible | – |
| Wrong dosage | – | – | Possible | – |
| False alarm | – | – | Possible | – |
| Sound perception | – | – | – | Possible |
| Permanent damage | Unlikely | Unlikely | Unlikely | Unlikely |

**Table 5.25.** *Types of damage for a few active medical devices exposed to electromagnetic fields*

### 5.7.3. *Passive medical devices*

The most common passive implants are hip and knee prostheses. The plates and screws used in orthopedics are also passive implants. These implants contain no electronics. Only metal and conductive implants are affected by electromagnetic fields. They are generally made of non-ferromagnetic materials, such as titanium or ceramic.

Under the action of strong high-frequency electromagnetic fields, heating and, in extreme cases, burning can occur around these implants. The same is true for jewelry, piercings, casings for implantable medical devices and other conductive objects worn on or inside the body. Alongside these effects, they distort the MRI image nearby, leading to loss of information for diagnosis.

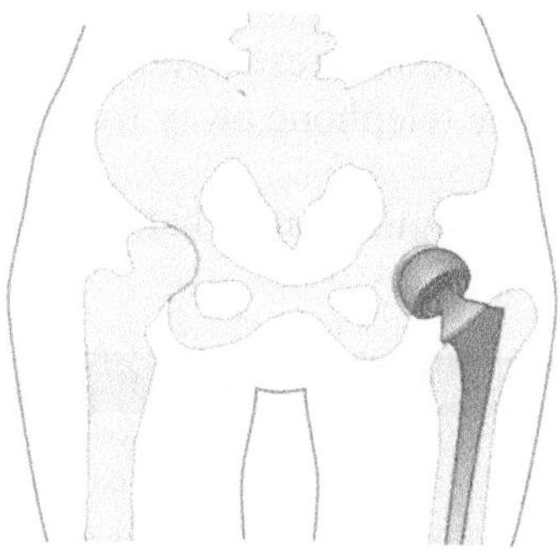

**Figure 5.16.** *Passive implant: hip prosthesis*

### 5.7.4. *Limits for active implants*

Active implantable or worn medical devices are not disturbed by the levels of electromagnetic fields found in daily life, although they exceed the standardized levels [STD 34], as progress has been made both in detection software and in the selectivity of connection filters.

Normally, these people are sufficiently protected by compliance with the reference levels specified in the 1999/519/EC Council Recommendation. However, in some situations, these levels are not sufficient to provide adequate protection, i.e. to guarantee that some devices will not malfunction during high occupational exposure (see Appendix 1, Tables A1.3 and A1.4) [ANS 16].

Malfunctioning should be understood in the non-nominal sense, since active devices switch to back-up mode in the presence of interference, returning to nominal mode when interference ceases. Strictly speaking, this is not a malfunction in the sense of operating safety, but a normal functioning mode. The consequences of extended unavailability of all the functions of such devices can be serious if they have a vital function.

The electromagnetic fields levels considered are those where the medical device and its sensors are implanted or worn. In many cases, using equipment that radiates in a very localized way poses no difficulty, provided that it is away from the close vicinity of the implant.

The most common example is certainly the radiation produced by a mobile telephone, which wearers of pacemakers are advised not to use or place near their device. Beyond this reservation, they can use it without risk. They simply have to keep the telephone away from their trunk, for example.

Some examples of configurations that may cause an implant to malfunction.

| Sources | Placement | Notes |
|---|---|---|
| Motorized device powered by the mains, such as a drill | Device kept in the implant area | Risk of malfunction when the device is in contact with the body next to the implant |
| Antitheft systems | Entrances and exits of shops, libraries, etc. | Avoid stopping in front of the gates (simply walk through) |
| Mobile telephones | Mobile telephones kept near the implant | Hold the device more than 15 cm from the implant area |

**Table 5.26.** *Situations where failure of electronic implants is possible*

NOTE.– The electromagnetic immunity of AIMD depends on the instantaneous value of the radiation, rather than on its rms value. This rules out any calculation of the time average for radiofrequency fields greater than 100 kHz [EC 04]. Consequently, directly using the conclusions of an assessment in relation to the exposure limits of the 1999/519/EC Recommendation is not always relevant.

### 5.7.5. *Pregnant women*

It has never been demonstrated that electromagnetic fields, whether of low or high frequencies, represent a risk for pregnant women if they are exposed up to the levels relating to workers (ALs or ELVs). Induced electric fields are considered to potentially affect the development of the fetus. We know that fetuses are sensitive to temperature elevations. However, there is no established mechanism, suggesting a pregnant woman and her unborn child are sensitive to the effects of electromagnetic fields, any more than any other adult is.

Although no recommendation or specific guideline exists, as a precaution, it may be wise and practical to limit the exposure of pregnant women at work to the reference levels of the general public, i.e. to apply the levels defined by the 1999/519/EC Recommendation. This precaution features in the practical guide that accompanies the EMF Directive [EC 04]. Some legislative texts use this advice [IEE 05, ARP 02].

# 6

# Exposure Indices

This chapter explains how calculated or measured electric, magnetic and electromagnetic fields should be compared with the exposure limits. How can we verify that a signal composed of many frequencies respects limits that are themselves dependent on frequency? The main methods used are presented here. Mathematical notation is required.

## 6.1. General introduction

The two principal ICNIRP Guidelines (1998 and 2010) and the documents that refer to them (Recommendation 1999/519/EC, the EMF Directive, etc.) define the basic restrictions, also known as exposure limit values (ELVs), established from proven biological effects. The majority of them are defined by quantities within the exposed organism (induced electric fields, current densities, absorbed power, etc.). Reference levels or action levels (ALs) have been inferred from these limits and they are defined in measurable quantities (ambient electric fields, magnetic fields, etc.). These limits depend on frequency to follow tissue sensitivity and variation in the coupling coefficient (see Chapters 3–5).

Exposure levels in relation to the basic restrictions are assessed through simulation based on digital humanoid models, phantom measurements or calculations when the models are simple. The result and the limits can be directly compared when the signals are sinusoidal. When the assessed quantities are lower than the basic restrictions or reference levels, the exposure is deemed compliant. At low frequencies, sinusoidal signals are mainly encountered close to the 50 Hz power distribution network (lines, transformers, etc.).

Today, direct comparison is no longer possible in most cases, since the signals radiated by electrical household devices and industrial and medical equipment have complex waveforms that result in a broad frequency spectrum.

The reference documents give very little indication as to how to assess exposure in these situations. Guidelines are often ambiguous, which does not make it easy to choose a method or to write standards. The EMF Directive [EC 01] is an exception, because the proposed method should not lend itself to interpretation.

A few of the signal definitions used to illustrate concretely different comparison methods are outlined.

## 6.2. Signals and definitions

The following low-frequency formulae and definitions apply to magnetic induction, since this is the most commonly assessed radiation. They are, nonetheless, valid for electric fields and for comparison with basic restrictions (or ELV). At high frequencies, the electric field is mainly measured and therefore used in equations.

### 6.2.1. *Sinusoidal signal*

A sinusoidal magnetic induction $B$ of infinite duration is expressed over time by:

$$B(t) = B_{pk}.\cos(2.\pi.f_0.t + \theta_0) \quad [6.1]$$

where $B_{pk}$ is the peak amplitude of the induction of frequency $f_o$ and of the initial phase $\theta_o$. $t$ conventionally represents time.

#### 6.2.1.1. *rms value*

The *root mean square* (rms) value or the quadratic mean of a signal $S(t)$ is the square root of the square of a signal averaged over a given period of time:

$$S_{rms} = \sqrt{\frac{1}{T}\int_T S(t)^2 .dt} \quad [6.2]$$

where $T$ is equal to the calculation time. The rms value is usual for a periodic signal, because for voltage, this corresponds to the value of a direct voltage that would produce identical heat in a resistor. The rms value is homogenous to the square root of a power. The interest of this value becomes clear when thermal effects are analyzed.

Calculating this value for sinusoidal magnetic induction [6.1] leads to:

$$B_{rms} = \frac{B_{pk}}{\sqrt{2}} \qquad [6.3]$$

whatever the frequency and phase.

The exposure limit levels given in the reference documents are generally rms values. The EMF Directive defines peak ELVs (internal electric field) at low frequencies. Of course, both lead to the same result, provided that the factor √2 is not omitted when necessary.

It is crucial to distinguish peak value and rms value when dealing with exposure to electromagnetic fields.

### 6.2.2. *Complex signal*

The concept of complex waves refers to non-sinusoidal signals in the context of low-frequency electric and magnetic fields. This designation covers mainly non-sinusoidal periodic signals (e.g. distorted sine signals), limited duration signals (e.g. pulses), modulated signals and random signals.

Any periodic signal can be represented as an infinite number of spectral components. Magnetic induction can be expressed as follows:

$$B(t) = \sum_{i=0}^{\infty} B_{i\,pk}.\cos(2.\pi.f_i.t + \theta_i) \qquad [6.4]$$

with $f_1$ signal frequency (inverse of its period), $B_{i\,pk}$ peak amplitude of the component at frequency $f_i$ and $\theta_i$ initial phase of this component. The $f_i$s, multiples of $f_1$, are called harmonics. There can be a lot of them, but in practice their numbers are limited. The conversion of a signal into this form is called Fourier series development[1].

Thus, a complex signal is a signal made up of at least two frequency components of significant amplitude.

The formula [6.4] indicates that knowing the amplitude and phase of each component makes it possible to recreate the original signal.

Any signal sampled regularly over a limited period of time can also be expressed as the sum of a spectral component. These are distributed across a band equal to the sampling frequency with an interval (spectral resolution) equal to this frequency divided by the number of samples. Spectral components are not necessarily harmonics, but their sum makes it possible to recreate the original sampled signal, provided that the precautions for use have been complied. The decomposition of a sampled signal into spectral components over a limited period of time is known as a discrete Fourier transform.

Direct and inverse Fourier transform techniques can be used to process signals in either the time or frequency domain.

If the $B_{i\,\mathrm{rms}}$, the rms value of each component $f_i$ , is used instead of the peak value, the previous relationship becomes:

$$B(t) = \sum_i \sqrt{2}.B_{irms}.\cos(2.\pi.f_i.t + \theta_i) \quad [6.5]$$

The resulting quadratic, or square root of the sum of the squares (*root sum square*), of the rms components can be used to calculate the rms value of the complete signal:

$$B_{RSS} = \sqrt{\sum_i B_{irms}{}^2} \quad [6.6]$$

1 Joseph Fourier (1768–1830) demonstrated that to facilitate the study of a signal, it can be broken down into a sum (finite or infinite) of sine and cosine. Each component is processed separately in time and the signal can ultimately be recreated by adding the components together.

Application of the Parseval theorem shows that $B_{RSS} = B_{rms}$, making it possible to indifferently calculate the rms value of a signal in the time or frequency domain.

The signal is periodic when its spectral components are multiples of a fundamental frequency [6.4]. The signal is known as coherent.

The signal is incoherent when its spectral components appear to be independent in frequency and phase. Their peak amplitudes may also vary in time. The relationship between the frequencies is not a rational number and there is therefore no apparent periodicity in the signal. Observation over a long period of time makes it possible to determine the peak amplitude at a given moment, which is difficult to predict. It appears when all the components are in maximum phase and is then equal to the sum of the amplitude of each component.

When the number of spectral components becomes significant, the signal may appear completely random, like noise. It is then difficult to observe any periodicity at all (see Table 6.1).

Such situations occur during exposure to multiple sources that are completely independent from one another or to a single source radiating on several uncorrelated frequencies or even emitting random signals (e.g. radiation from the brushes of an electric motor).

The result of an exposure assessment may depend on the form of the signal as well as on the calculation method (see Chapter 7 for examples).

Table 6.1 shows some simple and complex waveforms.

Signals modulated in radiofrequency, such as those discussed in Chapter 2 (OFDM modulation), are not shown. Their temporal representation cannot be used, because they look like those of random signals (noise). Statistical curves (probability densities, autocorrelation curves, etc.) are required [LET 13].

These various waveforms are combined in sophisticated systems or during exposure to multiple sources.

Electric, magnetic and electromagnetic fields come in diverse forms and sinusoidal radiation tends to be the exception at low frequencies, beyond the domain of the transport and distribution of power energy.

| Signals | Forms | Representations | Comments/ examples |
|---|---|---|---|
| Simple | Sinusoidal | 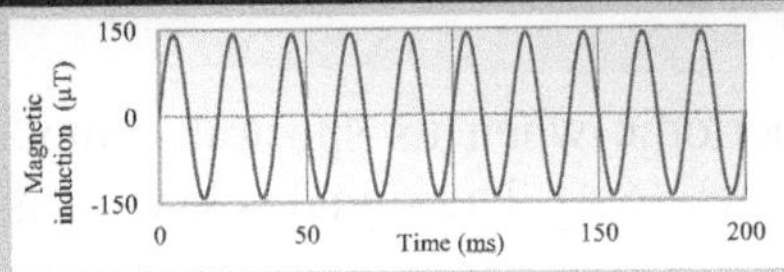 | Continuous wave (CW) Power 50/60 Hz Pure carrier |
| Complex | Rectangular | 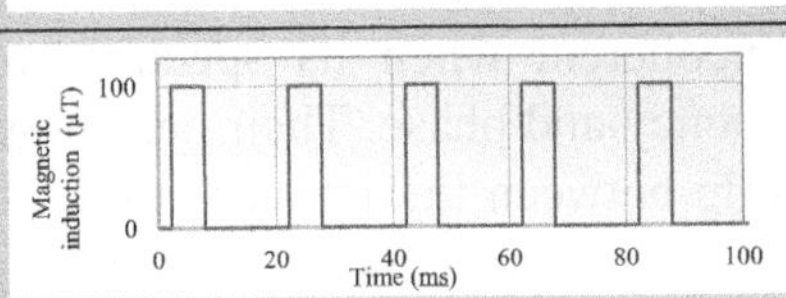 | Arc welding (pulsed current) |
| | Clipped | 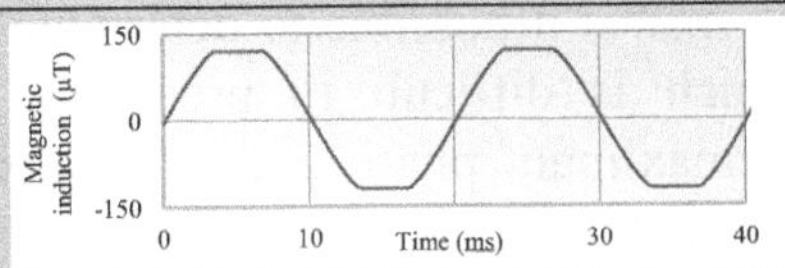 | Saturated signal |
| | Pulse or burst | 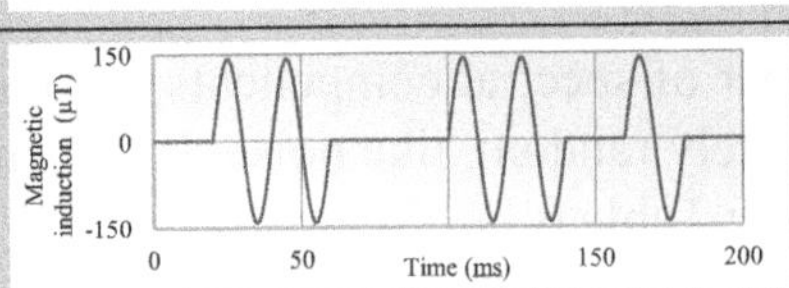 | Current On-Off regulation RADAR signal |
| | Rapid switching | 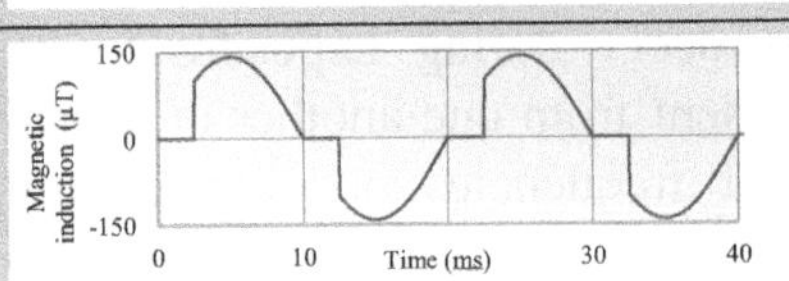 | Current regulation (power regulator, etc.) |
| | Long switching | 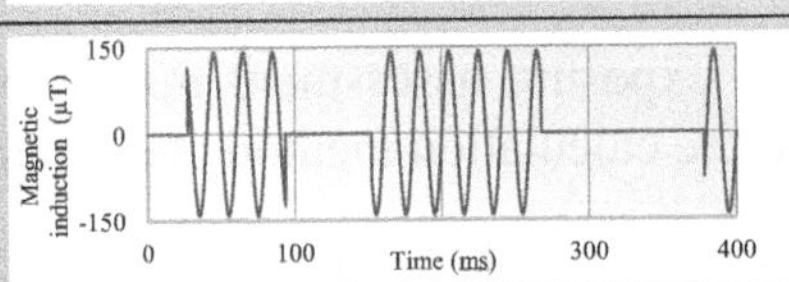 | Current On-Off regulation |
| | Coherent | 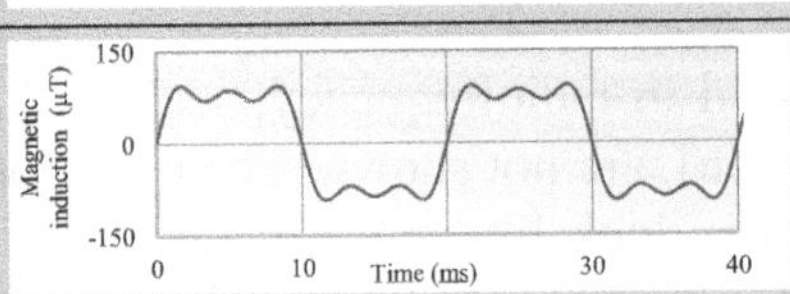 | Signal with harmonics |
| | Incoherent | 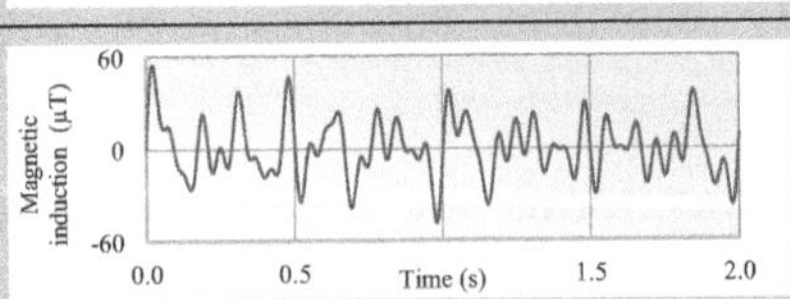 | Signal formed from 7 incoherent spectral components |

**Table 6.1.** *Examples of simple and complex waveforms*

### 6.2.3. *rms value of a complex signal (example)*

The rms value formula [6.2] applies to all signal types. The calculation may be exemplified by an *S*(*t*) signal composed of a series of bursts of four sinusoids of a Ton duration, repeated at T interval, as shown in Figure 6.1. The peak amplitude of the signal is $\sqrt{2}.So$. The rms value of the burst is equal to *So*.

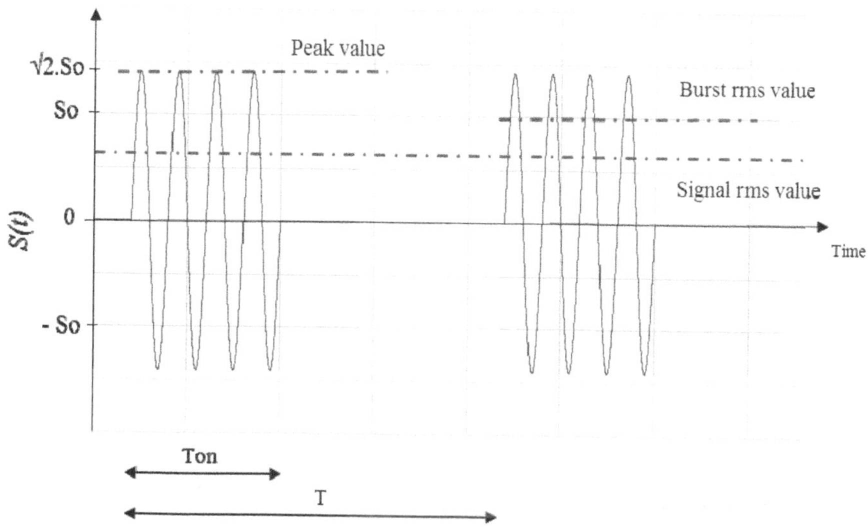


**Figure 6.1.** *Example of peak value, rms value of a burst*

The rms value of the signal is calculated by averaging the rms value of the burst over the repetition period, thus:

$$S_{rms} = \sqrt{\frac{1}{T}\int_{Ton} \sqrt{2}.So.\sin(...)^2\, dt} = So.\sqrt{\frac{Ton}{T}} \qquad [6.7]$$

The ratio *Ton* to *T* is defined as the duty cycle.

The rms value is proportional to the root of the relative time. The average power is proportional to this time.

## 6.3. Introduction to exposure indices

There are many methods for assessing the conformity of an exposure to a limit; the most relevant are outlined here. Regardless of the limits, results differ according to the form of the signal to be analyzed and the method used, unless this signal is sinusoidal and of infinite duration.

The most common exposure limits have been selected to illustrate the various methods:

– general public → reference levels according to ICNIRP 1998 [ICN 98];

– workers → EMF Directive ALs [EC 01].

## 6.4. Exposure index for high-frequency fields

In the past, a number of unfortunate health-related accidents caused by thermal effects made it necessary to establish the first recommendations aiming to protect people from high-frequency electromagnetic fields. These were published before those for low frequencies, although the stimulation of nervous tissues by an electric or magnetic field had been known earlier [ARS 1896].

To protect people from overheating, the first exposure limits issued by INIRC-IRPA (which later became ICNIRP) were signal powers or field rms values. Their temporal form was not considered, because the fundamental need was to be protected from thermal effects. The spectrum from 100 kHz to 300 GHz was covered.

### 6.4.1. *Exposure to a single frequency greater than 100 kHz*

Comparison is simple in the case of a single-frequency electromagnetic field where the quantity retained (electric field, magnetic field or induction, power density) is compared to the corresponding reference level.

The formalism for power density is as follows:

$$S \leq Sl \quad [6.8]$$

where $S$ is the measured or calculated power density and $Sl$ the reference level corresponding to the frequency of $S$. According to [1.16], when the measured quantity is the electric field, the relationship becomes:

$$E^2 \le El^2 \text{ (or } E \le El) \qquad [6.9]$$

where $E$ is the electric field and $El$ the corresponding reference level. Following the same principle, when the magnetic field is taken into account, the relationship becomes:

$$H^2 \le Hl^2 \text{ (or } H \le Hl) \qquad [6.10]$$

All these quantities are rms values, whatever the waveform or spectrum. They must be quadratically averaged over the defined exposure duration (6 min for frequencies lower than 6 GHz).

Most of the time, measuring probes can capture only a polarization of the field. At high frequencies, three measurements must be carried out simultaneously or sequentially to determine the total ambient field in a given position. These three components form an orthogonal tridimensional vector ($x$, $y$ and $z$). This aspect is discussed in Chapter 8.

These components are added up quadratically to obtain the total field intensity, the value that will be compared with the exposure limits. This produces the following for an electric field:

$$\sqrt{E_X{}^2 + E_Y{}^2 + E_Z{}^2} \le El \qquad [6.11]$$

$E$ is the most commonly used physical quantity in the high frequency area. The reasoning is also valid for magnetic fields and power density.

A source emitting a high-frequency signal is considered to be a source that radiates a sinusoidal signal, given that its bandwidth is generally narrow in relation to the central frequency, although it is modulated by the information to be transmitted.

A single vectorial component is used afterward to simplify the notation, bearing in mind that it is always the module that is taken into account for comparison:

$$\sqrt{E^2} \le El$$

or simply:

$$E \leq El \qquad [6.12]$$

### 6.4.2. *Exposure to multiple frequencies*

During exposure to multiple frequencies coming from a single source or from several independent sources, the previous process does not apply. Both the cumulative effect of the frequencies (the power absorbed at each frequency is added together) and the limit values that depend on the frequency must be taken into account.

When the limit value is identical, the powers or squared rms values can simply be added together:

$$E_1^2 + E_2^2 + \ldots \leq El^2 \qquad [6.13]$$

When the limit value is different at each frequency, the process involves adding up the unit exposures. Unit exposure consists of weighting each power by the corresponding exposure limit before carrying out the sum. The comparison is then performed with respect to 1:

$$\frac{E_1^2}{El_1^2} + \frac{E_2^2}{El_2^2} + \ldots \leq 1$$

thus, in short notation:

$$\sum_i \left( \frac{E_i^2}{El_i^2} \right) \leq 1 \qquad [6.14]$$

with $E_i$ module of the electric field of the *i*th frequency and $El_i$ the reference level at this frequency. $El_i$ depends on the frequency.

NOTE.– This formalism is used throughout this chapter and will not be re-explained each time. The components of the field ($E_i$) are generally obtained through high-frequency spectral analysis.

The ambient electric field to reference level ratio is a dimensionless number and is known as *Exposure Index* (*EI*) or sometimes exposure quotient:

$$Exposure\_Index \text{ or } EI^{SS} = \sum_i \left( \frac{E_i^2}{El_i^2} \right) \qquad [6.15]$$

SS denotes sum square.

*EI* is a dimensionless number. Conformity is guaranteed when it is equal to or less than 1, or 100% if expressed as a percentage. It also reveals the exposure margin. An index of 0.1 (10%) indicates that the exposure is 10 times lower than the limit. An index of 2 (200%) indicates that the exposure is two times higher. The index may be calculated from measured or calculated data. Measuring devices can show this value directly, according to the selected limits. In this case, the index is generally presented as a percentage.

Calculation of the index according to [6.15] produces the relative power margin. Its square root indicates the electric field margin. In both cases, the conformity limit remains equal to 1:

$$EI^{RSS} = \sqrt{\sum_i \left( \frac{E_i^2}{El_i^2} \right)} \qquad [6.16]$$

Root sum square (RSS) signifies the root of the sum of the squares. The 1998 ICNIRP Guidelines are the reference for calculating exposure at high frequencies, both for the general public and for occupationals. For electric fields, the sum is defined by limiting the frequency domain:

$$EI^{SS} = \sum_{100\,kHz}^{300\,GHz} \left( \frac{Ei}{Eli} \right)^2 \qquad [6.17]$$

This summation formula works for simultaneous exposure to many sources as well as to one source radiating on several frequencies.

NOTE.– The above relationship may present application difficulties. To simplify the processing or design of measuring devices, the most restrictive value limit in the band to be analyzed, or even in the entirety of the high-frequency domain, is sometimes only used (28 V/m for the general public and 61 V/m for workers). The result is therefore conservative:

$$EI^{SS} = \frac{1}{\min\left(El_i^2\right)} \sum_i \left(E_i^2\right) \qquad [6.18]$$

The weighting no longer depends on the frequency. The total level of rms field (or power) makes it possible to obtain *EI* directly.

## 6.5. Exposure Index for low-frequency fields

Historically, only radiations at 50 Hz or 60 Hz frequencies existed at low frequencies. These emanated from the production, transport, distribution and use of power. Controlling the power of electric devices by means of transformers, rheostats and slow switching systems, such as electromechanical contactors, did not alter the overall sinusoidal form of the current or the voltage.

The rms value of these sinusoidal fields was measured and then compared. This presented no difficulties and measuring devices were designed to meet this need. Standards were initially written primarily to take into account 50 and 60 Hz radiations and high-frequency radiations. Some of them are still in force [STD 04].

More recently, new technologies for converting and regulating electric power have appeared. They use voltage inverters with static relays, such as thyristors and power transistors (IGBT, etc.). The current delivered by the power network is rectified, chopped and sometimes rectified again, before being smoothed. This produces a complex radiated field with a spectrum rich in harmonic frequencies (multiple frequencies of 50/60 Hz or of the switching frequency). These spectra have become very broad because of the very short switching time of electronic components.

Standards and measuring devices followed these evolutions. The first devices covered a bandwidth of a few hundred hertz. Today, some devices can instantly measure the content from 1 Hz to 400 kHz.

The rms value is measured regardless of the waveform [STD 04]. Some devices also provide instantaneous values, since, as discussed in Chapter 4, at frequencies below 10 MHz and especially below 100 kHz, interaction between fields, whether electric or magnetic, and biological tissues produces phenomena related to their instantaneous values.

Based on this observation, many methods of assessing exposure levels in complex fields have appeared, with the feature that all produce the same result for a continuous sinusoidal field (CW).

### 6.5.1. *Exposure to a low-frequency sinusoidal signal*

As discussed in the previous section, the exposure level must be less than the desired reference threshold (reference level, action level, basic restriction or ELV). Comparison is easy in the context of sinusoidal radiation, such as magnetic induction [6.1]:

$$B_{rms} \leq Bl$$

where $B_{\mathrm{rms}}$ is the rms value of the magnetic field and $Bl$ the rms limit value of the field at the corresponding frequency. $Bl$ depends on the frequency.

The concept of *EIs* introduced for high frequencies [6.15] has also been adopted at low frequencies:

$$EI = \frac{B_{rms}}{Bl}$$

*EI* must remain lower than or equal to 1 (100%) to ensure conformity with the desired thresholds.

This calculation is performed in the frequency domain. The level and frequency of the signal must be known (to define the corresponding limit value).

A simpler way of determining this index is to shift into the time domain. This involves comparing a time-varying signal with a constant threshold. If it remains lower than this limit, the conformity of the exposure is guaranteed. This operation can be performed very simply, using a few electronic components.

The following is obtained for a sinusoidal signal [6.1]:

$$\max\left|Bpk.\cos(2.\pi.f_0.t+\theta_0)\right| \le \sqrt{2}.Bl$$

or:

$$\max\left|\frac{Bpk}{\sqrt{2}.Bl}.\cos(2.\pi.f_0.t+\theta_0)\right| \le 1$$

The absolute value makes it possible to take field intensity into account without considering its propagation direction.

The term on the left of this relationship is merely $EI$:

$$EI = \max\left|\frac{Bpk}{\sqrt{2}.Bl}.\cos(2.\pi.f_0.t+\theta_0)\right| \qquad [6.19]$$

$EI$ must be determined by taking into account the module of the field, which can radiate in any direction, as for high frequencies. For this, the field must be processed vectorially, i.e. processed separately on three channels, before the module or $EI$ is calculated. If the module of the field is expressed by:

$$B = \sqrt{{B_X}^2 + {B_Y}^2 + {B_Z}^2} \qquad [6.20]$$

with $B_X$, $B_Y$ and $B_Z$ the field values, respectively, on the $x$, $y$ and $z$ axes, $EI$ is written as follows:

$$EI = \sqrt{{EI_X}^2 + {EI_Y}^2 + {EI_Z}^2} \qquad [6.21]$$

$EI_X$, $EI_Y$ and $EI_Z$ are calculated separately before being added together. $B$ can be expressed as a peak value or an rms value. In this last case, the factor √2 should not be used in [6.19].

### 6.5.2. Exposure to a signal of complex form

Many exposure assessment criteria may be applied for complex radiations from a single source or from various sources. These assessments can operate in the time or frequency domain.

The best-known methods will be outlined here, starting with those proposed in the 1998 ICNIRP Guidelines.

#### 6.5.2.1. *1998 ICNIRP Guidelines (Recommendation 1999/519/EC)*

The formulation proposed in the 1998 ICNIRP Guidelines is relatively simple for non-thermal effects: weighted sum of each spectral component of a given field as an rms value. This formulation in the form of an index is expressed as follows for a magnetic field (general public exposure):

$$EI_{public}^{198} = \sum_{i=1Hz}^{65\,kHz}\left(\frac{B_i}{Bl_i}\right) + \sum_{i>65\,kHz}^{10\,MHz}\left(\frac{B_i}{6,25}\right) \qquad [6.22]$$

with $B_i$ magnetic induction in μT measured at frequency i, $Bl_i$ reference level at the corresponding frequency. The physical quantities of this relationship are rms values. This formulation is also used for electric fields [ICN 98].

This calculation is performed in the frequency domain when the amplitude and frequency of each component have been determined through spectral analysis, e.g. Fourier transform.

For the same frequency, the limits for thermal and non-thermal effects are different, and the quadratic value must be averaged over 6 min in the first case.

It should be noted that an algebraic sum of rms values, weighted or not, has no physical meaning (because it is equivalent to a root sum of powers), unlike the sum of peak values, which can be represented, despite the two operations producing the same result:

$$\sum_i\left(\frac{B_{i\ pk}}{\sqrt{2}\,.\,Bl_{i\ rms}}\right) = \sum_i\left(\frac{B_{i\ rms}}{Bl_{i\ rms}}\right) \qquad [6.23]$$

$B_{i\,pk}$ and $B_{i\,\mathrm{rms}}$ are, respectively, the peak value and the rms value of the *i*th frequential component. $Bl_{i\,\mathrm{rms}}$ is the rms limit value at the corresponding frequency.

Taking into account the three orthogonal components of the field, the index is written as follows:

$$EI^{I98}_{public} = \sqrt{\left(EI^{I98}_{public\,X}\right)^2 + \left(EI^{I98}_{public\,Y}\right)^2 + \left(EI^{I98}_{public\,Z}\right)^2} \qquad [6.24]$$

The three components have to be determined simultaneously.

Recommendation 1999/519/EC incorporates in full the calculation methods and reference levels for the general population from the 1998 ICNIRP Guidelines.

Determining parameters for spectral components is quite complex, because there are many restrictions. This calculation method is therefore used in practice only for specific applications.

#### 6.5.2.2. *RSS index*

For practical and historical reasons, the index based on the root of the quadratic sum of the weighted spectral components has also been established for low frequencies, according to this relationship:

$$EI^{RSS} = \sqrt{\sum_i \left(\frac{B_i}{Bl_i}\right)^2} \qquad [6.25]$$

As before, $B_i$ and $Bl_i$ are either rms values or peak values. The overall index is formulated using the three field orthogonal components, as follows:

$$EI^{RSS} = \sqrt{\left(EI^{RSS}{}_X\right)^2 + \left(EI^{RSS}{}_Y\right)^2 + \left(EI^{RSS}{}_Z\right)^2} \qquad [6.26]$$

where $EI^{RSS}{}_X$, $EI^{RSS}{}_Y$ and $EI^{RSS}{}_Z$ represent, respectively, the indices of components $x$, $y$ and $z$ of the magnetic field.

Equation [6.25] is identical to equation [6.16], which is used at high frequencies, the weighted sum of the powers of each frequency.

$EI^{RSS}$ is calculated in the frequency domain. This index always remains mathematically less than or equal to index $EI^{I98}$ [6.22].

### 6.5.2.3. *Weighted peak method*

The algebraic sum of the weighted values ($EI^{I98}$) can produce a very conservative estimate of the exposure, not scientifically justified for a signal rich in harmonics (e.g. a distorted signal). In theory, this sum could even diverge to infinity if the entirety of the harmonics were taken into account.

In its 2003 statement [ICN 03], ICNIRP proposed a slightly more advanced calculation method taking into account the waveform (temporal characteristics) of the radiation and the nature of biological interactions. Initially presented for basic restrictions, it can also be applied to reference levels. For exposure to magnetic fields, the following relationship is used:

$$W(t) = \sum_i \frac{B_i}{Bl_i}.\cos(2.\pi.f_i.t + \theta_i + \varphi_i) \qquad [6.27]$$

$W(t)$ corresponds to magnetic induction [6.5] weighted by the reference levels, $B_i$ is the rms value of the frequency component $f_i$, $Bl_i$ is the corresponding rms limit value, $\theta_i$ is its initial phase and $\varphi_i$ is the phase introduced by the weighting function at this frequency [JOK 00].

This sum depends on the contribution of the amplitude of each spectral component, as well as the combination of their phase. If we exclude phase information, this formula is comparable to that of ICNIRP 1998 [6.22].

$EI$ is calculated by taking the maximum of the absolute value of [6.27]:

$$EI^{WPM} = \max|W(t)| \qquad [6.28]$$

As before, $EI$ must remain lower that 1 so as not to exceed the exposure limits. This calculation method is called the "weighted peak method" (WPM) [EC 01]

This approach applies to any waveform, provided that it is observed over a sufficiently long period of time, in order to detect the instant at which the signal is at its maximum.

The index does not diverge for a periodic signal, whatever its form, and generally for all real signals. It remains identical to that of the 1998 ICNIRP index for a field composed of several sinusoids of non-multiple frequency with apparently random phases.

Relationship [6.27] can also be written as follows:

$$W(t) = \sum_i |WL_i|.B_{i\,pk} \cos(2.\pi.f_i.t + \theta_i + \varphi_i) \qquad [6.29]$$

where $B_{i\ pk}$ is the peak level of the field component at frequency $f_i$ and $WL$ depending on weighting equal to the inverse of the peak reference level:

$$|WL_i| = \frac{1}{\sqrt{2}.Bl_i} \quad \text{or} \quad |WL(f)| = \frac{1}{\sqrt{2}.Bl(f)} \qquad [6.30]$$

The curves in Figure 6.2 show the magnetic field reference level (Table 5.6) in peak value ($B_{pk} = \sqrt{2}.B_{rms}$, top curve) and its inverse function, which is simply the module of the weighting function (bottom curve). The reference level shown is that of the general public (ICNIRP 98). The curves are plotted from 8 Hz to 100 kHz.

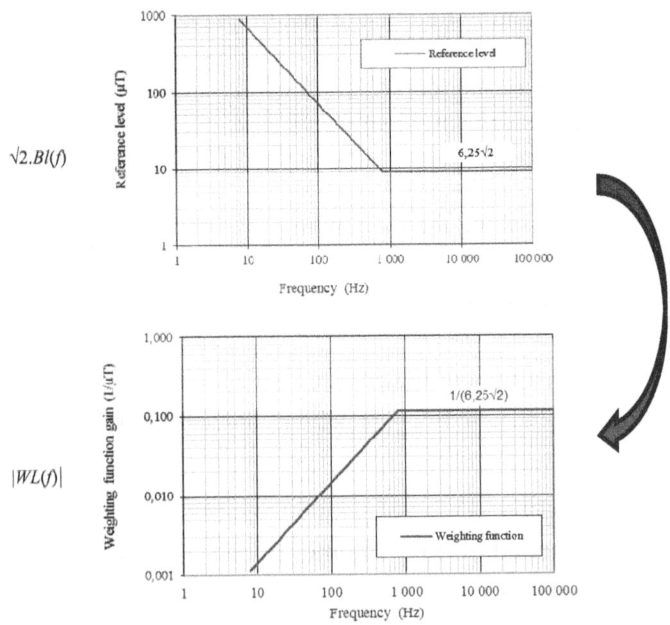


**Figure 6.2.** *Reference level (peak value) and module of the weighting function WL(f) depending on the frequency*

The curve in Figure 6.3 represents the phase of the weighting function ($\varphi_i$). This phase is 90° ($\pi/2$) up to 800 Hz and zero beyond this point.

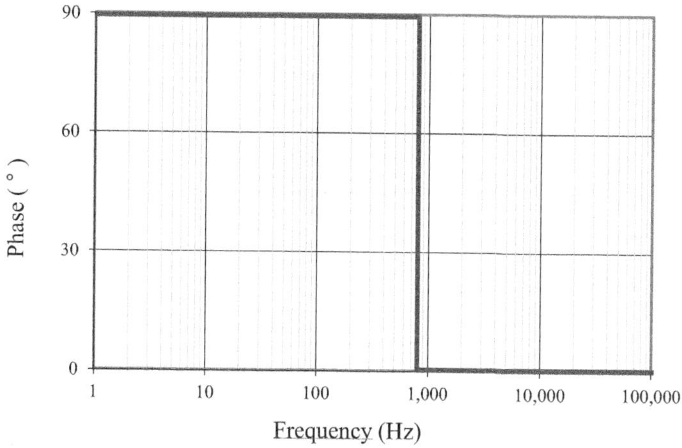


**Figure 6.3.** *Phase of the weighting function ($\varphi_i$) as a function of the frequency*

These curves are the response of a filter, a device whose transfer function (ratio between the output and input signals) depends on the frequency.

Thus, calculation of *EI* using the weighted peak method involves determining the spectrum of the waveform, then applying the relationship [6.27] or [6.29] and extracting the maximum from it. Calculation using the weighted peak technique in the spectral domain is shown in Figure 6.4.

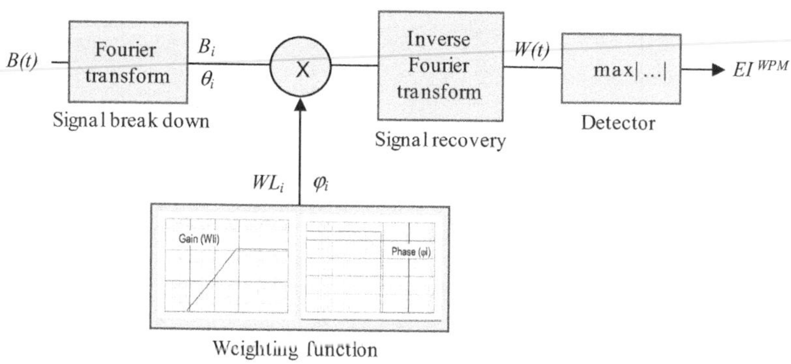


**Figure 6.4.** *Application of the weighted peak method in the frequency domain*

The filter must be approximate in order to be physically feasible. Its transfer function is obtained through smoothing of the theoretical weighting function represented by the piecewise linear curve, as shown in Figure 6.5.

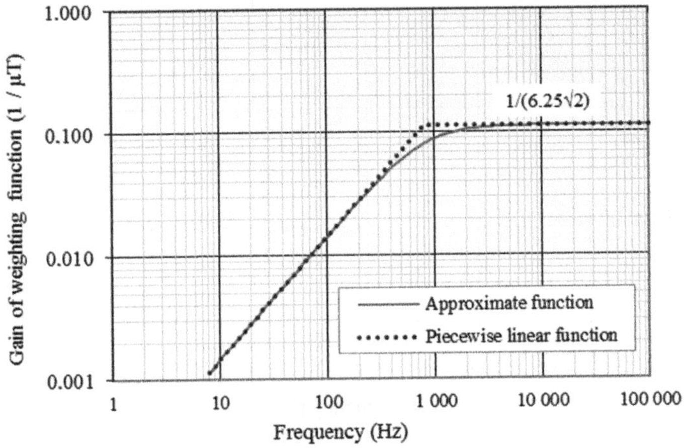


**Figure 6.5.** *Piecewise linear and approximate weighting (public reference level ICNIRP 1998 from 8 Hz to 100 kHz – peak values)*

This filter introduces a phase response that is also approximate (Figure 6.6).

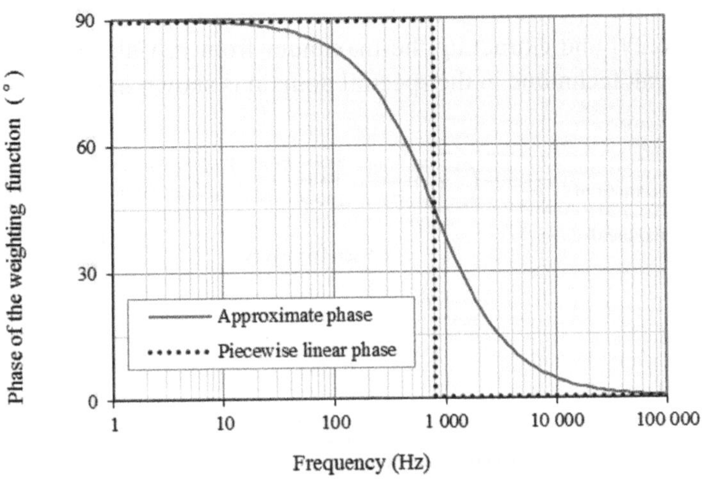


**Figure 6.6.** *Phase curve of the weighting filter*

The differences between the theoretical weighting function and the frequency response of the filter (approximate weighting) are recognized by ICNIRP [ICN 10].

The weighted peak method in the time domain can be mapped, as Figure 6.7 shows.

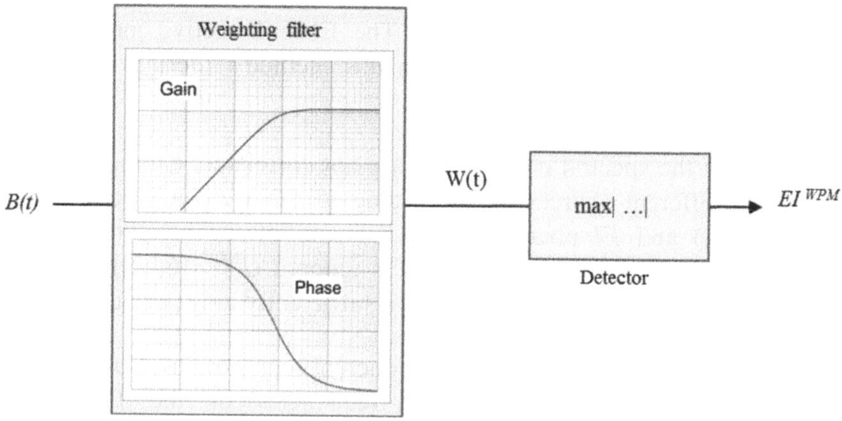


**Figure 6.7.** *Application of the weighted peak method in the time domain*

Calculating *EI* using the weighted peak method is applicable to all forms of low-frequency signal, both outside and inside the human body. Thus, it can be applied to the basic restrictions (or ELVs) defined in terms of electric field or current density and to the reference levels (or ALs) expressed in electric and magnetic fields and in magnetic induction. Appendix 2 gives instructions on the design of such filters.

Calculating the $EI^{WPM}$ index in the frequency or time domain produces the same result[2]. There may be significant differences, according to which either the linear weighting function or the approximate function is used (see section 6.5.3 for an example).

Calculating the index in the frequency domain (Figure 6.4) requires prior knowledge of the rms value, the frequency and the phase of each spectral

2 The theoretical demonstration can be justified using the Pancherel theorem: Fourier transform of the product of the convolution of two signals is equal to the product (ordinary multiplication) of the Fourier transforms of each one.

component of the signal. These parameters are generally obtained through Fourier transform, a technique that can pose numerous difficulties, in particular when signals are not stable during the analysis period. Furthermore, this operation must be handled delicately. It is wiser to process the signal directly in the time domain (Figure 6.7).

The weighted peak method was introduced by ICNIRP in 2003 [ICN 03] and in its 2010 Guidelines [ICN 10]. The EMF Directive indicates that exposure assessment must be based on this method (filtering in the time domain).

NOTE.– When the spectral components are not correlated, e.g. because they come from different sources, *EI* according to the weighted peak method ($EI^{WPM}$ [6.29]) and *EI* obtained from the sum of the weighted spectral components where the temporal aspect is ignored ($EI^{198}$ [6.22]) are equal, provided that the signal is observed over a sufficiently long period of time.

When the spectral components are coherent, e.g. coming from a source generating harmonics, i.e. multiple frequencies of the fundamental frequency, the weighted peak method produces inferior results to those obtained by way of an algebraic sum. These components are added together or subtracted as a function of their phase and can never all be completely added together. When these phases are ignored, all the components are added together, an operation that can greatly overestimate the exposure level.

The weighted peak method (see equations [6.28] and [6.29]) was presented for a single field component with the aim of simplifying notation. Measurements must usually be carried out with an isotopic measuring probe composed of three sensors positioned perpendicularly to measure the field, regardless of its propagation direction. The three channels must be processed simultaneously so as not to lose the phase information between them. Each component must be weighted separately. *EI* is then calculated as follows:

$$EI^{WPM} = \max\left(\sqrt{W_X(t)^2 + W_Y(t)^2 + W_Z(t)^2}\right) \qquad [6.31]$$

where $EI^{WPM}$ is $EI$ according to the weighted peak method and $W_X(t)$, $W_Y(t)$ and $W_Z(t)$ represent the weighted spatial components $B_X$, $B_Y$ and $B_Z$ of field $B(t)$. Figure 6.8 shows a complete representation of the processing.

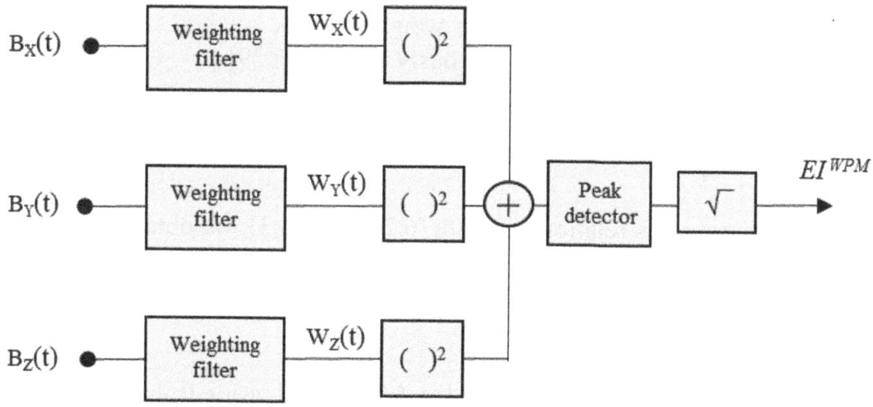


**Figure 6.8.** *Principle of processing using the weighted peak method ($EI^{WPM}$) in the time domain*

Each spatial field component is weighted separately through filtering in the time domain. The instantaneous square $W(t)$ modules are calculated then summated. The square root is extracted from the peak value to obtain a linear quantity. Information on the spatial orientation of the field has disappeared from the result.

### 6.5.2.4. *Exposure index in rms value*

*EI* in rms value is proposed in some standards such as [STD 04]. It is easily obtained in the time domain with weighting filters. Numerous commercially available fieldmeters incorporate this calculation method.

The $EI^{RSS}$ index, the square root of the sum of the rms values [6.25], is calculated in the spectral domain, but requires knowledge of the frequency decomposition of the radiation. Processing in the time domain, i.e. calculation of the $EI^{rms}$, produces the same results without this preliminary processing. This operation is formalized as follows:

$$EI^{rms} = \sqrt{\frac{2}{T}\int_T \left\{ \sum_i |WL_i|.B_{i\,pk} \cos\left(2.\pi.f_i.t + \theta_i + \phi_i\right) \right\}^2 .dt} \qquad [6.32]$$

with the previous notations, i.e. with $f_i$ the frequency of the *i*th component, $\theta_i$ its initial phase, $\varphi_i$ the phase of the filter at frequency i, $B_{i\,pk}$ its peak amplitude and $WL_i$ the weighting factor at the corresponding frequency [6.30].

The signal is filtered as for the weighted peak method. The notation [6.32] makes it possible to advantageously pick up [6.29]:

$$EI^{rms} = \sqrt{\frac{2}{T}\int_T W(t)^2 .dt} \qquad [6.33]$$

By combining this relationship with [6.28] and [6.3], we obtain:

$$EI^{rms} = \sqrt{\sum_i \left(|WLi|.Bi_{pk}\right)^2} \qquad [6.34]$$

The result of this index is identical to $EI^{RSS}$ [6.25], aside from the fact that its calculation operates in the time domain. This is an application of the Parseval theorem [6.6].

Taking into account the measurements according to three axes, the index becomes:

$$EI^{rms} = \sqrt{\frac{2}{T}\int_T \left(W_X(t)^2 + W_Y(t)^2 + W_Y(t)^2\right).dt} \qquad [6.35]$$

Similarly to the representation of the weighted peak method (Figure 6.8), this processing is implemented as shown in Figure 6.9.

The three spatial field components are weighted through filtering in the time domain. The instantaneous square $W(t)$ module is calculated and then generally averaged on one second. Finally, the square root is extracted.

The difference between calculating the $EI^{rms}$ index and the $EI^{WPM}$ index is related to the detector, which calculates an average over a determined duration rather than extracting the peak value.

A device displaying the weighted peak *EI* can show both results on a single hardware or software basis. When the weighting filters are disabled, either the peak field value or its rms value can be obtained directly according to the detector.

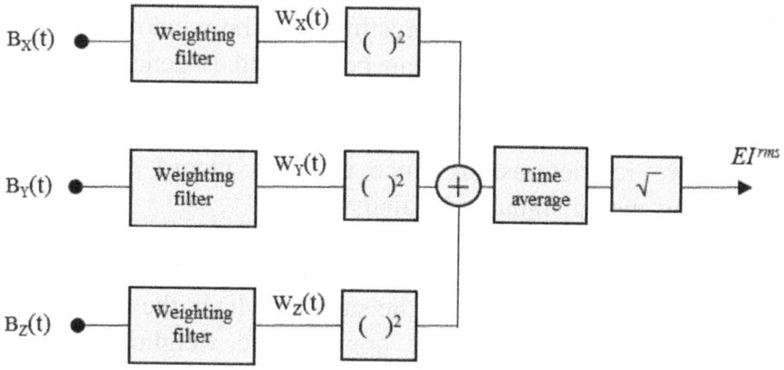


**Figure 6.9.** *Representation of the processing of the $EI^{rms}$ index*

This rms value calculation also applies to high-frequency electric and magnetic fields.

See Chapter 7 for a number of examples of *EI* calculations.

### 6.5.3. *Comparison of weighted peak techniques*

The weighted peak method can be applied in the time or frequency domain. In both cases, the piecewise linear weighting function per frequency band or an approximate function (Figures 6.4 and 6.7) may be used. The latter can be carried out physically using electronic circuits or digital filters with a simple structure.

The difficulty encountered in the frequency domain is due to the elaboration of a complex spectrum (amplitude and phase) through Fourier transform. This operation is reiterated for each calculation. The restrictions relating to this operation are as follows:

– the signal must be sampled rapidly to obtain a good frequency resolution;

– the number of samples is limited by the calculation power of the analysis tool;

– the signal must remain stable during data collection (unsuited to transient phases or installations starting-up);

– the signal must be processed over a whole number of periods to prevent artifacts from appearing (presence of spectral leakage or unreal frequencies) causing the exposure level to be overestimated. The windowing function by means of which this phenomenon can be reduced is generally unsuited to this type of application.

The first two restrictions influence the data record duration and the bandwidth of the analog signal to be analyzed.

If this is not entirely convincing, it should be noted that working in the time or frequency domain produces the same result. In the time domain, the restriction relates to producing the approximate weighting filter, which requires some elementary precautions. This filter is not redesigned for each processing. The temporal filter can also be produced using the piecewise linear weighting function. In this case, it must be digital, because it cannot exist physically. It must be reviewed at almost every processing, in particular when the sampling frequency changes. Using Fourier transform poses the same difficulties. This technique is rarely used.

In brief, the two weighted peak methods used in practice are the weighted peak method in the time domain with an approximate weighting function (analog filter or its digital equivalent) and the weighted peak method in the frequency domain with the piecewise linear weighting function (theoretical weighting).

The example shown in Figures 6.10 and 6.11 allows us to compare these two techniques for a trapezoid magnetic field from a real case. The magnetic field reaches a peak value of 100 mT in 20 ms (the electric current generating this field is controlled). The field remains constant for 40 ms, then diminishes in a regular slope for 20 ms (dotted curve).

$W(t)$, represented by the solid curve on Figure 6.10, is the field in the time domain weighted by the filter corresponding to the high action level applicable to magnetic fields in the EMF Directive (see Appendix 2).

*EI* is the maximum value that the filter response takes [6.28]. It can be read directly on the graph and can also be calculated analytically. It is worth approximately 1.9.

Figure 6.11 represents the same field. The signal is processed in the frequency domain, according to Figure 6.4: a Fourier transform has been applied, followed by piecewise linear weighting in amplitude and phase. The resulting signal has been recreated. *EI* corresponds to its maximum.

A weighting function that follows the reference levels (or AL) perfectly does not exist naturally and causes overshooting. The peak value reaches 2.1.

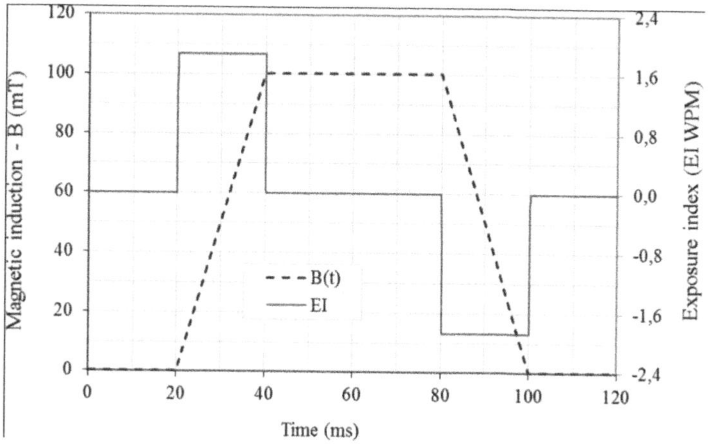


**Figure 6.10.** *Magnetic induction and exposure index (time domain)*

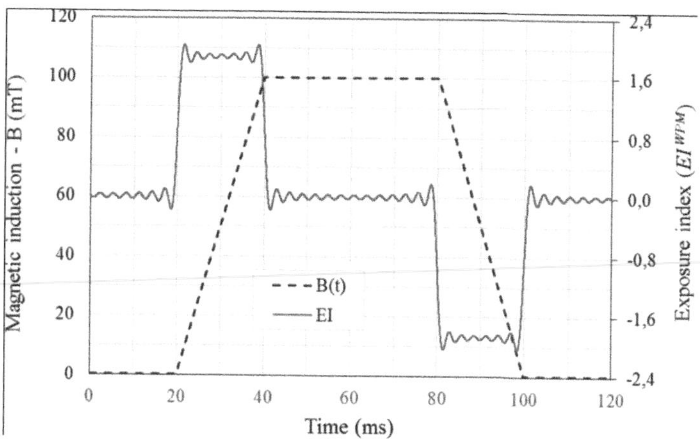


**Figure 6.11.** *Magnetic induction and exposure index (frequency domain)*

This case illustrates the difference between the two methods. The weighted peak method in the time domain is suited to processing in real time (Figure 6.7). It is relatively simple to perform using electronic circuits or a digital filter, although its concept may seem more complex.

The weighted peak method in the frequency domain requires expertise in Fourier transform techniques. It involves breaking down a temporal signal into its spectral components, weighting them, then recreating a signal in time through inverse Fourier transform (Figure 6.4).

It should be noted that this operation is not causal, because the amplitude of the signal issuing from the processing changes before the input signal changes. This operation can be performed only in non-real time (in physics, effect always precedes cause). Furthermore, the level of overshooting depends on the number of harmonics retained and on the slope of the signal. It is difficult to predict.

This example illustrates the importance of the weighted peak technique. It is, however, wise to choose the application domain (time or frequency) depending on the waveform. The time domain must be prioritized for complex signals. In the case of incoherent sinusoidal signals, processing in the frequency domain can be faster. Sometimes, the nature of the signals with regard to existing measuring device resources means there is no choice as to the method.

Chapter 7 applies the various calculation methods.

## 6.6. EIs: contact currents and induced currents

The 1998 ICNIRP Guidelines indicate that exposure must be assessed for contact currents in the same way as for electric or magnetic fields when they consist of several frequencies [6.22]:

$$EI_{Ic}^{I98} = \sum_{n=1Hz}^{110MHz} \left( \frac{I_n}{Ic_n} \right) \qquad [6.36]$$

with $I_n$ contact current component at frequency $n$ and $Ic_n$ reference level of the contact current at frequency $n$ (Table 5.20).

The weighted peak method can be used to assess the exposure of these currents, as noted in the 2010 ICNIRP Guidelines. The calculations are identical to those presented in the previous sections. The shape of the weighting filters is different.

Situations in which contact currents appear often involve sources radiating a continuous sinusoidal signal. In this case, the exposure assessment method is not of great importance, except for the first milliseconds.

Currents induced in limbs cause thermal effects. The formulas applicable to high-frequency exposure are used [6.15]:

$$EI_{Il} = \sum_{k=10MHz}^{110MHz} \left( \frac{I_k^{\,2}}{Il_k^{\,2}} \right) \qquad [6.37]$$

with $I_k$ current component in a limb at frequency $k$ and $Il_k$ reference level of the induced current (Table 5.19).

Given that the limit $Il$ is constant in the frequency domain (45 mA for the general public and 100 mA for workers), the calculation is simplified:

$$EI_{Il} = \frac{1}{Il^2} \sum_{k=10MHz}^{110MHz} I_k^{\,2} \qquad [6.38]$$

This is the rms-induced current to exposure limit ratio. Calculation performed using the rms method in the time domain [6.34] produces the same result.

## 6.7. Summary of exposure indices

By calculating the index, we can assess the exposure level in relation to a predefined limit. The effects at high frequencies are thermal. Thus, the method to be used must be based on a sum of powers or a quadratic sum of rms values. It must also be averaged over 6 min ($f < 6$ GHz).

At low frequencies, where the aim is to protect against non-thermal effects, such as sensorial stimulation or excitation of nerve tissues, calculating the index is more complicated and the instantaneous value of the signal has to be considered. The four most common methods encountered in this domain are:

– $EI^{I98}$: algebraic sum of weighted spectral components;

– $EI^{RSS}$: quadratic sum of weighted spectral components;

– $EI^{rms}$: rms value of the weighted signal;

– $EI^{WPM}$: weighted peak method (maximum value of the weighted signal).

These methods can be applied to all internal physical quantities (basic restrictions or ELVs) and to all external physical quantities (reference levels or action levels, including contact currents). These quantities can come from calculation or measuring.

The exposure indices $EI^{RSS}$ and $EI^{rms}$ are related to the thermal effects. Furthermore, they are the only ones that should be applied at high frequencies. $EI^{I98}$ and $EI^{RSS}$ are determined only in the frequency domain and $EI^{rms}$ works only in the time domain.

$EI^{WPM}$ takes into account the instantaneous value of the field. Its use is relevant at low frequencies. It can be calculated in either domain, depending on practical considerations, but processing in the time domain is generally simpler.

In the frequency domain, the first operation consists of determining the spectral components. The radiated signal must remain stable throughout the acquisition.

To determine the exposure level in the time domain, an analog or digital filter must be used to weight the signal. Calculation using the weighted peak method involves retaining the peak value of this signal after weighting as *EI*.

$EI^{rms}$ is obtained by calculating an rms value, generally over one second, whatever the waveform of the signal [STD 04].

Table 6.2 summarizes the various ways of determining the exposure levels with low-frequency radiations for magnetic induction (magnetic field).

The indices can be compared in the low-frequency band and with an identical reference level ($EI^{RSS} = EI^{rms}$).

Generally, whatever the amplitude and waveform of the radiation, we have:

$$EI^{rms} \leq EI^{WPM} \leq EI^{I98} \qquad [6.39]$$

This expression is simplified for some waveforms.

All the indices are equal for a sinusoidal signal:

$$EI^{\mathrm{rms}} = EI^{\mathrm{WPM}} = EI^{\mathrm{I98}} \qquad [6.40]$$

When two or more incoherent sinusoidal signals are present (without any particular phase or frequency relationship), provided that the signal is observed for a sufficiently long period of time, we have:

$$EI^{\mathrm{rms}} < EI^{\mathrm{WPM}} = EI^{\mathrm{I98}} \qquad [6.41]$$

The weighted peak method and the method proposed by ICNIRP in 1998 give same results. For a periodic signal containing one or several harmonic frequencies, we have a strict inequality:

$$EI^{\mathrm{rms}} < EI^{\mathrm{WPM}} < EI^{\mathrm{I98}} \qquad [6.42]$$

A theoretical signal may be an exception to this relationship.

There are other ways of calculating an index, such as dividing the rms value relationship by the reference limit at the frequency of the fundamental (generally at 50 Hz).

Assessment according to the 98 ICNIRP Guidelines ($EI^{\mathrm{I98}}$) can strongly overestimate the exposure. Performing it is not particularly practical when this is possible; it is therefore used only for certain applications. The $EI^{\mathrm{rms}}$ index should be reserved only for simple signals (sine). In other cases, it is difficult to justify (at low frequencies).

On the basis of current knowledge, the weighted peak method is the method that best represents low-frequency biophysical and biological phenomena (taking into account coupling and stimulation effects). It is processed preferentially in the time domain, where it can be performed by analog and digital filters. It can be applied to any waveform.

The 2010 ICNIRP Guidelines have recommended this method introduced in 2003. The EMF Directive indicates that the exposure assessment shall be based on it. It is beginning to be widespread in standards and legislation.

Chapter 7 compares the indices applied to several examples of theoretical and real signals.

| Domain | Indices | Exposure index calculations | References | Comments | Formula |
|---|---|---|---|---|---|
| Frequency | ICNIRP 98 | $EI^{I98} = \sum_i \frac{Bi}{Bli}$ | 1998 ICNIRP Guidelines<br>Recommendation<br>1999/519/CE | Algebraic sum of weighted rms values | [6.23] |
| | RSS | $EI^{RSS} = \sqrt{\sum_i \left(\frac{Bi}{Bli}\right)^2}$ | EN 63233<br>1998 ICNIRP Guidelines (HF) | Root of the sum of weighted powers (squared rms values)<br>Applicable at LF and HF | [6.16] |
| Time | rms | $EI^{rms} = \sqrt{\frac{1}{T}\int_T \left\{\sum_i \frac{B_{ipk}}{Bl_i}\cos\left(2.\pi.f_i.t+\theta_i+\phi_i\right)\right\}^2 .dt}$ | EN 63233 | Calculation of the rms value in the time domain after weighting | [6.34] |
| FREQUENCY OR TIME | Weighted peak method | $EI^{WPM} = Max\left\lvert\sum_i \lvert WL_i\rvert .B_{ipk}\cos(2.\pi.f_i.t+\theta_i+\phi_i)\right\rvert$ | 2003 ICNIRP statement<br>2010 ICNIRP Guidelines<br>EMF Directive | Determination of maximum (peak value) of a signal after weighting | [6.28] |

$B_i$: rms value of the frequential component i of the field.
$B_{i\,pk}$: peak value of the frequential component i of the field.
$Bl_i$ limit at frequency i (rms value).

$WL_i$: inverse of $\sqrt{2}.Bl_i$.

$\theta_i$: initial phase of the frequency component i of the field.
$\varphi_i$: phase of the weighting function at frequency i.

**Table 6.2.** *Reminder of exposure indices*

# 7

# Applications of Exposure Indices

## 7.1. Introduction

This chapter applies the *exposure index* calculation methods presented in Chapter 6 to low-frequency signals of varying complexity. The first examples are based on theoretical, but realistic signals. The subsequent examples come from data recorded on three radiation sources. These examples will help the reader familiarize themself with the concept of indices and compare the results of the various methods.

NOTE.– Chapters 8 and 9 do not depend on the examples given here. This chapter can therefore be read later.

Calculating the *exposure index* at high frequencies presents no particular difficulty, since it is the powers (or root mean square [rms] values) that are simply added.

Exposure indices are calculated on one axis. The reasoning remains valid on the field module (three axes). Three channels must then be used and the results added together, as relationships [6.24], [6.31] and [6.35] show. Magnetic induction is used in the formulas in this chapter. The reference levels aimed at protecting the general public and the action levels (ALs) aimed at protecting workers are applied.

The following indices are compared in reference to the general public limits as per the 1998 ICNIRP Guidelines (see Chapter 5):

– $EI^{I98}_{public}$ : algebraic sum of weighted spectral components;

– $EI_{public}^{rms}$ or $EI_{public}^{RSS}$ : rms value or resulting quadratic;

– $EI_{public}^{WPM}$ : weighted peak method.

The following indices are calculated for the ALs defined by the EMF Directive (workers) (see Chapter 5 and Appendix 2):

– $EI_{LAL}^{WPM}$ : weighted peak method in reference to low ALs (LALs);

– $EI_{HAL}^{WPM}$ : weighted peak method in reference to high ALs (HALs).

To obtain the WPM index in relation to limb ALs ( $EI_{lbAL}^{WPM}$ ), the index $EI_{HAL}^{WPM}$ must be divided by 3.

## 7.2. Theoretical signals

The first six examples deal with theoretical, but realistic signals. They range from the simplest to the most complex:

1) signal formed from two sinusoids with close amplitude and frequency (50 and 55 Hz);

2) signal formed from two coherent sinusoids (50 and 150 Hz in phase and in phase opposition);

3) signal formed from one sinusoid at 50 Hz and one sinusoid at 3.0 kHz;

4) burst formed from periods of sinusoids at 50 Hz;

5) sinusoidal signal at 50 Hz, chopped (by a fast switching circuit e.g. thyristors or IGBT);

6) square signal at 50 Hz.

The chosen waveforms allow the *exposure index* to be easily and analytically determined. This section aims to understand the importance of calculation methods in the establishment of exposure levels.

Magnetic induction $B(t)$ and weighted induction $W(t)$ [6.29] for the ALs as presented in Appendix 2 [A2.3] and [A2.5] are plotted as a function of time. The absolute peak value of $W(t)$ corresponds to the *exposure index* using the weighted peak method [6.28].

The indices are determined in the spectral or time domain. As the calculation of indices $EI^{RSS}$ and $EI^{rms}$ produces the same results, the choice is made according to the composition or waveform of the signal, to minimize the operations. This principle is applied to the weighted peak method with the slight difference that when the frequency domain is chosen, the theoretical weighting function (piecewise linear) is used, while approximate weighting is used for processing in the time domain. This can produce differences in the results, because approximate weighting may differ by 3 dB (–30%, +40%), as allowed by the 2010 ICNIRP Guidelines (see section 6.5.2.3).

The frequency, the corresponding reference level, the rms value and the peak-to-peak value are given for each signal.

The amplitude of the signals has been chosen to facilitate the calculation and interpretation of the results. In many cases, a quick calculation is adequate for determining the index. In some cases, they have been calculated digitally.

### 7.2.1. *Example 1: two incoherent sinusoidal signals with similar frequencies*

The waveform in this first example is composed of two sinusoids which can be considered incoherent in frequency. This signal is one of the simplest of those defined as complex.

The radiation is composed of two signals at 50 and 55 Hz with respective peak amplitudes of $100\sqrt{2}$ and $90.9\sqrt{2}$ µT. These amplitudes have been chosen because they are equal to the general public reference level [ICN 98]. Induction is expressed in the following form:

$$B_1(t) = 100\sqrt{2}\cos(2.\pi.50.t) + 90.9\sqrt{2}\cos(2.\pi.55.t) \quad [7.1]$$

$B_1(t)$ is represented by the solid curve in Figure 7.1 (axis on the left).

The peak value of the field is 270 µT ($\sqrt{2}[100+90.9]$). It is reached in each signal period, i.e. every 200 ms. Calculating the indices presents no difficulties.

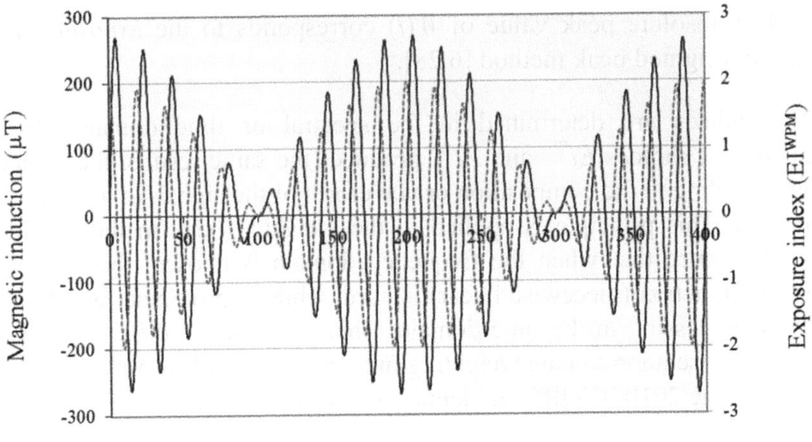


**Figure 7.1.** *Signal composed of two sinusoids with similar amplitude and frequency (solid curve)*

### 7.2.1.1. *Index as per ICNIRP 98*

The reference levels as per ICNIRP 1998 at 50 and 55 Hz are 100 and 90.9 µT respectively (see Table 5.6), meaning [6.22] can be used to calculate the index:

$$EI^{198}_{public} = \sum_i \frac{B_i}{Bl_i} = \left(\frac{100}{100}\right) + \left(\frac{90.9}{90.9}\right) = 2 \qquad [7.2]$$

The data are rms values. The index is the algebraic sum of the two weighted components.

### 7.2.1.2. *Index in rms value*

The *exposure index* in the frequency domain ($EI^{RSS}_{public}$) remains easier to calculate than the index in the time domain $EI^{rms}_{public}$:

$$EI^{RSS}_{public} = \sqrt{\sum_i \left(\frac{B_i}{Bl_i}\right)^2} = \sqrt{\left(\frac{100}{100}\right)^2 + \left(\frac{90.9}{90.9}\right)^2} = \sqrt{2} \approx 1.414... \qquad [7.3]$$

The index is the quadratic sum of the two weighted components.

#### 7.2.1.3. *Index as per the weighted peak method*

The weighted peak method [6.29] as per the reference level in the 1998 ICNIRP Guidelines can be applied to this signal. The weighted signal is expressed thus:

$$W_{1\text{public}}(t) = \frac{100\sqrt{2}}{100\sqrt{2}}.\cos\left(2.\pi.50.t + \frac{\pi}{2}\right) + \frac{90.9\sqrt{2}}{90.9\sqrt{2}}\cos\left(2.\pi.55.t + \frac{\pi}{2}\right) \quad [7.4]$$

$W_1(t)$ is represented by the dotted curve in Figure 7.1 (axis on the right).

NOTE.– In this chapter, $W_n(t)$ represents the signal $B_n(t)$ weighted by the general public reference levels or ALs. Its literal expression will not be restated.

The *exposure index* is obtained when the maximums (or minimums) of each weighted sinusoid coincide. In this example, the weighted sinusoids simultaneously reach a peak amplitude of 1:

$$EI_{public}^{WPM} = \text{Max}\left|W_{1\text{public}}(t)\right| = 2 \quad [7.5]$$

This index is also obtained by adding together the weighted peak indices of each frequency component.

#### 7.2.1.4. *Summary of previous results*

Table 7.1 summarizes the previously calculated values.

| | Sinusoid 50 Hz | Sinusoid 55 Hz | $B_1(t)$ |
|---|---|---|---|
| Brms (µT) | 100 | 90.9 | 140 |
| Bpeak (µT) | 100√2 | 90.9√2 | 190.9√2 |
| *General public reference level in ICNIRP 98 (µT rms)* | 100 | ~90.9 | – |
| $EI_{public}^{198}$ | 1 | 1 | 2 |
| $EI_{public}^{RSS}$ | 1 | 1 | √2 |
| $EI_{public}^{WPM}$ | 1 | 1 | 2 |

**Table 7.1.** *Exposure indices of $B_1(t)$ compared to the reference level (general public)*

The first two columns relate to the two sinusoidal signals (50 and 55 Hz). The final column summarizes the values of the example. This column shows the equality of the indices $EI^{I98}_{public}$ and $EI^{WPM}_{public}$ for a sum of incoherent signals. The index $EI^{RSS}_{public}$ remains the lowest.

#### 7.2.1.5. *Index as per the weighted peak method $EI^{WPM}_{LAL}$ for workers*

For workers, $EI^{WPM}_{LAL}$ is calculated by applying the LALs (sensory effects) of the EMF Directive (1,000 μT for both frequencies – Table 5.4) to the weighted peak method:

$$EI^{WPM}_{LAL} = \text{Max}\left|\frac{100\sqrt{2}}{1{,}000\sqrt{2}}.\cos\left(2.\pi.50.t\right) + \frac{90.9\sqrt{2}}{1{,}000\sqrt{2}}\cos\left(2.\pi.55.t\right)\right| \approx 0.191 \qquad [7.6]$$

Dephasing due to the weighting function has been omitted. It would not have altered the result.

#### 7.2.1.6. *Index as per the weighted peak method $EI^{WPM}_{HAL}$ for workers*

$EI^{WPM}_{HAL}$ is calculated by applying the HALs (health effects) of the EMF Directive (6,000 and 5,450 μT, respectively, at 50 and 55 Hz; Table 5.4):

$$EI^{WPM}_{HAL} = \text{Max}\left|\frac{100\sqrt{2}}{6{,}000\sqrt{2}}.\cos\left(2.\pi.50.t\right) + \frac{90.9\sqrt{2}}{5{,}450\sqrt{2}}\cos\left(2.\pi.55.t\right)\right| \approx 0.0333 \qquad [7.7]$$

Table 7.2 summarizes these values.

| | Sinusoid 50 Hz | Sinusoid 55 Hz | $B_1(t)$ |
|---|---|---|---|
| *Low ALs (μT rms)* | 1,000 | 1,000 | – |
| $EI^{WPM}_{LAL}$ | 0.1 | 0.0909 | 0.1909 |
| *High ALs (μT rms)* | 6,000 | ≈5450 | – |
| $EI^{WPM}_{HAL}$ | 1/60 | 1/60 | 1/30 ≈ 0.0333 |

**Table 7.2.** *Exposure indices for $B_1(t)$ compared to action levels (EMF Directive)*

Unsurprisingly, the index $EI_{LAL}^{WPM}$ is 10 times lower than that of $EI_{public}^{WPM}$. $EI_{HAL}^{WPM}$ is six times lower again (these values are ratios between the ALs and the reference level).

### 7.2.2. Example 2: signal made up of two coherent sinusoidal signals (influence of the phase)

In this example, the signal consists of two sinusoidal signals of 50 and 150 Hz frequency and peak amplitudes of 100√2 and 100/3√2 µT, respectively. The component at 150 Hz can represent the third-order harmonic. The influence of the phase between these signals is analyzed:

$$B_2(t) = 100\sqrt{2}\,\sin(2.\pi.50.t) + \frac{100}{3}\sqrt{2}\,\sin(2.\pi.150.t + \phi) \qquad [7.8]$$

The signals are compared with two differences in initial phase $\phi = 0°$ and $\phi = 180°$. These signals are plotted over two periods of 40 ms (solid curves in Figure 7.2).

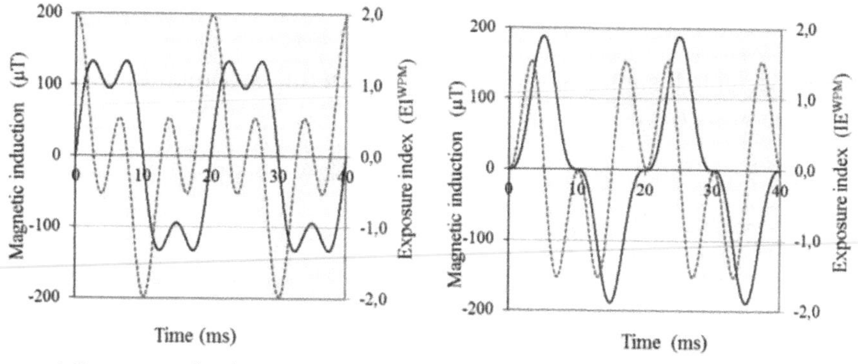


a) Components in phase (ϕ = 0°)

b) Components in phase opposite (ϕ = 180°)

**Figure 7.2.** *Waveform of a signal made up of two coherent sinusoids*

The signal is reproduced in each period of the 50 Hz component (20 ms). An observation period over this interval is sufficient to detect the maximum level.

The exposure indices $EI^{I98}$ and $EI^{RSS}$ are calculated as in Example 1. The levels selected for $B_2$ produce the same results (2 and √2, respectively).

#### 7.2.2.1. *Weighted peak method versus reference level (general public)*

Let us apply the weighted peak method to $B_2(t)$, as the $W_2(t)$ curves show in dots in Figure 7.2, by using the general public reference level.

#### 7.2.2.2. *Zero initial dephasing ($\phi$ = 0°)*

The *exposure index* takes its value when the maximum of each weighted sinusoid coincides. This occurs at $t = 0$ (modulo the half-period if the absolute value is taken into account). At this instant, $W(t)_{\text{public}}$ takes the value of 2, the weighted peak index cannot exceed the sum of the indices of each component (equal to $EI^{I98}$). This instant corresponds to the maximum derivative of $B_2(t)$.

#### 7.2.2.3. *Phase opposition ($\phi$ = 180°)*

The maximum slope of a signal is canceled out by that of the other signal, rather than being added together, as they were in the previous case. This inevitably produces a lower index (≈1.52).

The weighted signal $W_2(t)$ can be shown to be proportional to the derivative of $B_2(t)$.

Table 7.3 is obtained by following the process outlined in Example 1 (section 7.2.1).

| | Sinusoid 50 Hz | Sinusoid 150 Hz | $B_2(t)$ ($\phi$ = 0°) | $B_2(t)$ ($\phi$ = 180°) |
|---|---|---|---|---|
| Brms (µT) | 100 | 100/3 | 105.4 | 105.4 |
| Bpeak (µT) | 100√2 | 100/3√2 | 133.33 | 189, etc. |
| *General public reference level in ICNIRP 98 (µT rms)* | 100 | 100/3 | – | – |
| $EI^{I98}_{public}$ | 1 | 1 | 2 | 2 |
| $EI^{RSS}_{public}$ | 1 | 1 | √2 | √2 |
| $EI^{WPM}_{public}$ | 1 | 1 | 2 | ≈1.52 |

**Table 7.3.** *Exposure indices for $B_2(t)$ compared to the reference level (general public)*

This table proves that the phase difference between the spectral components disappears in results obtained through frequential methods $EI^{I98}_{public}$ and $EI^{RSS}_{public}$, unlike the weighted peak method. This reasoning applies regardless of the number of harmonics. Furthermore, when they are numerous, the index $EI^{I98}$ can lead to a significant overestimation of exposure (see Example 6, section 7.2.6).

Furthermore, this example shows that although we know the peak or rms value of a complex field, we cannot deduce the *exposure index* from it. Here, the field with the highest peak value produces the lowest index.

#### 7.2.2.4. *Index calculation in relation to ALs (EMF Directive)*

When the weighted peak method with the LALs and HALs of the EMF Directive is applied, the indices take on the values summarized in Table 7.4, on the one hand, for $\phi = -90°$ and, on the other hand, for $\phi = +90°$ (these dephasings have been chosen to maximize the difference between the results).

| | Sinusoid 50 Hz | Sinusoid 150 Hz | $B_2(t)$ ($\phi = -90°$) | $B_2(t)$ ($\phi = +90°$) |
|---|---|---|---|---|
| Brms (µT) | 100 | 100/3 | 105.4 | 105.4 |
| *Low ALs (µT rms)* | 1,000 | 1,000 | – | – |
| $EI^{WPM}_{LAL}$ | 0.1 | 0.0333 | ~0.13 | ~0.09 |
| *High ALs (µT rms)* | 6,000 | 2,000 | – | – |
| $EI^{WPM}_{HAL}$ | 1/60 | 1/60 | ~0.03 | ~0.03 |

**Table 7.4.** *Exposure indices for $B_2(t)$ in relation to action levels (EMF Directive)*

The index $EI^{WPM}_{LAL}$ (EMF Directive) is more than 10 times lower than the index $EI^{WPM}_{public}$ in relation to the reference levels defined for the general public by the 1998 ICNIRP Guidelines (or the EC Recommendation).

### 7.2.3. *Example 3: signal made up of two sinusoidal signals with very different frequencies*

For this third example, based on hypotheses similar to the first examples, the magnetic induction proposed consists of one sinusoid at 50 Hz and one sinusoid at 3 kHz, with respective peak amplitudes of $100\sqrt{2}$ and $6.25\sqrt{2}$ µT. The induction $B_3(t)$ is written in the following form:

$$B_3(t) = 100\sqrt{2}\,\sin(2.\pi.50.t) + 6.25\sqrt{2}\,\sin(2.\pi.3{,}000.t) \qquad [7.9]$$

Its representation is shown in Figure 7.3 (solid curve).

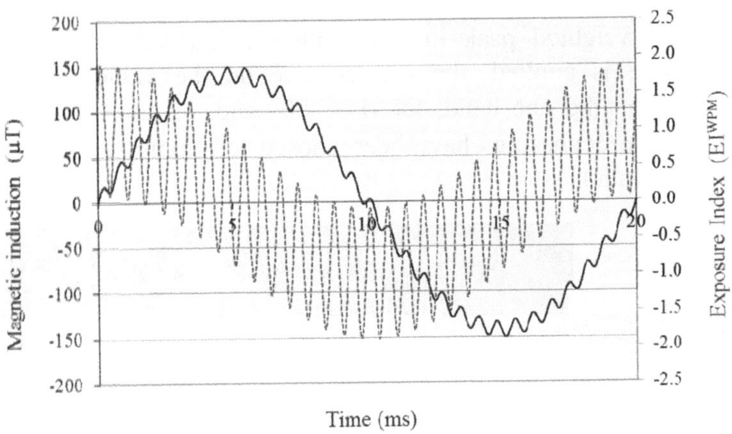


**Figure 7.3.** *Waveform B3(t) composed of two sinusoids*

#### 7.2.3.1. *Indices/reference levels (general public)*

Determination of $EI^{198}$ and $EI^{RSS}$ is immediate. In the time domain, analysis over a period of 20 ms is sufficient to be certain that the instant at which $W_3(t)$ reaches its maximum, as shown in Figure 7.3 (dotted curve), has been taken into account. The exposure indices calculated for the general public reference levels are summarized in Table 7.5.

Both frequency components reach the same level after application of the weighting function, although the amplitude of the high-frequency component is 16 times lower than that of the 50 Hz component (see Figure 7.3).

Furthermore, their contribution to the result is identical. The phase difference between the two sinusoids has no measurable influence on the result.

| | Sinusoid 50 Hz | Sinusoid 3 kHz | $B_3(t)$ |
|---|---|---|---|
| Brms (μT) | 100 | 6.25 | 100.2 |
| Bpeak (μT) | $100\sqrt{2}$ | $6.25\sqrt{2}$ | ≈150.3 |
| *General public reference level in ICNIRP 98 (μT rms)* | 100 | 6.25 | – |
| $EI_{public}^{I98}$ | 1 | 1 | 2 |
| $EI_{public}^{RSS}$ | 1 | 1 | $\sqrt{2}$ |
| $EI_{public}^{WPM}$ | 1 | 1 | 2 |

**Table 7.5.** *Exposure indices for $B_3(t)$ compared to the reference level (general public)*

### 7.2.3.2. *Index calculation in relation to ALs (EMF Directive)*

For this example, the indices depending on the LALs and HALs from the EMF Directive (magnetic field) are summarized in Table 7.6.

| | Sinusoid 50 Hz | Sinusoid 3 kHz | $B_3(t)$ |
|---|---|---|---|
| Brms (μT) | 100 | 6.25 | 100.2 |
| *Low ALs (μT rms)* | 1,000 (1,117)* | 100 (141)* | – |
| $EI_{LAL}^{WPM}$ | 0.100 (0.112)* | 0.0625 (0.044)* | 0.163 (0.156)* |
| *High ALs (μT rms)* | 6,000 (6,001)* | 100 (141)* | – |
| $EI_{HAL}^{WPM}$ | 0.0167 (0.0167)* | 0.0625 (0.044)* | 0.079 (0.046)* |
| *Result obtained with the approximate weighting function. | | | |

**Table 7.6.** *Exposure indices for $B_3(t)$ in relation to action levels (EMF Directive)*

Indices have been determined with theoretical and approximate ALs (values in brackets). At 3 kHz (cutoff frequency), the difference is $\sqrt{2}$ (3 dB).

For this signal, the indices are smaller with approximate values (with a 300 Hz rather than a 3 kHz component, $EI_{LAL}^{WPM}$ would have been larger).

The differences between indices $EI_{public}^{WPM}$ and $EI_{HAL}^{WPM}$ are due to the levels selected (general public reference levels and EMF Directive ALs).

### 7.2.4. *Example 4: burst*

Example 4 deals with magnetic induction formed of a burst of four periods of one 50 Hz sinusoid. The burst is repeated every 200 ms. The duty cycle (DC) of signal duration 80 ms (4 × 20 ms) over a period (200 ms) is 40%. The rms value of the signal during the burst is 1,000 μT (~1,414 μT peak):

$$B_4(t) = REC \; * 1000 \sqrt{2} \sin(2.\pi.50.t) \qquad [7.10]$$

with:

$$REC = \begin{cases} 1 & \text{for } 20 \text{ ms} < t < 100\text{ms (modulo 200ms)} \\ 0 & \text{elsewhere} \end{cases}$$

The burst begins and ends when the amplitude of the sinusoid is zero (absence of discontinuity in the phase). These signals are encountered in current regulation for thermal applications where the adjustment variable is DC.

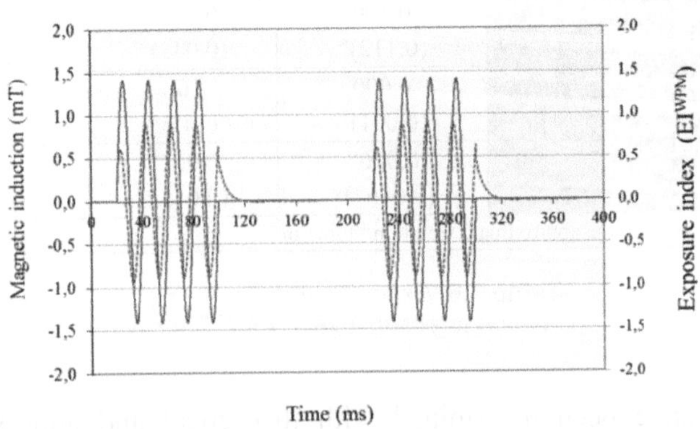


**Figure 7.4.** *50 Hz burst with a 80 ms duration repeating every 200 ms*

#### 7.2.4.1. *Index calculation in relation to the reference level (general public)*

The rms value of this signal is calculated from the rms value of the sinusoidal signal (1,000 μT) according to formula [6.7]:

$$B_{4rms} = \sqrt{\frac{1}{T}\int_T B_4(t)^2 .dt} = \sqrt{\frac{1}{200ms}\int_{80ms}\left(1000.\sqrt{2}.\sin(.)\right)^2 .dt} = \sqrt{\frac{80}{200}}.1000 \approx 632\ \mu T \quad [7.11]$$

The indices referred to in relation to the 1998 INCIRP reference level are summarized in Table 7.7. The first column of values acts as a reference for a continuous sinusoidal signal (CW) of the same amplitude.

| | Sinusoid at 50 Hz | Burst $B_4(t)$ |
|---|---|---|
| Brms (μT) | 1,000 | 1,000√0.4 ~632 |
| Bpeak (μT) | 1,000√2 | 1,000√2 |
| *General public reference level in ICNIRP 98 (μT rms)* | 100 | 100 |
| $EI_{public}^{198}$ | 10 | ~20 |
| $EI_{public}^{RSS}$ | 10 | ~6.32 |
| $EI_{public}^{WPM}$ | 10 | 10 |

**Table 7.7.** *Exposure indices for B4(t) in relation to the reference level (general public)*

The index $EI_{public}^{198}$ of the series of bursts is very high due to the numerous harmonics.

#### 7.2.4.2. *Index calculation in relation to ALs*

The indices as a function of the LALs and HALs in the EMF Directive are summarized in Table 7.8 for $B_4(t)$.

| | Sinusoid 50 Hz | Burst $B_4(t)$ |
|---|---|---|
| Brms (μT) | 1,000 | 1,000√0.4 ≈632 |
| *Low ALs (μT rms)* | 1,000 | 1,000 |
| $EI_{LAL}^{WPM}$ | 0.9 | 0.9 |
| *High ALs (μT rms)* | 6,000 | 6,000 |
| $EI_{HAL}^{WPM}$ | 1/6 | 1/6 |

**Table 7.8.** *Exposure indices for $B_4(t)$ in relation to action levels*

The index $EI_{LAL}^{WPM}$ is obtained by retaining the maximum of the curve $W_4(t)$ represented in dots in Figure 7.4. Its theoretical value is equal to 1. The approximate weighting function introduces a deviation of 1 dB (10%) on the signal. This deviation does not appear for $EI_{HAL}^{WPM}$ (refer to Figures A2.4 and A2.6).

The continuous sinusoid (CW) and the switched sinusoid as shown (with no phase discontinuity at the beginning or end of the burst) produce the same result when the weighted peak method is applied.

### 7.2.5. *Example 5: "chopped" sinusoidal signal*

The magnetic induction in this example is a 50 Hz sinusoid present over three quarters of its period; this corresponds to a delayed conduction angle α of 45°. This signal is found in the regulation of electric currents by thyristors or IGBT.

Over the first half-period, the signal is formulated as follows, depending on the phase, by placing $\theta(t) = 2.\pi.50.t$:

$$\text{for } 0 < \theta < \alpha = 45, \; B_5(t) = 0 \qquad [7.12]$$

$$\text{for } \alpha = 45° < \theta < 180, \; B_5(t) = 1000\sqrt{2}\sin(\theta(t)) \; \mu\text{T}$$

The peak amplitude of the signal is 1,414 μT and the full-wave rms value (without chopping) is 1,000 μT. In this example, the amplitude instantaneously takes the value of 1,000√2 sin (45°) or 1,000 μT (1 mT) at the switchover (plotted in solid lines in Figure 7.5).

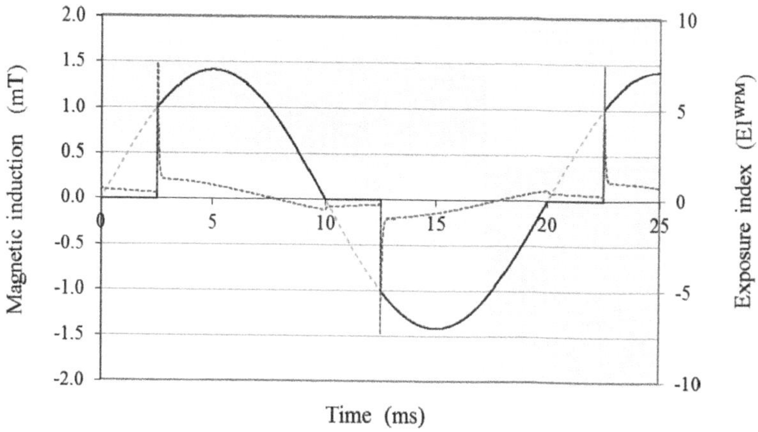


**Figure 7.5.** *Representation of $B_5(t)$ and $W_5(t)$ (chopped 50 Hz sinusoid)*

The rms value of the induction is, according to [6.2]:

$$B_{5\,rms}^2 = \frac{1}{\pi}\cdot\int_{\alpha}^{\pi}\left[1000\sqrt{2}\,\sin(\theta)\right]^2 .d\theta \qquad [7.13]$$

or:

$$B_{5\,rms} = 1000.\sqrt{1-\frac{\alpha}{\pi}+\frac{\sin(2\alpha)}{2\pi}}$$

let for $\alpha = \pi/4$ (45°), $B_{5\,rms} = 1{,}000.\sqrt{(3/4 + 1/2\pi)} \approx 953$ μT.

### 7.2.5.1. *Index calculation in relation to the reference level (general public)*

Exposure indices in relation to the reference level (general public) are given in Table 7.9.

The full-wave values are listed in the central column. The transition at the switchover must be analyzed as a high-frequency signal.

The reference level is therefore 6.25 μT rms for the general public.

The index $EI_{public}^{I98}$ is high, due to the many harmonics, as discussed in Example 4.

| | Sinusoid 50 Hz (full wave) | Sinusoid 50 Hz Chopped (45°) $B_5(t)$ |
|---|---|---|
| Brms (μT) | 1,000 | ~977 |
| Bpeak (μT) | 1,000√2 | 1,000√2 |
| *General public reference level in ICNIRP 98 (μT rms)* | 100 | 6.25 |
| $EI_{public}^{I98}$ | 10 | ~246 |
| $EI_{public}^{RSS}$ | 10 | ~77 |
| $EI_{public}^{WPM}$ | 10 | 160/√2 ~113 |

**Table 7.9.** *Exposure indices for $B_5(t)$ in relation to the reference level (general public)*

### 7.2.5.2. *Index calculation in relation to ALs*

The indices as a function of the LALs and HALs in the EMF Directive are summarized in Table 7.10. The AL to be taken into consideration is that of high frequencies, i.e. 100 μT rms. Indeed, the weighting filter is high-pass, which does not affect abrupt transitions. Only the filter gain at high frequencies should be considered.

| | Sinusoid 50 Hz (full wave) | $B_5(t)$ Chopped 50 Hz sinusoid (45°) |
|---|---|---|
| Brms (μT) | 1,000 | ~977 |
| *Low ALs (μT rms)* | 1,000 | 100 |
| $EI_{LAL}^{WPM}$ | 1 | ~10/√2 or ~7.1 |
| *High ALs (μT rms)* | 6,000 | 100 |
| $EI_{HAL}^{WPM}$ | 1/6 | 10/√2 or ~7.1 |

**Table 7.10.** *Exposure indices for $B_5(t)$ in relation to the action levels (EMF Directive)*

The signal weighted by the LALs is represented by the dotted curve in Figure 7.5. The signal weighted by the HALs is plotted in the same way, as shown in Figure 7.6.

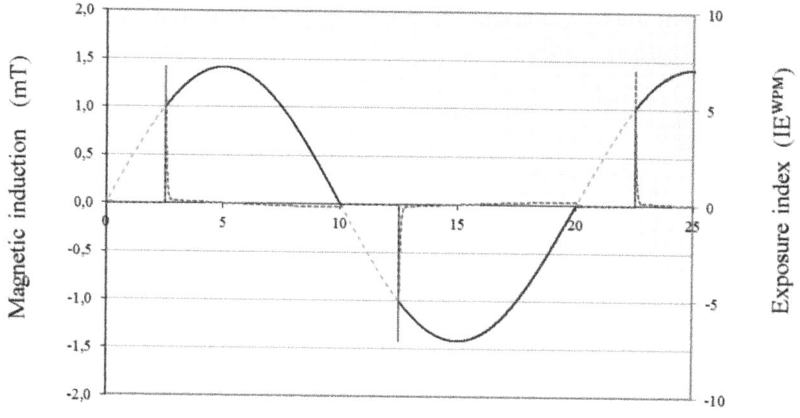


**Figure 7.6.** *Chopped sinusoid (ignition angle 45°) and signal weighted by the high action levels (HAL)*

The *exposure index* calculated in relation to LALs ($EI_{LAL}^{WPM}$) and the *exposure index* in relation to HALs ($EI_{HAL}^{WPM}$) are almost identical. The difference comes from the initial conditions (just before the switchover, the levels are not identical). By dividing the latter index by 3, we obtain the *exposure index* relative to the limb ALs ($EI_{lbAL}^{WPM}$), i.e. around 2.36 for this example.

In practice, a signal with such an instantaneous induction or current transition could not exist. The electric circuit is generally inductive, which limits the rise time (time variation). This results in a lower *exposure index*.

In this example, the index is given from the first milliseconds following the transition (switchover of electrical circuits).

### 7.2.6. *Example 6: square signal*

The magnetic induction in this example is a perfect square signal with peak amplitude $B_p = 100/\sqrt{2}$ ($100\sqrt{2}$ peak-to-peak) and a period of 20 ms

($f$ = 50 Hz) (see Figure 7.7). This type of signal, although it may be complex, has the advantage of being relatively simply formulated in both the time domain and the frequency domain. It is composed of harmonics that stretch out infinitely in theory, making it possible to emphasize the differences between the *exposure indices*.

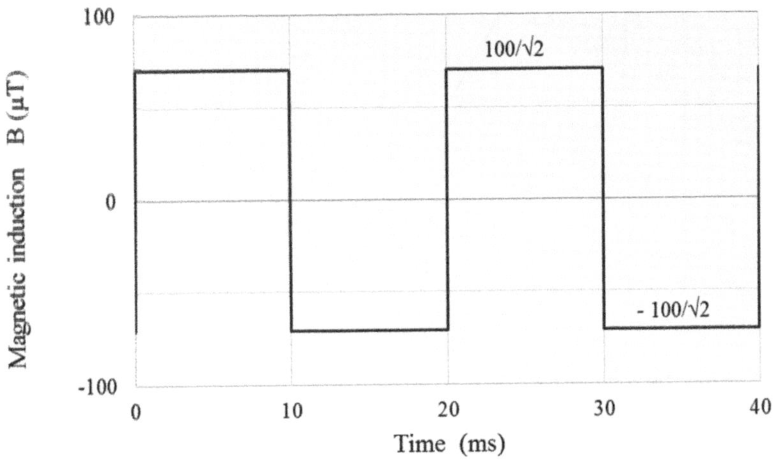


**Figure 7.7.** *$B_6(t)$: Square signal at 50 Hz with peak amplitude 100/√2 µT*

### 7.2.6.1. *Indices in the frequency domain*

This signal emphasizes the difficulty of using the index $EI_{public}^{I98}$. It can be shown that the *exposure index* increases with the number of harmonics taken into account, without converging on a limit value, which has no physical reality. Generally, this means first and foremost that the result depends on the bandwidth of the measuring chain when a signal changes state very rapidly.

In practice, $EI_{public}^{I98}$ broadly overestimates exposure. This is why its use is restricted to a few specific applications where only this assessment method is applicable.

On the contrary, the calculation of the index $EI_{public}^{RSS}$ (root of the sum of the squares) converges, regardless of the number of harmonics, but the *exposure index* is relatively low.

### 7.2.6.2. *Indices as per the weighted peak method in the time domain*

The *exposure index* as per the weighted peak method in the time domain ($EI_{public}^{WPM}$) is initially calculated as shown in Figure 6.7. This means the signal is weighted with the first-order filter (introduced in Appendix 2 and illustrated by Figure A2.2). This filter approximates the reference levels defined by the 1998 ICNIRP Guidelines (general public) between 8 Hz and 150 kHz. The weighted signal $W(t)$ is plotted in Figure 7.8 at steady state (dotted curve).

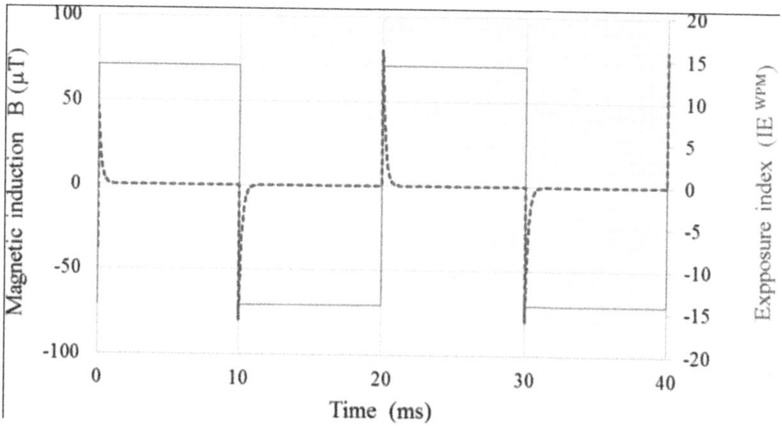


**Figure 7.8.** *Weighting applied to a square signal at 50 Hz*

The weighting filter derives the input signal. It allows only the rising and falling transitions to pass. On a rising transition, the expression of the output is an exponentially decreasing function:

$$W(t) = \frac{B_{pp}}{6.25\sqrt{2}}.\exp\left(\frac{-t}{\tau_2}\right) \qquad [7.14]$$

with $B_{pp} = 2.B_p = 100\sqrt{2}$ , $\tau_2$ is the time constant of the filter of value $\frac{1}{2.\pi.f_c}$ with $f_c$ = 800 Hz or $\tau_2 = 200$ µs. The detailed representation of $W(t)$ normalized over the first millisecond is given in Figure 7.9.

The peak value of $W(t)$ obtained at each transition fixes the *exposure index* at 16 in this example [$EI^{WPM} = \max|\ W(t)\ | = (100\sqrt{2})/(6.25\sqrt{2})$].

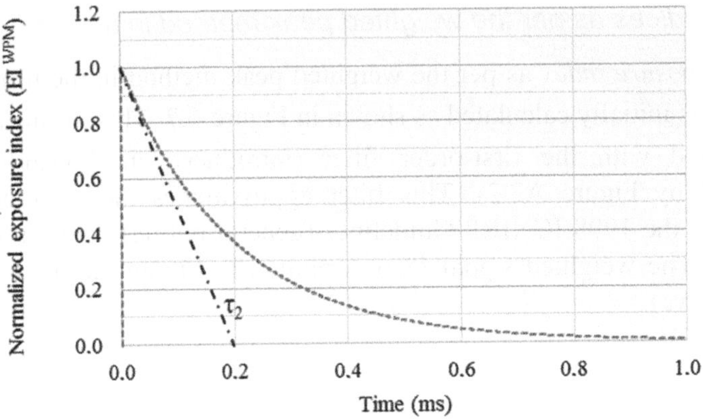


**Figure 7.9.** *Response of the weighting filter W(t) normalized with the change in induction level (square waveform)*

The index $EI_{public}^{rms}$ is calculated directly from [7.14] with relationship [6.2], a simpler operation than determining $EI_{public}^{RSS}$ from spectral components (remember these two indices are identical).

The index values are grouped together in Table 7.11 in comparison with a continuous sinusoidal signal with the same frequency and rms value.

| | Sinusoid (CW) at 50 Hz | $B_6(t)$ |
|---|---|---|
| Brms (μT) | 100/√2 | 100/√2 |
| B peak-to-peak (μT) | 200 | 100√2 |
| *Reference level GP (μT rms)* | 100 | 6.25 |
| $EI_{public}^{I98}$ | 100/(100√2) ≈ 0.7 | (*) |
| $EI_{public}^{rms}$ | 100/(100√2) ≈ 0.7 | 1.6 |
| $EI_{public}^{WPM}$ | 100/(100√2) ≈ 0.7 | 16 |
| (*)Index linked to number of harmonics considered (could exceed 100). | | |

**Table 7.11.** *Indices in relation to the limit for the general public (ICNIRP 98)*

Referring to the same exposure limits, index $EI^{rms}$ is 16 times lower than index $EI^{WPM}$. $EI^{I98}$ can be 10 times higher than the latter.

Let us apply the same reasoning by using the weighting filter corresponding to the LALs of the EMF Directive (section A2.2.1). The response $W(t)$ to the defined square signal is shown in Figure 7.10.

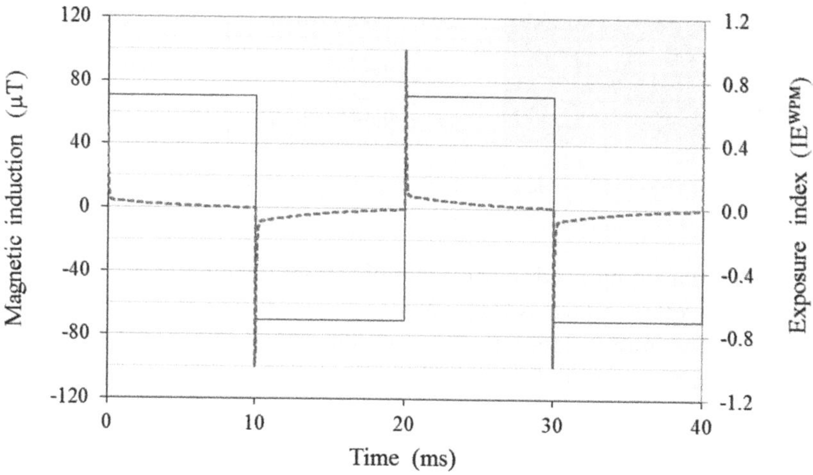


**Figure 7.10.** *Response of the LAL weighting filter to a square signal at 50 Hz*

The convergence of $W(t)$ is practically reached at 10 ms. The *exposure index* is calculated as follows (supposing that $W(t) = 0$ before each transition):

$$EI_{LAL}^{WPM} = \max|W(t)| = \frac{B_{pp}}{\sqrt{2}.Blh} = \frac{100\sqrt{2}}{100\sqrt{2}} = 1 \qquad [7.15]$$

The index for HALs is identical, because the limit values at frequencies over 3 kHz are identical ($Blh = 100\sqrt{2}$).

The indices as a function of LALs and HALs (EMF Directive) are summarized in Table 7.12 for this square signal and a sinusoidal signal.

| | Sinusoid (CW) at 50 Hz | $B_6(t)$ |
|---|---|---|
| Brms (μT) | 100/√2 | 100/√2 |
| B peak-to-peak (μT) | 200 | 100√2 |
| *Low ALs (μT peak)* | 1,000√2 | 100√2 |
| $EI_{LAL}^{WPM}$ | 100/(1,000√2) ≈ 0.07 | 1 |
| *High ALs (μT peak)* | 6,000√2 | 100√2 |
| $EI_{HAL}^{WPM}$ | 100/(6,000√2) ≈ 0.01 | 1 |

**Table 7.12.** *Iindices* $EI_{LAL}^{WPM}$ *and* $EI_{HAL}^{WPM}$ *(EMF Directive)*

NOTE.– The reasoning used in this example is applicable when the repetition period of the square signal is long. When it is short (less than a few ms), the calculations are more elaborate, because the initial state of the weighting filters at steady state (before transitions) must be taken into account to determine the index in the time domain. The most effective methods are therefore digital.

In practice, such results are not obtained for a measured square signal, because it has a limited bandwidth, for the following reasons:

– the rise and fall times of the current (magnetic field) are limited by design;

– the series inductance created by the wires limits the current variations;

– a current limiter can be placed on the power installations to avoid tripping the installation or reduce the radiated disturbances;

– the charge can be inductive (motor, etc.);

– the fieldmeter has a limited measurement bandwidth.

This means the exposure indices are lower than the value obtained on a perfect signal. This can be used as upper limit.

## 7.3. Real signals

The signals presented in the examples in this section have been measured from sources radiating a magnetic field. The three examples used are as follows:

– hand-held electric drill connected to the mains;

– resistive welding gun;

– transcranial magnetic stimulation (TMS) device.

A triaxial broadband field meter (1 Hz–400 kHz) with analog outputs was used to capture the radiation (see Chapter 8). The three axes ($B_X$, $B_Y$ and $B_Z$) were recorded simultaneously. This makes it possible to keep the phase information required for calculation using the weighted peak method (Figure 6.8). The data files were digitally processed. The data collected need no corrective preprocessing, because they directly represent the magnetic field (induction) picked up.

Processing was performed using a spreadsheet tool (Excel or similar). More elaborate calculation software is required for large files. Sampling frequencies were chosen that were sufficiently high to not influence the results.

The measuring device also supplies the rms value, the peak field value, $EI^{WPM}$ or $EI^{rms}$ every ¼ of a second (250 ms) on a display unit and serial link. These quantities are not simultaneously available.

The indices presented in this section come from calculation in the time domain with filters that approximate the weighting functions. Processing in the frequency domain is more complex and is of only limited interest in these three examples.

### 7.3.1. *Example 7: hand-held electric drill*

The first source of radiation used as a real example is a hand-held electric drill supplied by the mains. Its maximum power is 500 W. The product was acquired in a DIY shop.

The rotation speed and minimum couple of the tool were adjusted. No resistive couple was applied. The fieldmeter probe was immobilized, fixed to the back of the tool.

Figure 7.11 shows the magnetic field when the tool is switched on. The highest level measuring axis is plotted to preserve the field sign.

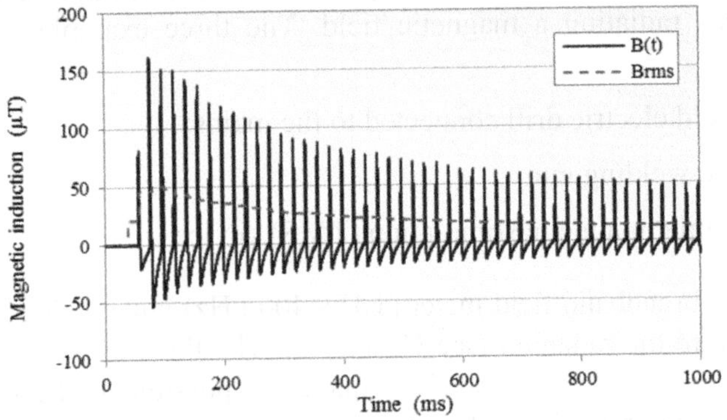


**Figure 7.11.** *Recorded magnetic induction and its rms value calculated over 20 ms when the tool is activated*

The current, and therefore the magnetic field when the tool is switched on, is more significant in overcoming the moment of inertia of the rotating parts. The radiation level stabilizes in less than 1 s. The peak value reaches 161 µT. The field module calculated over the three axes [6.20] reaches 208 µT. The rms value is calculated over a mains period, i.e. over 20 ms. The maximum rms value is 50 µT.

Figure 7.12 shows a detail of the magnetic field at a stabilized rotation speed.

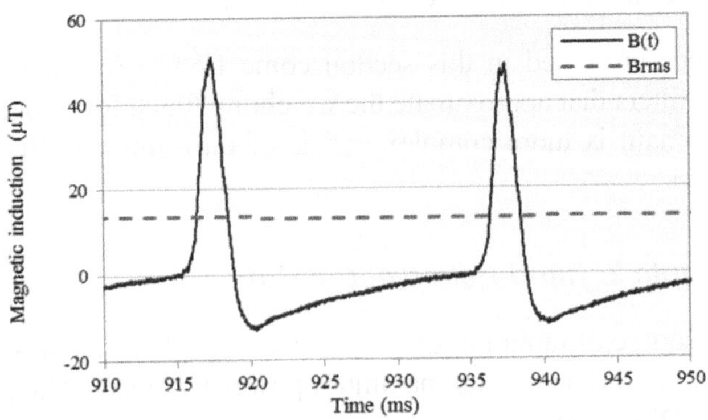


**Figure 7.12.** *Magnetic induction and rms value calcuated over 20 ms in a stabilized regime*

The peak value of the induction rises to 52 μT in Figure 7.13. The rms value holds steady at 14 μT. The field module calculated over the three axes ($x$, $y$, $z$) has a peak value of 65 μT and an rms value of 16 μT.

#### 7.3.1.1. *Calculation of the index in relation to the reference level (general public)*

Figure 7.13 shows the filtered signal $W(t)$ leaving the weighting filter referenced for the general public (ICNIRP 98), superimposed on the induction.

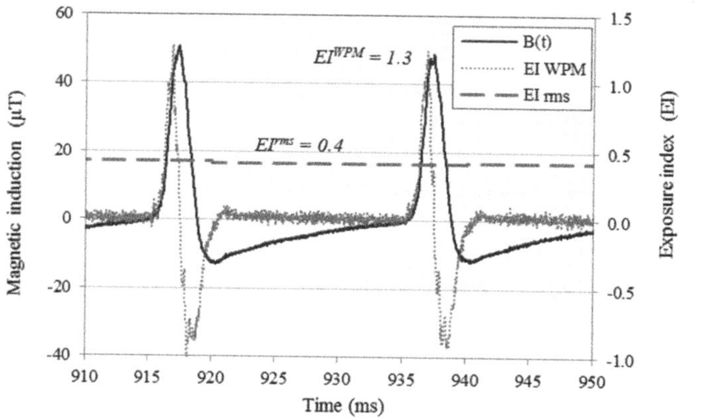


**Figure 7.13.** *Radiated B(t) and weighted W(t) magnetic induction/"general public"*

$W(t)$ corresponds to the derivation of $B(t)$. In this time sample, the *exposure index* $EI_{public}^{WPM}$ has a value of 1.3, using the weighted peak method with the general public reference level [ICN 98]. The index $EI_{public}^{rms}$ is 0.43, applying the rms value calculation. The rise in noise level is observed on $W(t)$.

The field level delivered at 4 Hz (250 ms) by the fieldmeter has also been collected with the detector in peak mode, then in rms mode (Figure 7.14). The rms value is averaged over the last second (sliding window average), in compliance with standard EN62233 [STD 04].

The curves produce a peak when the device is switched on (208 μT), as seen previously (Figure 7.12), followed by a stabilization of the field. The rms value is around five times lower than the peak value (which also means that the signal is not sinusoidal). These data collections are not strictly comparable, because they were carried out at different instants.

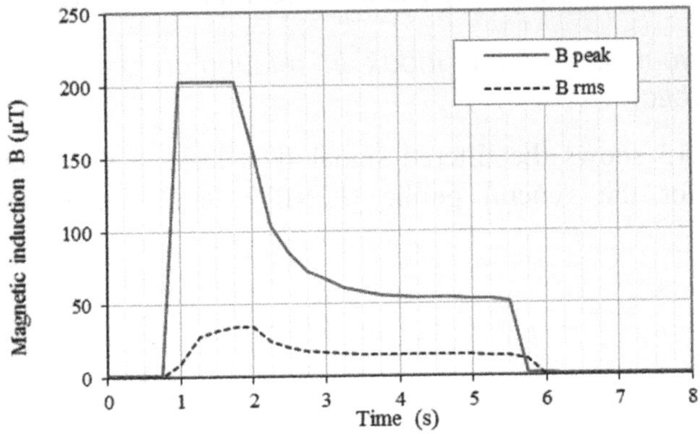


**Figure 7.14.** *Module of magnetic induction: peak and rms values*

The exposure indices $EI_{public}^{WPM}$ and $EI_{public}^{rms}$ provided by the measuring device were also recorded every 250 ms. They are plotted in Figure 7.15.

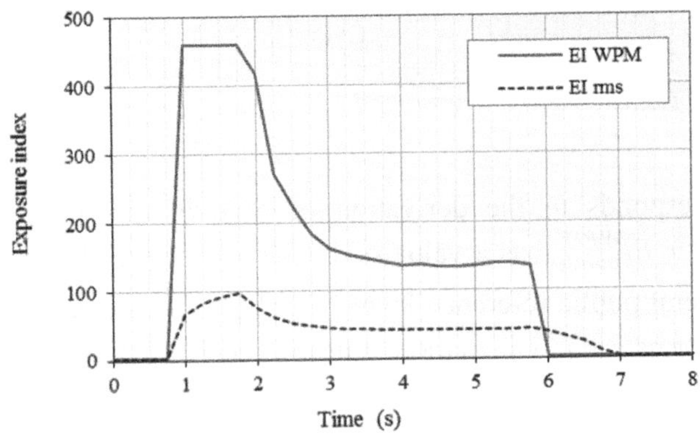


**Figure 7.15.** *Exposure indices $EI_{public}^{WPM}$ and $EI_{public}^{rms}$ measured by the fieldmeter on the tool and recorded*

The index $EI_{public}^{rms}$ is lower than 1 (100%); the drill complies with standard EN 62333 in the configuration tested.

It is possible to determine the indices in the frequency domain ($EI^{I98}_{public}$ and $EI^{RSS}_{public}$) in this case, but the calculation is lengthy. The waveform changes continuously in the starting phase of the motor (Figure 7.11). Furthermore, a Fourier transform synchronized with the mains voltage must be performed to obtain a precise result.

#### 7.3.1.2. *Calculation of the index in relation to ALs*

Figure 7.16 shows the result of calculating the *exposure index* $EI^{WPM}_{LAL}$ in relation to LALs (EMF Directive) in the time domain.

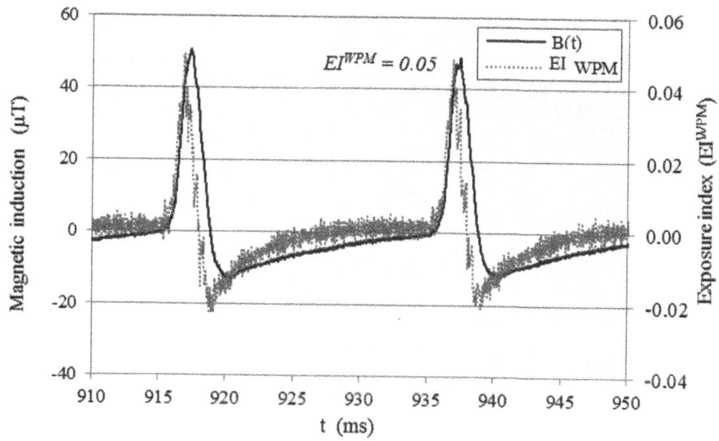


**Figure 7.16.** *B(t) and weighted W(t) magnetic induction/LALs*

Index $EI^{WPM}_{LAL}$ is 0.05, a value close to index $EI^{WPM}_{HAL}$ (0.03). Although the relevant signal is weak compared to the noise level, the conclusions are not affected (the tool complies with the EMF Directive). If necessary, background noise can be reduced, either before logging or through specific digital processing.

### 7.3.2. Example 8: welding gun

The second real source used is a resistive manual welding gun as described in section 2.2.2.4.2.

Resistance welding is performed by injecting a strong electric current between two electrodes clamping the metal sheets to be assembled. The sheets are welded at the point of contact through the Joule effect when the temperature is sufficiently high. When the current has flowed through and the solder has solidified, the pressure exerted by the electrodes is released. Currents with a strength of 6 kA to over 40 kA are commonly used for this assembly process in the metallurgy industry.

Figure 7.17 shows the gun from the side (left) and top (right).

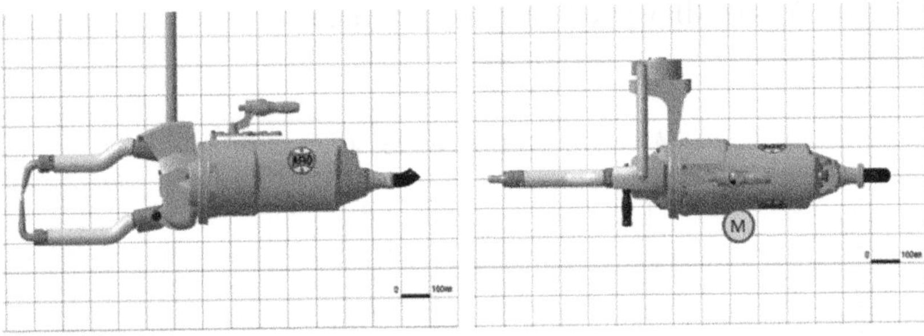


**Figure 7.17.** *Manual welding gun (source: AR0 Welding Technologies)*

The gun is supplied by the mains between two phases (400 V). A voltage-lowering transformer built into the body of the tool delivers the current during welding. This design allows strong magnetic radiation to remain localized around the gun arms. The primary current of the transformer, which is much lower than the welding current, is controlled by a switched electronic circuit (IGBT).

The principal characteristics of this equipment are as follows:

– portable welding gun with built-in transformer;

– reach (length): 80 cm; width = 25 cm (arm size);

– low-frequency alternative welding current (50 Hz);

– two-phase 400 V voltage supply;

– peak power 45 kVA (full wave);

– maximum continuous power 30 kVA.

The adjustment of the tool is as follows:

– welding current: two bursts of 160 ms (2 × 8 main periods) of 10 kA rms.

In this example, exposure is assessed on the side of the gun body. This area is located at the operator's waist height. The field measuring probe is positioned against the tool. The measuring point (*M*) is indicated in Figure 7.17.

The welding current and radiated magnetic field have been recorded. This recording makes it possible to calculate the exposure indices. The electric voltage supplying the two arms of the gun is lower than 10 V (very low voltage), the electric field is negligible, since it is largely below the limits (10 kV at 50 Hz/EMF Directive).

Only the first burst is shown, because the second is identical.

Figure 7.18 was obtained by recording the current. Its peak value reaches 17 kA and its rms value is 9.8 kA.

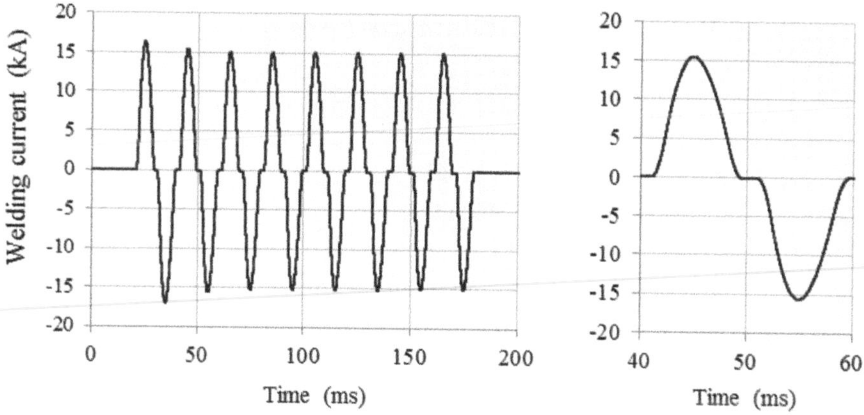


**Figure 7.18.** *Welding current (zoom on the second period on the right)*

Figure 7.19 shows the magnetic induction recorded, which, unsurprisingly, reproduces the current waveform. An induction component is shown. The absolute peak value is 1.1 mT (the peak module of the three components is 1.2 mT).

The recording of the induction makes it possible to calculate the exposure indices using the weighted peak method in the time domain.

The induction filtered by the weighting function referenced to the LALs is plotted in Figure 7.20. The index $EI_{LAL}^{WPM}$ is 0.64 (64%).

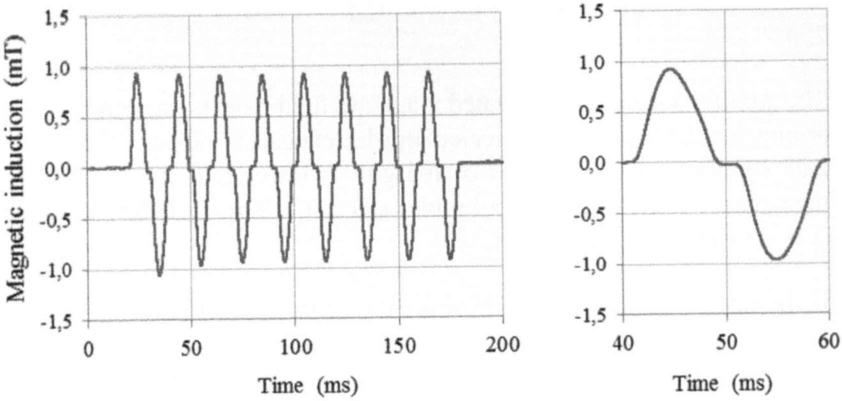


**Figure 7.19.** *Magnetic induction at the measuring point (zoom on one period on the right)*

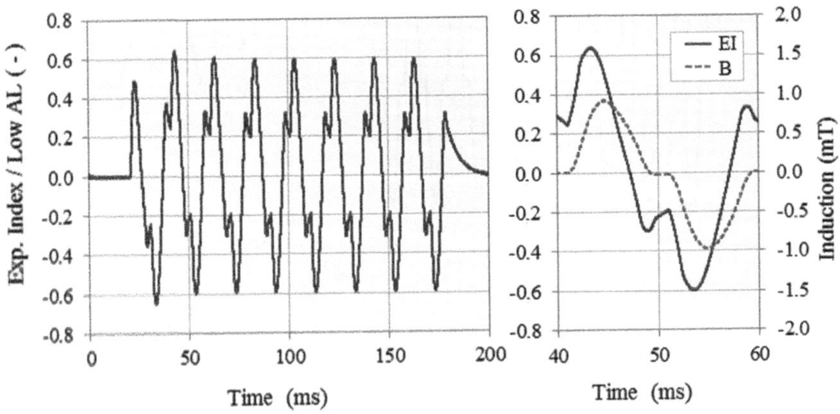


**Figure 7.20.** *Weighted induction/LAL(zoom on one period on the right with B plotted as a dotted line)*

In the same way, the induction weighted in relation to HALs is plotted in Figure 7.21. The index $EI_{HAL}^{WPM}$ is equal to 0.18. The trained eye will recognize the time derivative of the induction in this signal.

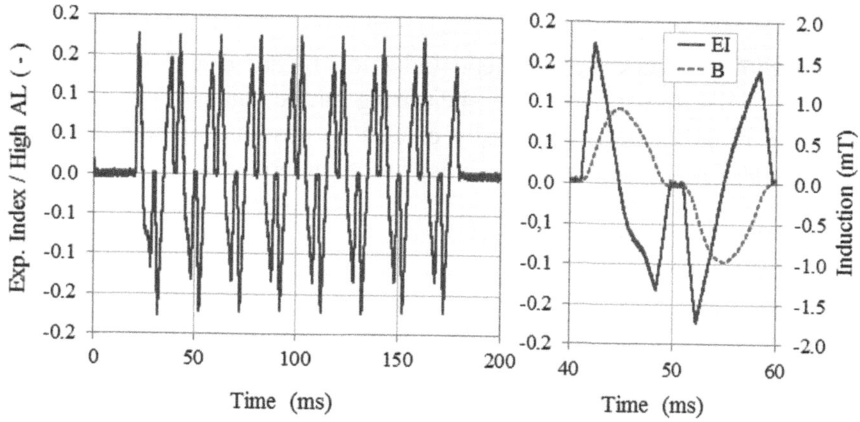


**Figure 7.21.** *Weighted induction/HAL (zoom on one period on the right with B as a dotted line)*

NOTE.– The index $EI_{LAL}^{WPM}$ is given as an indication; it does not need to be calculated at this point, since the LALs aim to protect the operator from sensory effects that could cause transient disturbed sensory perceptions to the central nervous system of the head only. Here, the measuring point is located at the operator's waist height, where only the index relating to HALs ($EI_{HAL}^{WPM}$) must be considered.

The weighted peak method is simple and quick to use. It is built into some measuring devices. They directly display the exposure indices with regard to LALs, HALs or ALs associated with limbs. These indices cannot be calculated simultaneously by a measuring device, unlike the possibilities offered by a recording of the magnetic field.

When the ALs are exceeded, the exposure level must be verified in relation to the exposure limit values, unless preventive or protective measures are implemented (see Chapter 9).

### 7.3.3. *Example 9: transcranial stimulation device*

The third real source used is a TMS device, such as that presented in Chapter 2 (section 2.2.3.2.2). These devices are designed with one or two circular coils (Figure 2.23) connected to a pulse generator.

Figure 7.22 shows the magnetic induction recorded a few centimeters from the coils (positioning the measuring probe against the coil risks damaging it, as the levels generally exceed its absolute maximum rating). The generator is adjusted in single pulse mode (monophasic).

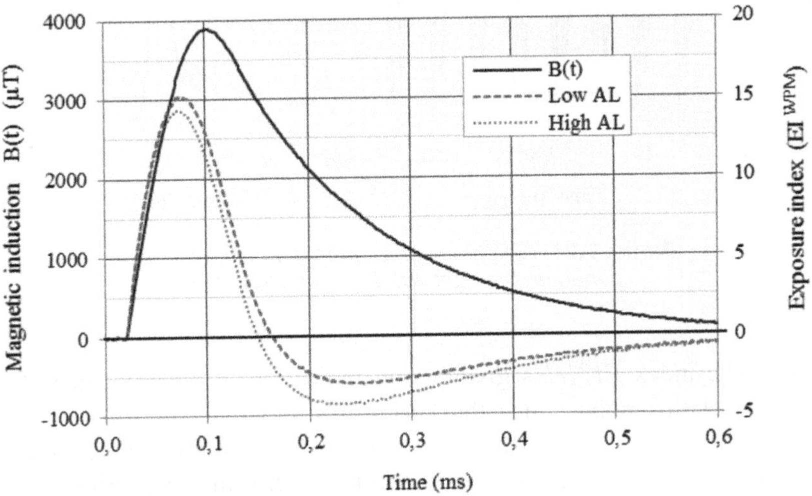


**Figure 7.22.** *Magnetic induction as a function of time and weighted inductions*

Peak induction reaches 3,900 µT. The signal is characterized first and foremost by a very short rise time. Induction variation reaches 100 mT/ms, inducing high levels of electric fields in cerebral tissues. The exposure indices in relation to LALs and HALs, $EI_{LAL}^{WPM}$ and $EI_{HAL}^{WPM}$, reach 15 and 14, respectively (exposure indices are necessarily lower than the peak-to-peak value of the induction divided by $100\sqrt{2}$, or, for this example, lower than 28).

The two indices are similar, because the signal is very fast, which means its spectrum is located mainly above 300 Hz, a domain where LALs and HALs are identical.

The pulse width at half-height is 1.7 ms. Measuring the rms value (generally over 1 s) or the rms index is irrelevant.

Prevention measures should be implemented to protect both the practitioner and the patient from the (unwanted) direct and indirect effects of magnetic fields [EC 05].

## 7.4. Conclusion on the index calculation examples

The examples show that the *exposure index* proposed by the 1998 ICNIRP Guidelines $EI_{public}^{I98}$ can greatly overestimate the exposure level in the case of complex signals. It can be shown that this index is fairly sensitive to ambient radiation noise, measuring noise and artifacts produced in particular by Fourier transform. This means additional processing of the signal is sometimes necessary.

The index $EI_{public}^{rms}$ (or $EI_{public}^{RSS}$ ) tends to underestimate the exposure level when the signal is discontinous, particularly when the peak factor (ratio of the peak value of the signal to its rms value) is significant.

The differences between the $EI_{public}^{rms}$, $EI_{public}^{WPM}$ and $EI_{public}^{I98}$ indices are more accentuated when the signal is richer in harmonics.

The weighted peak method ($EI^{WPM}$) is required, since it seems to better protect against the effects of the stimulation of excitable tissues. It has generally featured in regulations and standards since 2010.

The implementation of this method in the time domain presents no difficulties when some elementary precautions are in place, such as respect of Nyquist's theorem, which states that the sampling frequency must be more than twice the highest frequency of the signal. More significant limitations should be considered for analysis in the frequency domain. However, this domain of calculation remains interesting in a few specific situations.

A high noise level (noise from the device or ambient magnetic background) can significantly influence the weighted peak index.

It is possible to precisely calculate the *exposure index* when the signal or part of it is formulated mathematically, which is often possible. This operation allows the exposure level to be optimized in the design of equipment.

Graphic analysis of a signal, even a complex one, can be used to quantify maximum values of the weighted peak indices.

EXPOSURE TO MAGNETIC FIELDS QUIZ.–

– At low frequencies, is it possible to find an exposure situation that would be compliant for the general public and non-compliant for workers, considering that the exposure limits for the former are lower than for the latter?

– If so, under what conditions? What are the waveforms?

The information required to answer these questions can be found in this chapter and in Chapter 6.

# 8

# Exposure Assessment

## 8.1. Introduction

It is essential to estimate the electromagnetic field level in order to check whether the space occupied by people complies with standards, recommendations and regulations, or whether it poses any potential health risk.

Situations where the field exceeds the limits (reference levels or basic restrictions) are exceptional for the general public [THU 10] and most workers [ALA 11]. As a general rule, people are exposed to radiation levels well below these limits. In this case, the interest of an exposure level assessment is first and foremost to reassure. However, this process is required in situations where people, generally workers, are positioned near sources of strong radiation, such as radiofrequency antennae or equipment through which very strong currents flow (see Chapter 2). There is a potential risk of being exposed beyond the limits.

Separately from the *exposure index*, which determines whether or not the limit thresholds are being exceeded, it is also interesting to assess the electric, magnetic and electromagnetic field strength. Indeed, their physical quantity remains less abstract than a dimensionless index, in particular when it is low. A power density of 1 $mW{\cdot}m^2$ or 0.1 $\mu W{\cdot}cm^2$ equivalent to an electric field of 0.6 $V{\cdot}m^{-1}$ is a more concrete result than an *exposure index* of 1%.

Regardless of the required physical quantity, the exposure level can be assessed indifferently through measuring, calculation (analytical or numerical) or a combination of these two methods. The choice of the most relevant method will be dictated by the type of radiating source, environment configuration, availability of measuring equipment, simulation tools, available data and skills of the assessor. The complexity of the process also depends on the required precision level.

The methods can be validated by comparing the results of an analytical or numerical approach (simulation) with those of a metrological approach, which allows to conclude with a certain level of confidence.

### 8.1.1. *Theoretical approach*

The theoretical or analytical approach produces precise results when the configurations are simple. The results obtained are often conservative due to the precautions applied. In complex situations, the theoretical approach allows an order of magnitude to be estimated.

The calculations also make it possible to select the most suitable measuring device. They may also prevent unnecessary investigations. A source may be declared compliant (for the use intended by the manufacturer) if, for example, its electrical consumption in the worst cases produces a radiated power lower than the limits. Indeed, it would be inappropriate to measure the radiation of a remote control supplied by a button battery.

On the other hand, predictive results can be used to warn the assessor of potential hazard in the area to be measured.

The radiofrequency radiation level can also be predicted, as presented in section 1.3 (wave propagation), when the input data, such as maximum permitted output power and antenna characteristics, are available. Far-field predictions are more reliable the more incorporated the propagation phenomena. Less attention can be paid to the uncertainty of the results when the radiation levels are well below the limits, i.e. far from the transmitter.

The theoretical approach is sometimes the most useful method, with regard to the risk and the technical and human resources to be implemented. In some cases, for example insulated electric wires, this approach is more precise than measuring the magnetic field itself.

### 8.1.2. *Numerical approach*

In complex configurations, numerical simulation of the source can be used to refine upper and lower values, which were set analytically. The entire exposure chain may be simulated from the cause of the radiation to the biophysical mechanism. The main operation consists of modeling the source from 2D/3D plans or mapping of the ambient field. The human model most representative of the situation is then selected from a digital library (member of a virtual family: adult man or woman with various morphologies and postures, children, etc.). These human models are representative, with dielectric characteristics associated with each tissue (see section 3.2.1). This approach is introduced at the end of this chapter.

### 8.1.3. *Metrological approach*

The purpose of the metrological approach is to characterize the main properties of the ambient field, such as strength and frequency, or the exposure level (index) directly.

## 8.2. Measurement: general

### 8.2.1. *Process*

The purpose of measuring is to determine the parameters of an electric, magnetic or electromagnetic field in terms of strength, and frequency or waveform in order to calculate whether the exposure is compliant with the appropriate legislation. Taking these measurements around radiating sources is generally the first approach proposed, because it seems practical.

There is a wide range of exposure situations, producing a significant diversification of measuring systems and methodologies. The intrinsic characteristics of the radiated field and its spatial characteristics in the environment to be analyzed must be considered. The most suitable system and method for the required precision can then be determined.

Information may also be provided by the manufacturers. This is generally based on standards relating to their products (electric household appliances [STD 04], occupational equipment [STD 11]). It is often useful for comparing them, but takes into account only simple exposure configurations

[SCE 15]. However, these standards are an interesting basis for preparing a procedure representative of real exposure conditions.

As well assessing the potential risk relating to exposure to electromagnetic fields, measurements help validate a radiation source model.

Measurements must be able to be performed by a proficient person, but not necessarily an expert. To this end, they must be standardized, adapted for operational conditions and based on a compact, robust and reliable acquisition chain.

A rigorous and scientific process must be well considered before each measurement. Physical and practical sense is required. The results should not be unexpected or lead to an analysis error. An error in the operating of the device or an artifact[1] must be detected before the measuring campaign ends. To obtain reliable results, the approach may be divided into five phases.

#### 8.2.1.1. *Preparatory phase*

Before analysis begins, it is useful to have a little knowledge of the exposure area to be assessed. The following information is required as a minimum, although it may not necessarily be precise:

- identification of the relevant field source/s;
- estimate of the worst functional case producing maximum exposure;
- characteristics of the voltages, currents, powers, etc.;
- temporal and spatial characteristics of the radiation;
- exposure conditions, including the most negative situations;
- duration of exposure (for high frequencies [HFs] only);
- layout of elements that might influence radiation;
- proximity of other radiation sources;
- type of population exposed;
- environmental conditions;
- purpose of measuring.

1 An artifact is an aberrant signal due to the measuring conditions.

#### 8.2.1.2. *Choice of measuring device*

A definition of the objectives, including the information required for the measuring report, and an analysis of the procedures, reveal the technical requirements of the measuring chain.

Generally, as well as the nature of the field, the proximity of the source and the spatial and temporal distribution of the radiation are among the important points to be considered. The device is chosen during the preparatory phase. The choice may also be dictated by a standard. Before the measuring campaign, the correct functioning of the devices, their calibration date, the state of the batteries and the availability of the required accessories should be verified.

#### 8.2.1.3. *Pilot phase*

A pilot phase is essential when not much information is available (for example limited technical data or absence of procedure). It allows the choice of measuring device to be confirmed and the measuring points identified, as well as the moments when the radiation peaks. It also makes it possible to decide on a measuring protocol, supplement it and even adapt it to a specific situation.

During the measuring campaign, a procedure (which may be standardized) must be followed to make the results reproductive, even for a third party. This procedure minimizes operational errors, which may be greater than the measuring device uncertainties.

When a new technology emerges or when the radiating equipment is of limited or even unique use, it may be appropriate to write a dedicated procedure formalizing the latest developments in the domain. This can then be used as the basis for the writing of a standard, if many parties are involved, or to authorize an independent laboratory. Formalizing such a document internationally takes several years. Examples include the procedure for measuring magnetic fields under high voltage power lines [STD 13], the procedure for measuring transmission levels from mobile telephony and radio broadcasting base stations [ANF 15] and the procedure for measuring magnetic fields in the vehicle environment [STD 31].

A pilot phase is particularly recommended when the time allotted for measurement is very short. It may be performed on a dedicated piece of equipment, to avoid disturbing a production line, for example.

#### 8.2.1.4. *Measuring*

The process used is already well known by the time work gets underway, since it has been studied in the pilot phase. The identified measuring points may be adjusted when the position of the people has been observed.

Exposure measurements arouse interest as well as questions from people present on the site. Answering them lends the process and expertise credibility.

#### 8.2.1.5. *Assessment report*

Comparing the measurement results with the exposure limits produces the *exposure index*, the value recorded in the assessment report or compliance report. These results are correlated with previous analysis to confirm the conclusions.

### 8.2.2. *Measuring device requirements*

#### 8.2.2.1. *General specifications*

The general specifications of a field measuring device or chain must satisfy the relevant technical standards. For example, standards IEC 61786 [STD 01, STD 02] for measuring static magnetic fields or alternating electric or magnetic fields in the frequency band from 1 Hz to 100 kHz relating to human exposure. The specifications of these documents include:

– type of physical quantity required;

– frequency domain;

– spectral span;

– magnitude range (minimum level, maximum level);

– dimensions of the probes;

– electromagnetic immunity (electromagnetic compatibility);

– response to a modulated field;

– disturbances in the measurement of the magnetic field by an electric field (and vice versa);

– operating method;

– uncertainties of the device including measuring probes.

These final pieces of information are divided between the technical specifications of the device and the calibration certificate. These specifications are deemed to be met in the domain of use intended by the manufacturer.

Laboratory test benches can be used to precisely generate radiation in order to practice using the devices, verifying their state or calibrating them. Confidence in the results is related to understanding the measuring chain and to the precision it is expected to have.

A field measuring chain (broadband) consists of a probe, a detector, which processes the signal coming from the probe, and a display.

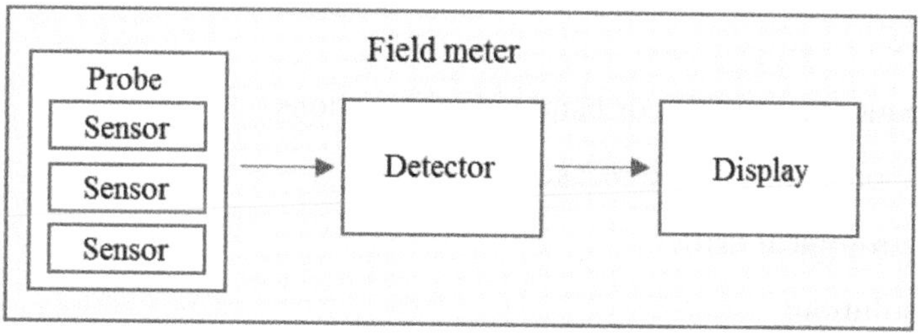


**Figure 8.1.** *Functional synoptic view of a three-axis field meter*

The sensor(s) are the basic elements of a measuring probe. They transform the intangible electromagnetic field into an easily measurable electric signal. These transformations are based on well-established physical principles. The sensors are arranged orthogonally to determine the module of the ambient electric, magnetic or electromagnetic field, which may be oriented in any direction (flux or polarization).

Several probes are required to cover the whole domain of exposure to electromagnetic fields, from static fields to 300 GHz electromagnetic fields, since none is able to cover such a range. The detector converts the electric signal into a physical root mean square (rms) or peak quantity, or into an *exposure index*. These quantities are adjusted from the calibration data. The display device is generally digital.

Measurements are carried out onsite in the position of the exposed people, but in their absence, since their presence might perturb the fields, making the repeatability or reproducibility of the measurements difficult.

#### 8.2.2.2. *Measurement uncertainty: general*

The uncertainty associated with a measurement is the parameter that characterizes its dispersion with reasonable probability. The purpose of measurement is, therefore, to estimate the physical quantity and the uncertainty affecting it. It is due (non-exhaustive list, see [JCG 10]) to various parameters, such as:

– calibration uncertainty;

– repeatability of the measurement (due mainly to variations in radiation level);

– reproducibility of the measurement (due mainly to the operating method);

– frequency response or bandwidth limitations

– response time of the measuring device;

– metrological drift;

– resolution;

– proximity of objects or obstacles;

– temperature and humidity (only for electric fields);

Uncertainties are explained afterward. Some of them can be reduced to negligible levels using a detailed measurement protocol.

In general, in the context of exposure, as for electromagnetic compatibility, the values measured are declared compliant as long as the actual uncertainty is lower than the maximum permissible measurement uncertainty. In this case, uncertainty is included in the determination of assessment quality, but must not be used in comparison with the limit values [STD 30, HSE 08].

The maximum uncertainties in Table 8.1 are permissible during assessment of exposure to electromagnetic fields by means of measuring or digital calculation [STD 11, STD 15].

| Frequencies | Measurements | Digital modeling |
|---|---|---|
| <10 kHz | +58%, –37% (±4 dB) | ±50% |
| 10 kHz–1 MHz | +41%, –30% (±3 dB) | ±50% |
| 1 MHz–30 MHz<br>1 GHZ–30 GHz | +41%, –30% (±3 dB) | ±40% |
| 30 MHz–1 GHz | +100%, –50% (±6 dB) | ±40% |
| 1–30 GHz | +100%, –50% (±6 dB) | ±50% |

**Table 8.1.** *Total permissible uncertainties in exposure to electromagnetic fields*

If uncertainty is confined within these limits, the standards generally recommend not including it in the results, since it was taken into account when the exposure limits were established. However, if the uncertainty is greater than the values indicated, its differential value must be added to the measurement before it is compared to the limits. Examples of the measurement uncertainty budget calculation of high-frequency electromagnetic fields are given in [STD 24].

## 8.3. Measuring low-frequency fields

### 8.3.1. *General*

The following devices are used to measure low-frequency fields:

– fieldmeter adapted to measuring static magnetic induction (up to a few hundred hertz or kilohertz);

– fieldmeter adapted to measuring time-varying magnetic induction;

– fieldmeter adapted to measuring time-varying electric fields.

There is a wide range of devices distinguished by their sensitivity, ranges and bandwidth, but also by their detector type. Some devices are able to measure the electric field and the magnetic field, but generally not simultaneously. Commercially available products without a calibration certificate cannot be considered reliable measuring devices.

The devices process the signals, either in the time domain or in the frequency domain, theoretically producing the same result (Chapter 6). In practice, each processing domain has advantages and disadvantages.

### 8.3.2. *Measuring low-frequency magnetic fields*

#### 8.3.2.1. *Magnetic field measuring sensor*

The usual technique for measuring a magnetic field consists of applying Faraday's law [1.5] [TAN 95]. A voltage is produced at the ends of an open loop placed in a time-varying magnetic field.

The standardized loops are generally circular coils with surface areas of 100 $cm^2$ and 3 $cm^2$. Coils of 100 $cm^2$ are used for better sensitivity. In general, coils of 3 $cm^2$ are reserved for measuring non-uniform fields encountered at a very short distance from the source.

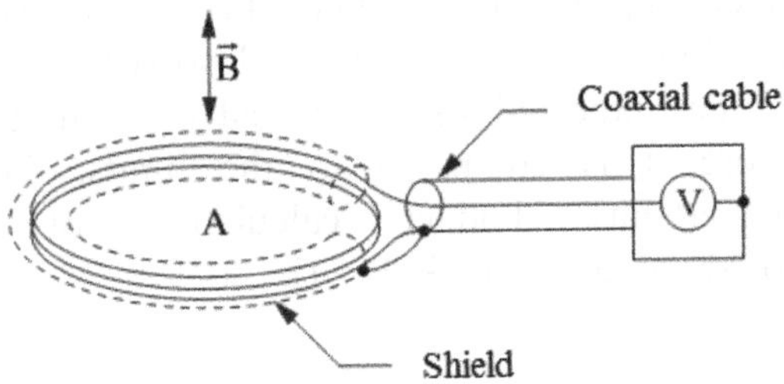


**Figure 8.2.** *Diagram of a magnetic fields measurer equipped with a coil*

In order to prevent the measurement from being influenced by the ambient electric field, the coil is protected with an electric shield, which is transparent to magnetic fields, a function fulfilled by a thin thickness of aluminum, for example.

Determining the *exposure index*, whether in the time domain (weighted peak technique, see Figure 6.7) or in the frequency domain (weighted vectorial spectrum, see Figure 6.4), involves using identical, concentric orthogonal coils, forming a spherical measuring probe.

**Figure 8.3.** *Triaxial probe composed of three orthogonal coils for measuring magnetic fields (source: Narda STS GmbH)*

The orientation of the probe in a quasi-uniform field is of no importance, because the probe, with its three sensors, is isotropic. Isotropic means the measurement result does not depend on the probe orientation.

The three signals are processed simultaneously, whether in real time or using software after recording, according to the principle presented in Figure 6.8.

In theory, a single coil might suffice over a periodic signal which is stable during acquisition, but this requires reorienting it carefully after each measurement and putting the signals back into phase before they are processed. Furthermore, a single measurement is possible, provided the coil is oriented perpendicular to the direction of magnetic flux.

The value of each field component is required for dosimetric assessment, because the direction of the field in relation to the person must be taken into account. The orientation of the coils must therefore be precisely known.

#### 8.3.2.2. *Device for measuring low-frequency magnetic fields*

A functional synoptic view of a magnetic field measurer is shown in Figure 6.8, to which the measuring probe and equalizers should be added. The display unit is not shown.

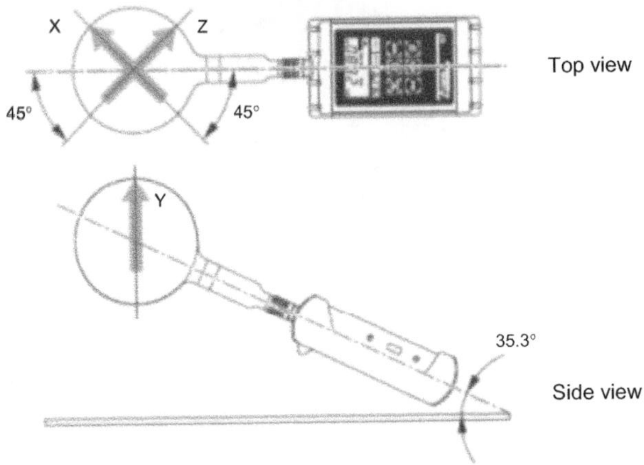


**Figure 8.4.** *Orientation of the coils of a probe for measuring magnetic fields (source: Narda STS GmbH)*

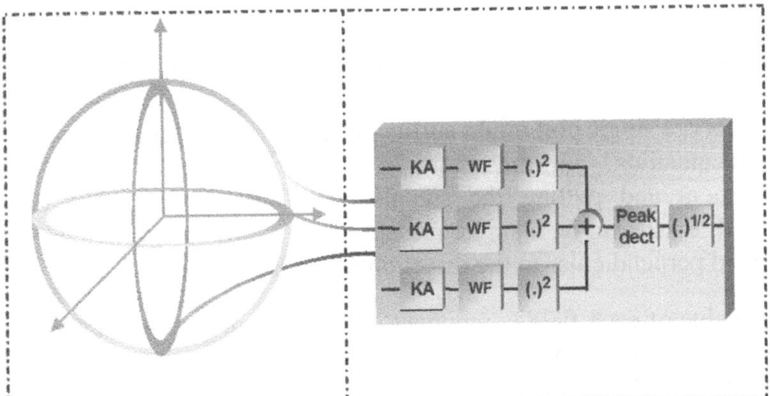


**Figure 8.5.** *Functional synoptic view of a magnetic fieldmeter*

The functions of the device are successive: the sensors convert the magnetic field into an electric signal, the equalizers (KA) flatten the frequency response of the sensors, the filters (WF) weight the signal. The module is then calculated and the detector extracts the maximum value over the measuring interval. Its square root corresponds to the *exposure index*.

This type of device may have a low cutoff frequency, adjustable between 1 Hz and 40 Hz, and a high cutoff frequency, between a few kilohertz and a few hundred kilohertz.

The device is equipped with a display, potentially with analog outputs for visualizing or recording the magnetic field (induction), and a digital interface enabling the recorded measurements to be downloaded and the software and calibration data to be updated.

The noise level of the device must be lower than 5% of the selected limit value. Measurements under this threshold must not be taken into account [STD 04].

#### 8.3.2.3. *Measuring low-frequency magnetic fields*

The device may be hand-held. The proximity of non-metal materials does not disturb the readings; the same is true for small non-magnetic metal parts. The presence of ferrous structures may significantly alter the spatial distribution of the magnetic field.

Repeatability, i.e. the realization of successive measurements in identical conditions, reduces uncertainties. Measurements in different positions may reveal field anomalies or a measuring problem (results differing significantly from the prediction).

#### 8.3.2.4. *Calibration of a device for measuring magnetic fields*

The calibration procedure consists of a set of operations carried out in specific conditions, with the aim of establishing a relationship between the values indicated by the device and the corresponding quantities generated by a calibration system with defined uncertainties. These procedures are standardized and performed by a duly accredited organization [STD 16].

Thus, calibration reveals the uncertainties of an instrument (random measurement errors) and makes it possible to compensate the trueness errors due, for example, to the tolerance of the components or their long-term drift.

Although calibration at regular intervals is regulated in standards or by a quality control procedure (generally every 2 years), the correct operating of a device must be verified before and after every measuring campaign. For example, coils subjected to strong fields may be damaged without warning (refer to the manufacturer's instructions for the absolute non-destructive ratings). Calibration is also necessary after every repair.

For the calibration, the probe or device is placed in a quasi-uniform sinusoidal field with known amplitude and direction. This field is produced using Helmholtz coils (see [STD 01]). Devices are calibrated at a minimum of three frequencies distributed in the measuring bandwidth. The 50/60 Hz frequency is always considered. The field level must be high to ensure that the results are not influenced by the background noise from the device and the ambient magnetic noise. Otherwise, additional uncertainty calibration must be taken into account.

If the device has automatic range selection, each of these should be calibrated.

A device without its probe may also be calibrated by directly injecting a voltage representative of the field into the detector or comparing it with a reference device. In general, uncertainties relating to the device's probe are greater than uncertainties relating to its detector.

The calibration certificate outlines the data relating to the device that should be included in the uncertainty budget.

#### 8.3.2.5. *Uncertainty sources in measurements of low-frequency magnetic fields*

Uncertainties are due to the residual difference between the information delivered by a measuring chain corrected of its errors after calibration and the actual corresponding quantities. Calibration makes it possible to quantify those relating to this chain.

The measuring uncertainties of a strong non-uniform field are related just as much, if not more, to the operating mode as to the precision of the devices. Far from the sources, where the fields become homogeneous, the overall uncertainty is essentially due to the device.

The measuring uncertainty is less critical when exposure levels are low. When the field to be measured is strong, far above the limits, the operator must not put himself in a hazardous position. The following sections summarize a few sources of uncertainty in measurements of low-frequency magnetic fields.

#### 8.3.2.5.1. Long-term drifts

Long-term drifts come from gradual changes in the components of the instrument. This ageing process alters the response of the device. Periodic calibration helps to correct these drifts.

#### 8.3.2.5.2. Electromagnetic immunity

The measuring device may be disturbed by the source to be measured or by the radiation of another source (electromagnetic compatibility). The measurement may be considerably influenced when its shielding is insufficient.

#### 8.3.2.5.3. Non-uniform magnetic field

Measuring devices are calibrated in a quasi-uniform field while the fields measured, close to equipment, decrease quickly with the distance. Specific uncertainties should be taken into account in this case.

*Measurement distance*

The uncertainty linked to the distance between the source and the probe is increasingly significant as this distance decreases. Indeed, the magnetic field transmitted by a source decreases with the distance *d* according to a law that is often as follows:

$$B = \frac{1}{d^{\alpha}} . B_c \qquad [8.1]$$

where $B_c$ is a constant dependent on the source, α is a constant between 1 and 3 in most cases (Figure 1.6). The standard deviation associated with distance uncertainty of rectangular distribution takes the form of [STD 02]:

$$\sigma_d = \pm \frac{\alpha}{\sqrt{3}} \frac{\Delta d}{d} . B \qquad [8.2]$$

The uncertainty introduced by an uncertain tangential position may be zero.

In practice, the probe is positioned in relation to its outer casing, while the reference point is its center.

*Concentricity*

The three field measuring loops of a triaxial probe do not necessarily have a common center. This is the case when the fieldmeter is fitted with coils with ferrite cores to increase its sensitivity. Furthermore, the proximity of the cores perturbs the ambient field. The manufacturer should find a compromise between this proximity and the error due to the decentering of the sensors. The sensors may also not be concentric when the device is multifunctional.

*Average*

A loop measures the average of the normal (perpendicular) component of the field crossing it. In a non-uniform field, the value measured is greater than the value in its center, considered the reference point. The difference is significant near the source, then decreases with the distance [MIS 93].

Figure 8.6 shows the relationships between the value at the center of the loop and the value of the magnetic field averaged over its surface area as a function of the distance for three field decreases ($1/d$, $1/d^2$ and $1/d^3$). $d/r$ is the distance relative to the radius of the loop. This is oriented perpendicular to the magnetic flux. These curves were obtained through calculation.

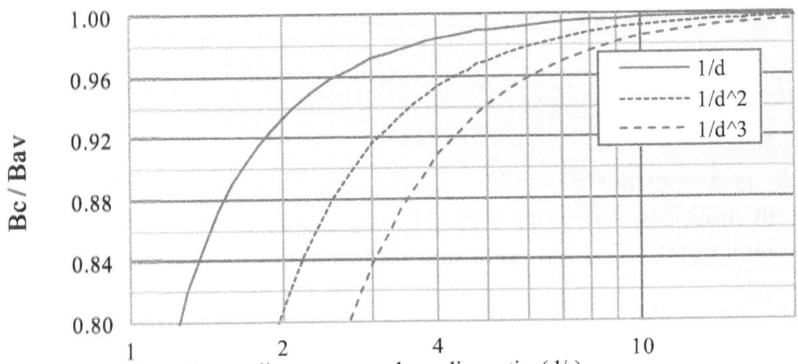


**Figure 8.6.** *Relationship between the field at the center and the field measured (averaged) by a loop for three decreases in the magnetic field*

NOTE.– Uncertainty is greatly reduced by using small loops. Standardized loops have a surface area of 3 $cm^2$ and 100 $cm^2$. Their respective theoretical radii are 1 cm and 5.6 cm. In practice, it not recommended to measure the magnetic field for conformity assessment when the distances in relation to the sources are smaller than the diameter of the measuring loops.

*Orientation*

There is an isotropy error relating to measurement of a non-uniform field by a loop system. It is of the order of magnitude of the error due to the average.

#### 8.3.2.5.4. Proximity of batteries

When the device is compact, its batteries are close to the measuring loops. The device has been calibrated with a certain type of batteries. Calibration may be distorted by a change in their chemical composition.

#### 8.3.2.5.5. Noise

*Noise of the device*

Electronic components generate unwanted background noise in the device, independently from the environment. This noise generally corresponds to the measurements displayed in the absence of fields. It determines the sensitivity of the device. Sensitivity is the smallest possible significant measurement. It is given for a quality criterion, which is often the signal to noise ratio.

*Ambient noise*

The ambient magnetic noise may be greater than the background noise of the device, especially in an industrial environment.

#### 8.3.2.5.6. Bandwidth limitation

A limited bandwidth can contribute significantly to measurement uncertainty, particularly for magnetic fields rich in harmonics, such as those radiated by power inverters.

The harmonics decrease with the frequency. The reference levels or action levels also decrease with the frequency, meaning a high-order harmonic can have as much weight in an *exposure index* calculation as the fundamental harmonic (see Example 3, section 7.2.3).

#### 8.3.2.5.7. Climatic environment

Calibration is performed at a controlled ambient temperature (~20°C). Devices should be used in the temperature range recommended by the manufacturers.

Air humidity poses no particular problem for magnetic fields, except when steam seeps in and condenses in the casing of the device.

#### 8.3.2.5.8. Peak factor

The calibers of a device are given in rms field value, which poses no difficulties during measurement of sinusoidal fields. These calibers are not suited to measuring signals with a high peak factor (peak value to rms value ratio), because the electric circuits may become saturated. The resulting non-linearity errors are difficult to quantify. Some devices alert the operator (overflow).

#### 8.3.2.5.9. Instantaneous dynamics

Some devices have automatic gain control, which adapts the measuring range to the field strength for optimum sensitivity. A short-duration signal with strong dynamics will not give the device time to adapt. In the best case, it will not deliver information. A permanent range must be selected manually for a correct measurement.

### 8.3.3. *Measuring a static magnetic field*

Static and quasi-static fields (<1 Hz) are measured using the Hall effect or magneto-resistive sensors. Their functioning domain extends from the continuous (0 Hz) to several kilohertz. Their range is between 100 μT and 10 T. They are not very sensitive. They are disturbed by the Earth's magnetic field (≈50 μT) and by the fields produced by the 50 Hz grid. Hall effect sensors are also sensitive to temperature, as they are made of semiconductive material. Regular calibration is necessary. These features mean this type of sensor is not generally used for low frequencies (ELF), but their millimetric size and mechanical ruggedness have advantages.

Static magnetic fields can be measured using a magnetometer with a saturable core. This type of device is used to measure fields in the range of 0.1 μT (and less) to 10 mT.

There are also other measuring techniques based on sensors with superconducting quantum devices (SQUID magnetometer).

## 8.3.4. *Measuring low-frequency electric fields*

Measurements of low-frequency electric fields are not common, since the maximum voltages surrounding us in our daily environment are relatively low, 240 V or 400 V at 50 Hz. With such values, electric fields remain far below the reference levels (5 $kV \cdot m^{-1}$ for the general public).

### 8.3.4.1. *Measuring sensor for electric fields*

The operating principle of a measuring device for electric fields is based on Gauss's theorem: a variable field produces a variable charge in a capacitive dipole (capacitor) [TAN 95, STD 01]. An impedance connecting the two electrodes enables charges to flow. The current thus created is proportional to the time variation of the electric field. The voltage between the electrodes is measured.

This kind of sensor allows "floating" measurements to be performed, i.e. with no connection to a ground. Here are two design principles.

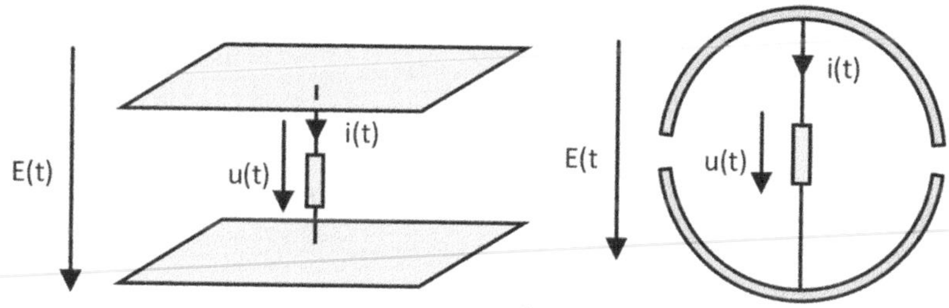


$$u(t) \propto i(t) = \frac{dQ}{dt} \propto \frac{dE}{dt}$$

**Figure 8.7.** *Examples of monoaxial sensors for electric fields*

The measuring cables must not disturb the electric field. They are shielded and insulated *a minima*, but fiber-optic solutions are well suited.

Three orthogonal dipoles (triaxial sensor) measure three components of an electric field in order to calculate the field module or determine the *exposure index* according to the principle presented for the magnetic field. The sensor may be cubical or spherical.

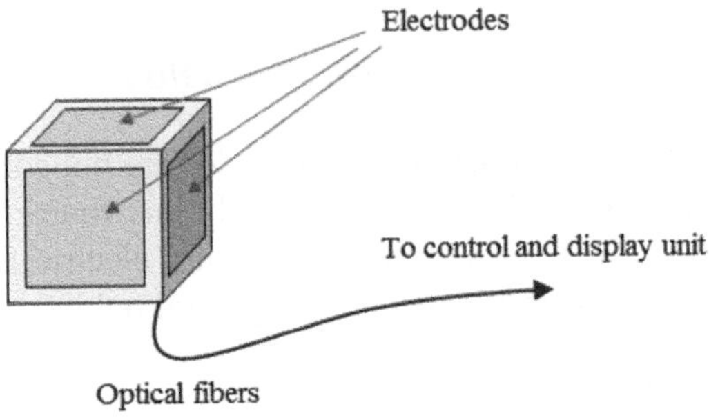


**Figure 8.8.** *Triaxial isotropic probe composed of three pairs of electrodesfor measuring an electric field*

Sensors connected to the ground can measure a field on or close to the ground.

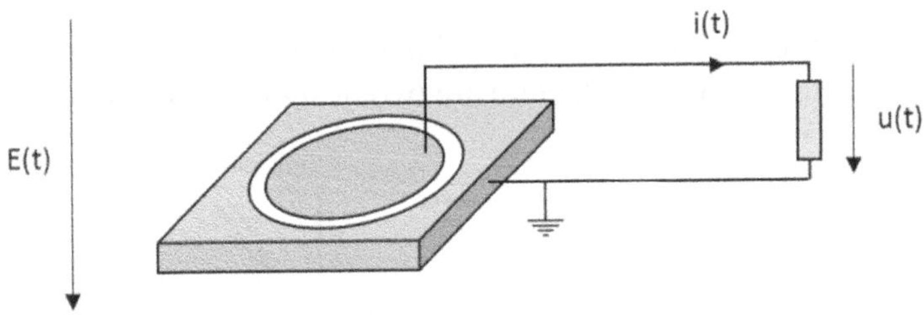


**Figure 8.9.** *Sensor for measuring electric fields in relation to the ground*

Sensors made up of optical fiber sensitive to electric fields, using the Pockels effect, are sometimes employed.

#### 8.3.4.1.1. Measuring sensor for Electric and magnetic fields

Some compact sensors are able to measure the electric and magnetic fields. A functional solution is presented in Figure 8.10. The coil picks up the magnetic field according to the principle outlined in the previous section and the two shield electrodes pick up the electric field. These must be transparent to the magnetic field.

The association of three of these sensors forms a triaxial spherical probe.

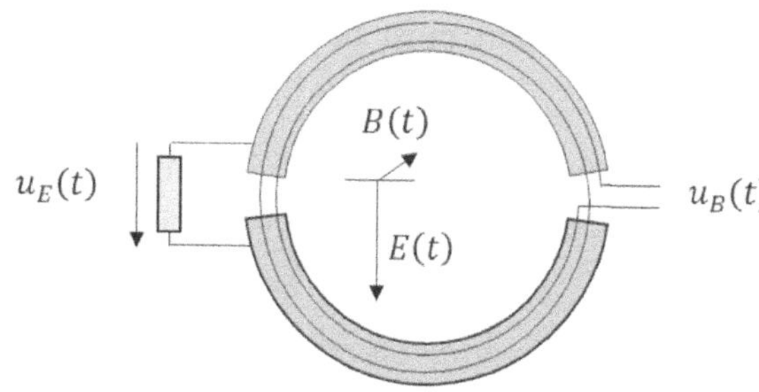


**Figure 8.10.** *Example of a monoaxial sensor for electric fields and magnetic fields*

### 8.3.4.2. *Device for measuring low-frequency electric fields*

Measuring devices are functionally composed of an isotropic probe sensitive to a low-frequency electric field, and a detector, incorporating an analog and digital signal processing. The general design is identical to that of a magnetic field measurer.

The detector can extract the rms value, the peak field value in V·m$^{-1}$, and the spectrum or *exposure index* of the electric signals delivered by the sensors. This processing may be built into the probe, between the electrodes, forming a compact unit. This integration guarantees a high level of electric insulation of the measuring chain in relation to the environment. To perfect this insulation, the user interface (display, controller keyboard and digital ports) is remote using an optical fiber. As an additional precaution, the sensor cannot be activated when its battery is being loaded.

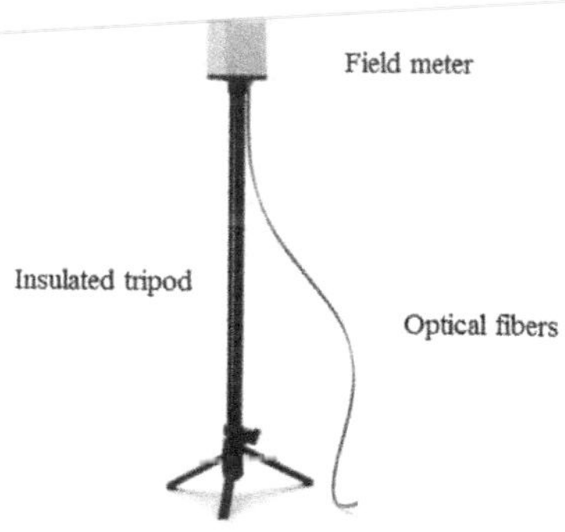


**Figure 8.11.** *Electric fieldmeter on its support (source: Narda STS GmbH)*

The measuring probe support must be made up of synthetic insulating materials.

A device must be carefully designed when its bandwidth exceeds around 10 MHz (the shorter the wavelength, the more sensitive the devices to the mechanic).

#### 8.3.4.3. *Measuring low-frequency electric fields*

Low-frequency electric fields are measured mainly near very high voltage lines and electrical distribution equipment. Measurements are also taken near radio transmission antennae in amplitude modulation for frequencies between 145 kHz and 2 MHz (section 2.2.3.1.1). More recently, the electric field (and the magnetic field) has been verified around wireless battery charging systems and inductive data transfer systems, although the magnetic field is more significant. The frequencies are typically between a few kilohertz and 100 kHz.

The devices are relatively simple for 50/60 Hz field measurements and become sophisticated as the frequency increases, providing field values and exposure indices ($IE^{I98}$, $IE^{RSS}$, $IE^{WPM}$, etc., see Chapter 6), as well as spectrums.

The voltage, frequency, geometry of the source and configuration of the environment should be considered in choosing the measuring device. The number of commercially available devices is small, since electric fields are not often measured.

The measuring method is very important, because the electric field is much more sensitive to the environment than the magnetic field. The movement of objects within a perimeter of several meters around the source perturbs the field, thus influencing the results. To produce reproducible measurements, the operator must stand at a sufficient distance from the measuring area.

The dimensions of the sensor must be small to prevent the device from significantly altering the charge distribution on the surface of nearby conductors. The use of an insulating tripod is mandatory.

#### 8.3.4.4. *Calibration of a device for measuring electric fields*

The general remarks relating to the calibration of a magnetic field measurement device also apply to electric field measuring devices.

The device to be calibrated is placed in a quasi-uniform electric field, generated by two parallel panels of large dimensions in relation to their gap and connected to a sinusoidal voltage generator (Figure 1.1). The sensor must remain far from any surface in order to reduce the proximity effect, which causes a change in the distribution of surface charges and therefore in the local electric field [STD 01]. The device is calibrated on each of its three sensitive axes at 50/60 Hz and at a few frequencies covering its bandwidth. Isotropy is verified, along with the cross-talk between the axes.

The signal to noise ratio must be sufficiently high to not add uncertainty. The voltage injection method used for magnetic field measuring devices also applies to the calibration of electric field meters.

A measuring sensor for electric fields is highly sensitive to mechanical dimensions. Calibration is mandatory when the device has been dismantled (repairs, etc.). Regular calibration does not mean a device does not need to be checked before a measuring campaign.

#### 8.3.4.5. *Sources of uncertainty in measurements of low-frequency electric fields*

Sources of uncertainty particular to measurements of low-frequency electric fields are presented hereafter. They substitute or complete the sources of uncertainties present when magnetic fields are measured.

##### 8.3.4.5.1. Proximity of the operator

The position of an assessor in relation to the field source during measurement of an electric field can affect the results by several tens of percentages, since the electric potential of the human body can be considered close to the ground.

Figure 8.12 [STD 02] illustrates this error under an overhead power line. When undisturbed, the electric field is vertical and can be considered uniform.

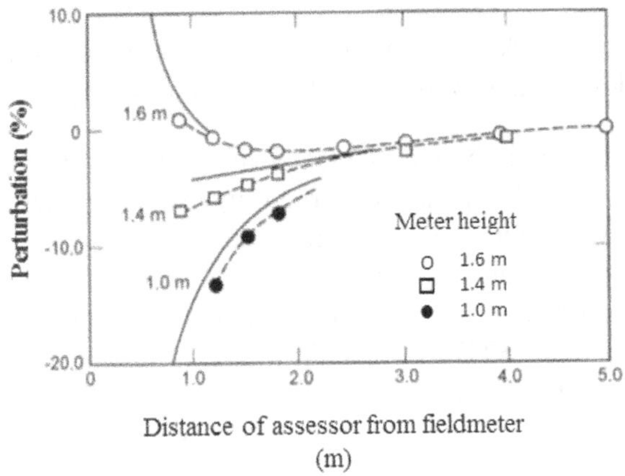


**Figure 8.12.** *Effects of proximity of the assessor during measurements of an electric field*

The measuring error is represented as a function of the distance between the assessor and the probe for various measuring heights in relation to the ground. The dot points come from the measurements and the curves are theoretical predictions. The graph shows that a minimum distance of 2 m must be respected for this application. It may be greater in other cases. The effect of the proximity of a body can be observed and is easily quantifiable experimentally.

### 8.3.4.5.2. Proximity of conductors

Measurement of the electric field may also be significantly disturbed if the probe is too close to objects of varying conductivity, such as infrastructures or vegetation. The reason is always a change in the distribution of surface charges.

It is necessary to remove mobile objects (or wait for them to be still) as far as possible to ensure the reproducibility of results.

If the distance between the probe and the surfaces exceeds two to three times the greatest dimension of the sensor (diameter or diagonal), the measuring error is not significant.

#### 8.3.4.5.3. Electrode asymmetry

Measuring differences occur when there are asymmetries in the design of the measuring probe. The result depends on the measuring axis of the device. The most sensitive axis and the geometric axis cannot also coincide.

#### 8.3.4.5.4. Anisotropy

There is a theoretical anisotropy error of a few percentages when the sensor is cubic in shape. This error is often omitted, because the devices are calibrated using a field perpendicular to the electrodes. Anisotropy is calculated from minimum and maximum values, determined through rotation of the probe in the field.

#### 8.3.4.5.5. Magnetic field

Measurement of the electric field may be disturbed by an ambient magnetic field, which is more difficult to control.

#### 8.3.4.5.6. Influence of the handle

Some devices are portable and have a handhold. This introduces an electric leakage impedance toward the ground through the assessor's hand and body. In order to limit this leakage, the handle must be oriented perpendicular to the field lines. This phenomenon may be observed during measuring. One reliable solution consists of taking measurements at a distance via a fixed probe, which is remote and connected by an optical fiber.

#### 8.3.4.5.7. Non-uniform electric field

Supposing the absence of proximity effects, the uncertainty due to the non-uniformity of the electric field is negligible in a large number of use-cases, whether in terms of the dimensions of the probe, its orientation or the gaps in distance.

#### 8.3.4.5.8. Climatic conditions

Air humidity alters the measured value of the electric field. Measurement is not recommended when the humidity level exceeds 70%. Furthermore, the measuring uncertainty risks may be aggravated by steam condensation. The variations in ambient temperature create air overpressures and depressions in the casing, which suck droplets or steam. When they reach the electrodes, leakage currents increase, producing an erroneous measurement.

If measurements are to be undertaken despite a significant level of humidity, use of a remote tripod is recommended [STD 01].

## 8.4. Measuring high-frequency electromagnetic fields

### 8.4.1. *General*

The physical quantities measured for quantifying high-frequency electromagnetic fields (100 kHz–300 GHz) are the electric field and the magnetic field. These quantities are related beyond a distance of a wavelength from the source [1.15] and make it possible to extract the power density from it [1.16]. The measuring device is chosen depending on the frequency domain to be analyzed, the characteristics of the radiation (reactive field, radiative field, etc.) and the number of radiation sources.

The basic constituents of a measuring chain are as follows:

– the probe, formed from one or several sensors;

– coaxial cables, which transfer electric signals from the probes to the processing unit, if it is remote;

– the processing unit, which converts the signals into quantities that can be used to quantify exposure.

Furthermore, broadband measurements are distinguished from narrowband measurements.

### 8.4.2. *Measuring sensors for electromagnetic fields*

The sensors used are radioelectric antennae, which deliver an electric power proportional to the received electric or magnetic field strength. The antenna factor, an essential element from the calibration certificate, makes it possible to directly associate a field value to it. An alternative basic technique consists of estimating the power through calorimetric measurement (bolometer). This power is then compared with the exposure limit converted into the same unit.

*Monopolar antennae* are distinguished from *dipole antennae.* These antennae pick up the electric component ($E$) of a high-frequency electromagnetic field:

– the monopolar antenna is composed of a conductor referenced to an electric ground (conductive surface, earth, metal chassis, etc.);

– the dipole antenna consists of two insulated conductors.

There are also loop antennae (current loop), which pick up the magnetic component ($H$) of the field, according to the principle presented for inductive measurements at low frequencies (LFs).

a) Broadband monopolar (conical antenna)

b) Broadband dipole (biconical antenna)

c) Magnetic loop

**Figure 8.13.** *Examples of antennae picking up the electric or magnetic field*

Antennae are also differentiated by their directivity. A directional antenna must be oriented toward an identified transmitter to characterize the field it emits. An antenna with low directivity captures the ambient fields coming from a large number of directions. A measurement method is associated with each type of antenna.

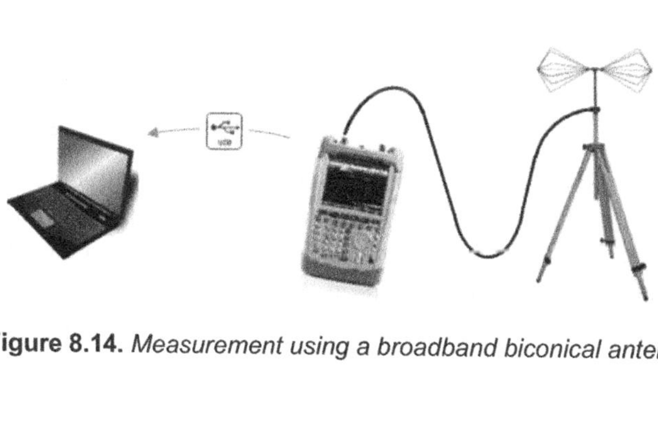

**Figure 8.14.** *Measurement using a broadband biconical antenna.*

The electric field must be measured on two orthogonal planes, generally oriented vertically and horizontally. The resulting field is the quadratic sum of the measurements:

$$E_{VH} = \sqrt{E_V^{\,2} + E_H^{\,2}} \qquad [8.3]$$

where $E_v$ is the field value in the vertical plane and $E_H$ is the field value in the horizontal plane. The overall field is the quadratic sum of the fields from each transmitter and their polarizations:

$$E = \sqrt{\sum_{i=1}^{N} \left( E_{V_i}^{\,2} + E_{H_i}^{\,2} \right)} \qquad [8.4]$$

$N$ is the number of transmitters. The *exposure index* is obtained by the sum of the weighted fields ($EI^{RSS}$ [6.16]).

A directional antenna can received only one spatial component of the field in a known direction. If the propagation or wave polarization direction is not known, measurement on three perpendicular axes will be carried out, keeping the electric center of the sensor in the same position. The quadratic addition of the three values gives the overall field value [6.11].

The resulting field is calculated thus:

$$\mathrm{E} = \sqrt{\sum_{j=1}^{M} \left[ \left(\mathrm{E}_{Xj}\right)^2 + \left(\mathrm{E}_{Yj}\right)^2 + \left(\mathrm{E}_{Zj}\right)^2 \right]} \qquad [8.5]$$

where $E_{Xj}$ $E_{Yj}$ $E_{Zj}$ are components $x$, $y$ and $z$ of the electric field ($V$ and $H$) at frequency $j$.

The process is facilitated by probes composed of three orthogonal antennae, forming an isotropic sensor (Figure 8.15).

For HF field measurements, the transmission must remain stable during the acquisition of the three components (the transmitter must remain immobile, for example). The advantage of HF compared to LF, where peak levels are sought, is that the result is not altered by a phase difference between the three measurement axes, meaning measurements can be taken successively.

The small sizes of the antennae used in the exposure domain mean they do not significantly affect the ambient field and the *maximas* of fields with significant spatial variations can be measured. This allows the requirements of the measuring procedures to be met. Small size also means low gain, which is generally not a constraint, since the levels to be measured are very high, a minimum of a few tenths of volts per meter compared to a few microvolts per meter required in telecommunication.

In summary, the measuring probe for electromagnetic fields:

– has one or several antennae, which transform the electric or magnetic field picked up into an electric signal representative of its strength;

– must not distort the radiation;

– must be small in order to perform localized measurements;

– must remain insensitive to interfering electromagnetic components (other polarizations, etc.);

– must be calibrated (known antenna factor and uncertainties).

### 8.4.3. *Measuring device for high-frequency electromagnetic fields*

There are devices dedicated to broadband measurements and others dedicated to narrowband measurements. These devices are complementary.

#### 8.4.3.1. *Broadband measurements*

Equipment designed to measure broad bands directly provides the total value of the electric field present across its entire bandwidth. This must be sufficiently wide to capture all the relevant frequencies. Some broadband devices have a weighted frequency response, which complies with a given exposure limit. They therefore display the *exposure index* directly.

This type of measurement allows a site to be rapidly analyzed, particularly if there are many radiation sources and if the site is large.

The device has few adjustments. Filters flatten the frequency response of the probe. A detector extracts from the signal the rms value of the measured electric field. The weighting function is not always used. The field level must then be compared with the lowest reference level in the measuring band, 28 V·m$^{-1}$ for example [6.18].

The global analysis therefore takes into account all the contributions over a considered frequency range, with no possible distinction between the various transmissions. It is often used during a preliminary study to quickly obtain an exposure level. This method is appropriate for all field types and there is no need to worry about their particular time and frequency characteristics.

A broadband device generally takes measurements in the time domain. The principal characteristics of a broadband measuring instrument are as follows:

– the instrument has an "isotropic" antenna. A directional antenna can be used to characterize an identified transmitter;

– the instrument measures the rms value of the total electric field in real time;

– the instrument is capable of averaging the rms value of the measured field in time in compliance with standards and directives (e.g. 6 min);

– the device is able to display the electric field level (V·m$^{-1}$), the equivalent magnetic field level (μT or A·m$^{-1}$) or the corresponding power density (W·m$^{-2}$);

– the instrument can display the maximum measured field level and possibly record the results;

– the instrument is accompanied by a calibration certificate specifying its contributions to the overall measuring uncertainty.

Broadband measuring devices are suitable when the field to be measured is lower than the exposure limits by an order of magnitude (for example 6 V·m$^{-1}$) [ANF 15]. However, they are not generally suited to measuring very low fields, due to lack of sensitivity, or measuring fields close to the limits, due to lack of precision. Furthermore, they are not able to discriminate between the signals received. For better precision, assessment must be performed in narrowband.

#### 8.4.3.2. *Narrowband or selective measurements*

Narrowband measurements are performed with a spectrum analyzer (SA) or receiver equipped with an antenna. The instantaneous bandwidth of the devices is limited, making it possible to precisely measure radiation levels, regardless of frequency. Furthermore, knowledge of the field frequency makes it possible to precisely calculate the *exposure index* corresponding to the reference level and identify the telecommunication service. The transmission time (e.g. the duty cycle) of the radiation can also be studied.

Measurements in narrowband require adjustments adapted to the time and frequency characteristics of each radiation source. A measuring chain dedicated to a transmission source or service contributes yet more precision to the field characterization.

The reduction in background noise provided by narrowband is not generally a criterion, since the levels measured are generally high (the background noise is proportional to the width of the analysis filter).

A narrowband measuring chain must:

– include an isotropic or directional antenna;

– contain antenna factor data (allowing the power measured to be converted into an electric field);

– have a calibration certificate for each element;

– allow its measuring parameters to be adjusted to the signals (resolution bandwidth [RBW], video bandwidth [VBW], etc.).

Figure 8.15 shows a conventional detailed measurement chain for determining an electromagnetic field level.

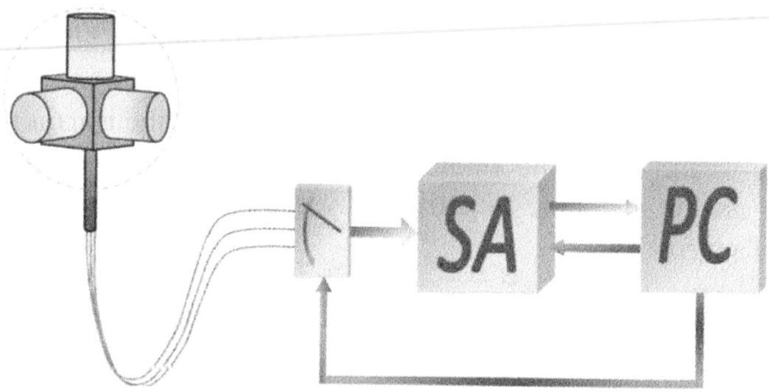


**Figure 8.15.** *Example of narrowband triaxial measuring chain based on three monopolar antennae*

This measuring chain is made up of a compact probe formed from three monopolar antennae to ensure isotropic reception, a RF three-channel multiplexer, a SA and a personal computer (PC). The multiplexer may be built into the probe. Coaxial cables guarantee the connection between the antennae and SA. The role of the multiplexer is to connect one of the three antennae to the analyzer, which has only one processing channel. The analyzer displays the spectrum of the signal of the selected antenna and provides detailed information of its content in terms of level and frequency, as well as its transmission cycle. The analyzer has good sensitivity, since although it covers a very wide frequency range (more than seven decades), its instantaneous bandwidth remains low (RBW filter).

The laptop (PC) records the spectrums, performs the sums to obtain the total field level and weights each frequency to establish the *exposure index*. The PC can also control the equipment, partially automatizing the measurements. Most of the equipment measures the electric field. The magnetic field and power density are obtained through calculation.

There are also some narrowband devices measuring the electric and magnetic components of an electromagnetic field. These devices are designed to measure up to a few tens of megahertz in the vicinity of the radiation source, a near-field zone where the two components are not simply correlated. Rejection of the unmeasured field must be sufficient to obtain a reliable result.

#### 8.4.3.3. *RF exposimetry*

Radiofrequency exposimeters, improperly called "dosimeters", are proposed to prevent the risk of exposure to electric fields. These devices are used by occupationals working in close proximity to very powerful transmission antennae. They are worn on the body and their main function is to alert the wearer when the ambient field level exceeds a preset threshold. These devices are also able to periodically record the measurements over a duration of several hours.

Their precision is fairly modest, because on the one hand, the level measured depends on the position of the equipment on the wearer and, on the other hand, his body can absorb or diffract part of the ambient radiation. This results in an underestimation of exposure [PIC 12]. These devices are not suited to exposure assessment in close proximity to sources (near field).

### 8.4.4. *Measuring high-frequency electromagnetic fields*

Electromagnetic fields must be measured in an unperturbed environment, i.e. in the absence of bodies. For his safety, the operator must ensure the exposure limits are not exceeded. If his presence affects the results, the probe or measuring device must be mounted on a tripod made of materials neutral to HFs (non-conductive composites, etc.) and he must stand at a certain distance.

HF electromagnetic fields are harder to measure than low-frequency magnetic fields, but the exposure level is simpler to calculate. The difficulty depends on the nature of the transmission and the field meters available. Complex modulations requiring increasingly wide bandwidths make it difficult to configure devices [NEU 03].

As a general rule, the electromagnetic field is not uniform, since the waves are reflected by conductive surfaces (ground, reinforced concrete or metal structures, etc.). Incident waves and reflected waves are added together or subtracted. Fading therefore occurs, causing strong variation in the field strength in time and space. These situations are particularly found in confined environments.

Measurements must be performed in various parts of the occupied area. Furthermore, for assessment of whole body exposure, measurements must be taken at various points to determine the average spatial value, as shown in Figure 8.16 [STD 15].

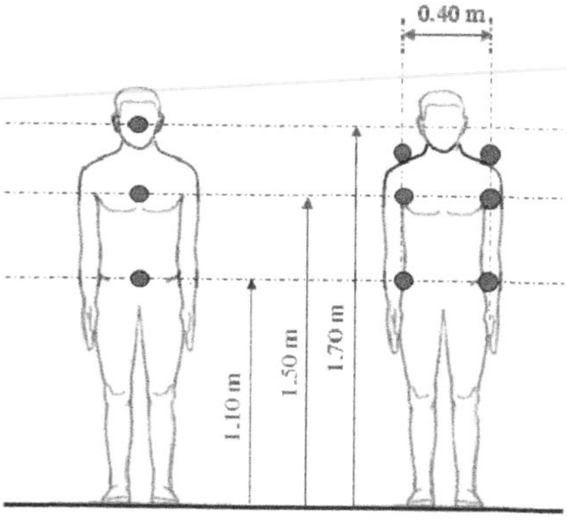


**Figure 8.16.** *Examples of measuring point positions for calculating an average spatial value*

The spatial average is calculated quadratically according to the following formula for $N$ electric field measuring points:

$$E = \sqrt{\frac{1}{N}\sum_{i=1}^{N} E_i^2} \qquad [8.6]$$

where $E_i$ is the module of the electric field at point $i$.

Three measuring points are generally recommended (e.g. at heights of 1.1 m, 1.5 m and 1.7 m from the ground) when the transmitter is far away. In some environments, additional measuring points may be required, depending on the size and posture of the body.

The number of measurements also depends on the confidence in the result, which must be reproducible. In some situations, this number may be significant.

The measured field must also be averaged over a duration in compliance with the standardized or regulatory requirements (e.g. 6 min for frequencies between 100 kHz and 10 GHz, see section 5.4).

A probe must be located at a minimum distance of several times its size from any conductive element, including the ground, to prevent coupling effects, which might significantly alter its performance.

Measurements at the beginning of the far-field zone and within direct visibility of the transmission source are easy, since reflection phenomena are weak. Pointing a directional antenna toward this source is sufficient. The results are reliable and reproducible [CNR 12]. They make it possible to extrapolate field levels and check the regulatory aspects of transmission are respected [EC 09].

There are many protocols and multiple standards dealing with the risk of exposure to HFs, such as [ANF 15]. They are suited to far-field exposure and depend on the type of antenna. Near-field exposure, especially in the reactive area (section 1.3.3.2), must be assessed through dosimetry (specific absorption rate [SAR]) to be more accurate. Currents induced in the limbs may sometimes be measured under these conditions (see sections 5.4.3 and 8.5.2).

Narrowband measurements require more attention, since each signal has its own characteristics, which do not need to be known when using a broadband device. For example, the RBW filter and VBW filter should be configured for each service [STD 28]. The detector to be used is the rms value detector.

### 8.4.5. *Calibration of the measuring chain for electromagnetic fields*

As with low-frequency measuring equipment, high-frequency measuring systems must be calibrated at regular intervals defined by standards or quality control procedures. This interval depends on the drift in the measuring chain response between the calibrations, and also on the conditions of use. The recommended interval is generally 2 years.

Calibration must comply with national or international technical standards and be performed by means of a calibration chain, each of whose element uncertainties are fully defined. Calibration follows a set of operations for quantifying them.

The antennae are calibrated at various frequencies in an anechoic chamber (Faraday cage lined with absorbent materials). This device is itself calibrated according to the standards in force. SA must also be calibrated. The results are incorporated into an overall assessment of measurement uncertainties.

These operations do not impede the verification of the correct working of the measuring chain, following a simplified procedure. This verification includes, as a minimum, visual inspection, checking of the battery, accessories and the calibration date. Coaxial cables are sometimes mistreated, often rolled up, unrolled and stressed. Their electrical state must often be checked, since defects may not be visible.

### 8.4.6. *Sources of uncertainty in measurements of HF electromagnetic fields*

The three main types of uncertainty in measurements of HF electromagnetic fields are as follows:

– uncertainties of the measuring chain;

– disturbances of the field by the operator;

– temporal and spatial variations of the field.

Without a preliminary study, it is difficult to predict which will be the most significant. The precision of the measuring chain corresponds almost entirely to that of the probe for narrowband measurements. The typical total uncertainty (expanded) of a narrowband measuring chain is 3 dB, which may seem high. The uncertainty at HFs is given in dB (1 dB corresponds to an electric field difference of ±10%, 3 dB corresponds to a difference of –30/+40% and 6 dB to a difference of –50/+100%).

For broadband measurements, uncertainties relating to equalization and weighting filters, which are less precise than the filters used at LFs, must also be considered. Environmental and operational uncertainties must be taken into account.

Measurement precision is less important when the field level is weak. Assessment must be rigorous when the field level is close to the limits. For example, environmental conditions must be verified (proximity of the operator, distance between probe and surface, etc.).

The radiation level coming from the source may vary greatly in time and space, due to propagation fading. Many measurements must then be carried out over a significant duration in order to correctly characterize the exposure.

#### 8.4.6.1. *Proximity of the operator*

The same precautions taken at LFs should be taken for measurements of electric fields of up to a few tens of megahertz. Disturbances induced by the operator and measuring device must be avoided. The operator may absorb or mask the field, reducing the value measured. He may also reflect or diffract it, increasing the level. A distance of 3 m may be recommended around an isotropic probe in order to avoid these phenomena.

#### 8.4.6.2. *Connection cables*

Uncertainty due to coaxial cables affects only measuring chains made up of separate elements. The cables may act as antennae and directly pick up a signal, adding it to that coming from the probe. Exposure is therefore overestimated. In order to minimize this effect, use of high-quality cables, orientated in the direction of electric field propagation when the field level is high, is recommended. The cables must not form a loop.

#### 8.4.6.3. *Probe direct coupling with the environment*

The proximity of even weakly conductive or dielectric elements around the probe may alter its properties significantly through direct coupling. A minimum distance of three times the largest dimension of the probe should be respected. This also makes the measurements reproducible.

#### 8.4.6.4. *Obstacle in the environment*

This involves removing non-permanent elements that might attenuate the signal. Mobile objects must be immobilized to guarantee the repeatability of the measurements.

#### 8.4.6.5. *Out-of-band rejection*

The probe and measuring filters cannot sufficiently reject out-of-band signals. Characterization of a weak signal is affected in a strong-field environment. It should be ensured that out-of-measuring-band radiation cannot disturb the measurements.

#### 8.4.6.6. *Intermodulation*

Intermodulation phenomena occur when the measuring device's electronic components are saturated. This may cause artifact frequencies to appear in the measuring band.

#### 8.4.6.7. *Peak factor*

Measuring instruments are calibrated with sinusoidal signals. Thus, particular attention must be paid during analysis of signals in pulse regime (e.g. radar and Wi-Fi or 4G digital transmission), for which the peak factor (or crest factor), the peak value to the rms value ratio, is significant. As well as the possible saturation of the reception circuits, the signals could be too short, leaving the detectors no time to assess them accurately.

#### 8.4.6.8. *Electromagnetic compatibility*

Measuring instruments must have significant electromagnetic susceptibility (immunity) performances in order to not be disturbed by strong radiated fields. Electromagnetic compatibility problems may also induce exposure non-conformities, although the ambient field level may be low.

It is sometimes necessary to insulate the device from radiated and conducted disturbances (propagating in the power cable, for example).

## 8.5. Measuring the contact current and induced current

Contact currents and induced currents generally cannot be assessed through calculation. Measuring them is often the only alternative.

These currents are not produced under the same conditions and produce different effects. They may be present simultaneously.

### 8.5.1. *Measuring the contact current (indirect effect)*

A contact current is established between a conductor and a person when they have different potentials (see section 5.5.1). A strong current stimulates the nervous system and may cause an electric shock or superficial burns. These are indirect effects. Currents are measured in a very wide frequency band (from a few hertz up to 110 MHz).

One method for measuring contact currents consists of holding one end of an amperometer and touching the conductive object with the other end. In practice, a conductor with a calibrated contact surface is held, then it is brought into contact with the conductive object to be analyzed. An amperometric clamp allows the current to be recorded or measured. Standard electromagnetic compatibility sensors can be used for one–off measurements (Figure 8.17). For recurrent measurements, a dedicated device is more practical. For safety reasons, measuring may also be performed on an electric model with the same dielectric properties as the human body.

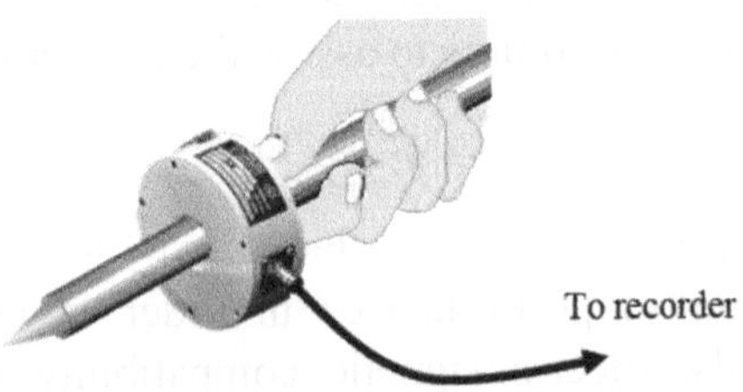


**Figure 8.17.** *Measuring the contact current with a 20 Hz 10 MHz probe*

### 8.5.2. *Measuring an induced current (direct effect)*

Measuring the electric field, in order to determine the exposure level of a person working in the immediate vicinity of a source (reactive zone), such as a radio transmitter from the FM band (section 2.2.3.1.1) or a HF welding device (section 2.2.3.2.1), is barely possible due to the coupling of the antenna with this source or with nearby conductive elements. Modeling the environment and calculating the SAR is tedious. The practical alternative is to measure the currents induced in the limbs of the exposed person, which may be significant at frequencies between 10 MHz and 110 MHz (these currents may induce thermal effects). Reference levels and action levels have been defined to facilitate this assessment (section 5.4.3).

The available measuring devices use an amperometric clamp. The plate system, resembling a bathroom scale, placed on the ground, is difficult to use and therefore not often employed [ICN 09b, AND 07].

The clamp is based on a transformer with an openable magnetic circuit. It is sized to be comfortably adjusted above the ankles or around the arms (Figure 8.18). The worker can move freely. The measuring sensor is connected to a display through an optical fiber in order to prevent the assessor from perturbing the electric field.

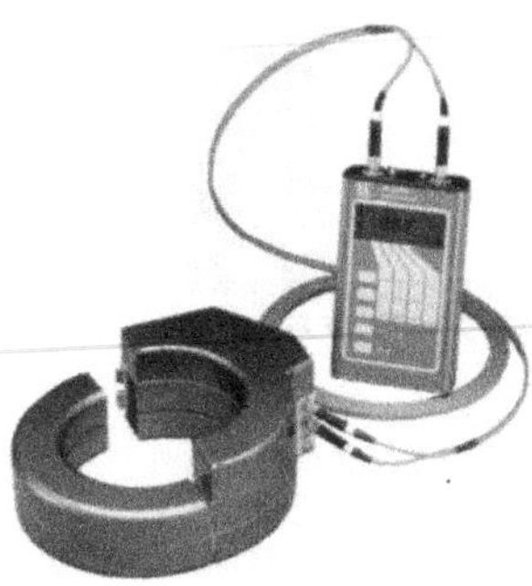

**Figure 8.18.** *Amperometric clamp (source: ETS.Lindgren Ltd.)*

The induced currents vary considerably depending on the size of the worker, his corpulence and the orientation of his arms. These measurements are also sensitive to the type of sole in his shoes. The composition of the materials of his clothes should also be considered. Accessories (jewelry, wristwatch, etc.) may influence the results.

The use of an antenna equivalent to a person from a dielectric point of view improves the reproducibility of the results. This antenna also makes it possible to assess exposure in the presence of intensive fields (of strengths beyond those which cause a critical effect). The signal may be processed in the time or frequency domain.

Due to the strong sensitivity of measurements to the configuration, a strict procedure must be followed and several measurements performed, in order to take into account the most negative situation with certainty. The rms value of the current should be determined over a duration of 6 min, for example.

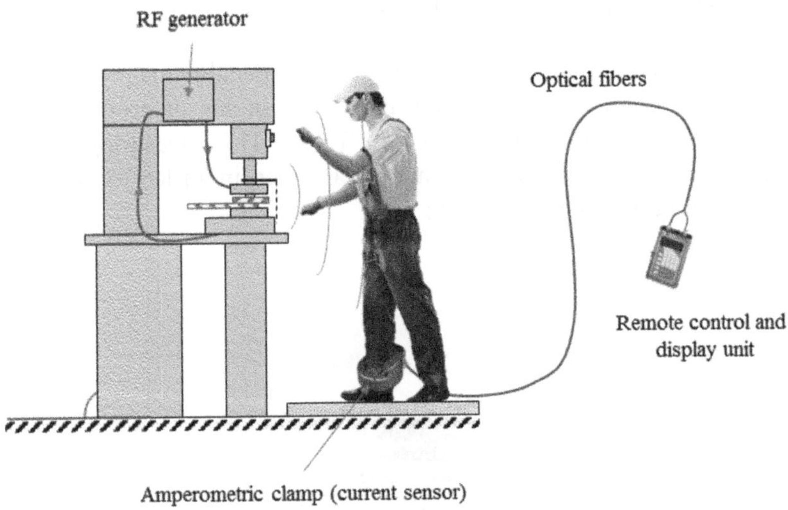


**Figure 8.19.** *Measurement of the current induced during HF welding. For a color version of this figure, see www.iste.co.uk/staebler/electromagnetic.zip*

Measurements with an ankle clamp generally produce reproducible results and are not greatly disturbed by radioelectric interference. They also allow the effectiveness of protective clothing made up of conductive materials to be verified. The clothing can be arranged on the clamp.

An amperometric clamp for induced currents may be used to characterize contact currents, if it is compatible.

## 8.6. Introduction to dosimetry

Assessment of exposure in relation to the basic restrictions (or exposure limit values) belongs to the domain of dosimetry, which will be merely introduced here. Its development lies beyond the scope of this book.

### 8.6.1. *Definition*

Physical quantities (electric fields, currents, heat, etc.) are induced through coupling in biological tissues immersed in an electromagnetic field and may generate adverse biological health effects when they are intensive. To prevent these effects, basic restrictions or exposure limit values have been established. Measuring them in most of the frequency domain requires invasive techniques that are difficult from a metrological point of view, and even more so from an ethical point of view. Reference levels (or action levels), which should be compared with the measured or calculated fields in the location of the human body, have been derived from this observation.

Electromagnetic dosimetry consists of quantifying the internal physical quantities depending on the distribution and characteristics of the external electromagnetic field, taking into account the characteristics of the organism (dielectric properties of tissues, size, etc.). Dosimetry therefore involves analyzing the coupling mechanisms between the external fields and the internal quantities of an exposed organism. It determines the mathematical relationships (coupling coefficients) between these quantities.

More generally, dosimetry makes it possible:

– to deduce reference levels (action levels) from the basic restrictions (exposure limit values);

– conversely, to quantify the internal quantities from the external fields in real conditions;

– to carry out biological and medical research for therapeutic and diagnostic purposes (knowledge of field strength within the exposed body).

The second point can be used to verify that the basic restrictions (or exposure limit values) are not exceeded in strong exposure situations.

The three main coupling phenomena have been presented previously (see section 3.3): electric field coupling, magnetic field coupling and electromagnetic field coupling:

1) a low-frequency electric field ($f < 10$ MHz) induces charges on the surface of the exposed body. The movement of these charges creates a current within the body associated with an induced electric field;

2) a low-frequency magnetic field ($f < 10$ MHz) produces an electric field that induces a current within the body when it is variable (Faraday's law and eddy current). These quantities increase as the time variation of the field and the exposed surface become more significant;

3) a high-frequency electromagnetic field ($f > 100$ kHz) is absorbed by the tissues. This absorption is transformed into heat.

The reference levels have been established for the most negative coupling conditions, especially considering whole body exposure in a uniform field. In many situations, strong fields are highly non-uniform and only one body part is exposed to these levels. In these conditions, the reference levels may be exceeded, while the basic restrictions are respected. The interest of dosimetry lies in verifying this.

To carry out an assessment through dosimetry, the environmental electromagnetic characteristics, such as frequency, waveform, propagation direction, polarization and spatial distribution of the fields must be known in the location of the exposed body. This information should be supplemented by the exposure conditions, posture and morphology of the person for a precise calculation.

The basic restrictions within the body at LFs are established in terms of the electric field $E$ ($V \cdot m^{-1}$) or current density $J$ ($A \cdot m^{-2}$).

The main basic restriction at HFs is SAR ($W \cdot kg^{-1}$). It is supplemented by the specific absorption or SA ($J \cdot kg^{-1}$) and the power density $S$ ($W \cdot m^{-2}$).

The exposure domain at frequencies higher than 6 GHz is not affected by dosimetry because the quantities are outside the body. It is the same for frequencies lower than 1 Hz.

Two electromagnetic dosimetry techniques are differentiated:

– experimental dosimetry;

– theoretical dosimetry, which is divided into analytical dosimetry and numerical dosimetry.

### 8.6.2. *Experimental dosimetry*

Experimental dosimetry belongs to the measurement domain. A measuring chain enables the relevant quantities to be directly determined. Measurements are performed on laboratory samples, synthetic materials of simplified form [STD 33], animals or artificial models of whole or partial bodies, electrically representative of various morphologies. These models are called phantoms.

Metrological laboratory conditions and specific devices are required to perform these measurements. Detailed procedures have been written [STD 09, STD 10]. As these investigations are difficult, they are limited to generic configurations. The results are then extrapolated for other situations.

The physical quantities measured by experimental dosimetry are the internal electric field, temperature elevation and the absorbed power.

The SAR, the power absorbed per unit of mass, is determined by the electric field strength, the speed of temperature increase or the absorption of radiated power. At LFs, the current density is deduced from the electric field.

#### 8.6.2.1. *Measuring the internal electric field*

Electric field strength is measured inside the exposed biological system, using implantable probes. The number of measuring points is limited in manual operation. Many measurements can be taken automatically, by a robotized arm, in a phantom containing a homogeneous physiological solution or gel (Figure 8.20).

The measuring probes must be very small and have dielectric characteristics close to the surrounding materials in order to not disturb the existing field.

*Triaxial* measuring probes allow each electric field component to be measured (vectorial measurement). They simplify the procedure by preventing its reorientation in orthogonal directions, according to the principle presented in section 8.4.3.2 (ambient electromagnetic field measurement). The probe must be rigid in order to be automatically positioned.

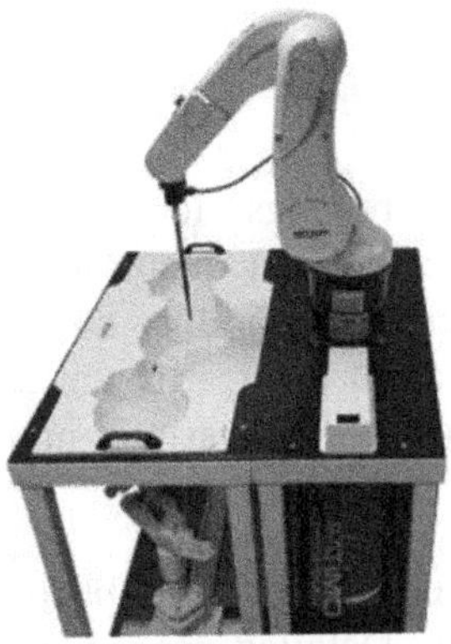

**Figure 8.20.** *Automatized measurement of the electric field in a phantom representative of a body part, to determine the mobile telephone SAR (source: Microwave Vision Group)*

There are probes made up of dipole antennae and electro-optical sensors. Dipole measuring probes incorporate a detector (e.g. a rectifier), filter and high impedance transmission line in order not to disturb the medium (Figure 8.21). The rectified voltage representative of the field strength is transmitted by the line toward the measuring device (voltmeter). The spectral content of the field is lost.

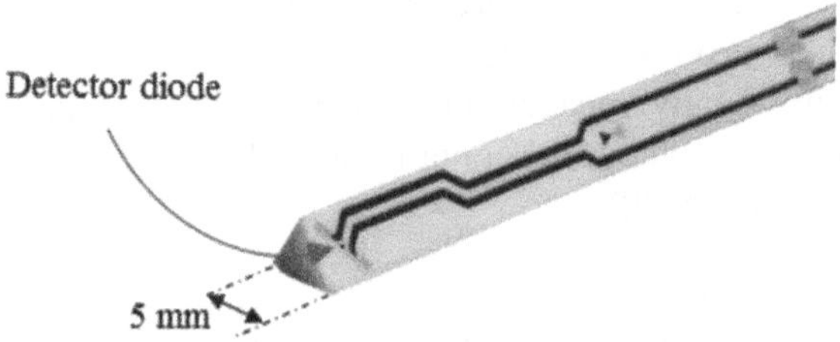


**Figure 8.21.** *Ceramic triaxial probe for measuring an electric field of 30 MHz–3 GHz (source: Microwave Vision Group)*

Electro-optical sensors make it possible to determine the amplitude, phase and direction of the field in order to obtain vectorial (3D) information in the time domain. These sensors use the Pockels effect (modification of the optical crystal properties by an electric field) or implement optical interference phenomena using a laser beam [JAR 12]. Fiber optics are often associated with these sensors in order to transmit information without disturbing the surrounding electric field. The measuring system and laser generator are remote. These probes cover a very large spectral band, theoretically from the continuous (DC) to several GHz (~20 GHz).

Measuring the electric field is the only technique possible at LFs. The three spatial components are extracted from the probe. The current density, which is not measurable, is deduced from the electric field:

$$J = \sigma . E \qquad [8.7]$$

where $J$ represents current density ($A{\cdot}m^{-2}$), $E$ the electric field in $V{\cdot}m^{-1}$ and $\sigma$ tissue conductivity in $S{\cdot}m^{-1}$.

#### 8.6.2.2. *Obtaining the SAR through the electric field*

The SAR is calculated when the rms value of the electric field $E$ has been measured at HFs according to formula [3.10]:

$$SAR = \sigma . \frac{E^2}{\rho_m} \qquad [8.8]$$

where $\sigma$ represents the conductivity of the medium ($S{\cdot}m^{-1}$) at the field frequency and $\rho_m$ its volumic mass ($kg{\cdot}m^{-3}$).

#### 8.6.2.3. *Obtaining the SAR by measuring temperature variation*

The SAR can also be obtained by measuring the temperature variation. This variation is due to the radiation absorption. The SAR is obtained with [3.9]:

$$SAR = C . \frac{\partial T}{\partial t_{(t=0)}} \qquad [8.9]$$

where $C$ is the specific heat capacity of the biological medium ($J{\cdot}kg^{-1}{\cdot}K^{-1}$) and $\frac{\partial T}{\partial t_{(t=0)}}$ the slope at the origin of the temperature elevation. The SAR is determined in the absence of thermal diffusion phenomena. It is the slope at the origin of the curves in Figure 4.2.

Probes can be used to measure the temperature of a laboratory animal at a few local points. The use of a phantom is required when the investigation needs to be more thorough. This technique can be used to assess the distribution of the SAR in a materially economical way, but it takes a very long time and requires perfect control of experimental conditions.

As with the previous technique, the probe must be prevented from disturbing the local electric field. Fiber-optic temperature sensors, thermistors and thermocouples are used. The probe must present a low thermal inertia and be able to measure a temperature variation lower than a hundredth of a degree Celsius.

Thermistors have the advantage of a small size, good temperature resolution and low cost, but the disadvantage of being made from metal elements, which disturb the electric field. Temperature sensors based on thermo-optic effects do not have this disadvantage.

Temperature elevation must be sufficiently significant to be measurable and fast to render losses due to the thermal conduction of the tissues negligible. Furthermore, the exposure duration must be very short in order to avoid a rise in the average temperature of the model.

Calorimetric and thermographic techniques are not discussed here.

#### 8.6.2.4. *Obtaining the SAR through differential power*

The general principle of this technique consists of placing a biological sample in a shielded chamber, such as a Faraday cage or waveguide equipped with an input and output connection. Calibrated electromagnetic radiation is diffused through the chamber and the output power is measured. The absorbed power is obtained through the difference, taking losses, mainly due to reflection toward the source, into consideration. Many power measurers and directional couplers are required. The power absorbed (Pa) is given by:

$$Pa = Pi - (Po + Pr) \quad [8.10]$$

with Pi input power, Po output power and Pr reflected power.

The power actually absorbed by the sample is the difference of the powers absorbed with or without the sample. The average SAR is obtained by dividing this result by the sample mass.

The temperature stability of the couplers and power measuring probes should be controlled to ensure the results are reproducible [DUR 86, ROA 09]. This technique can be used only to obtain the average whole body SAR.

### 8.6.3. *Theoretical dosimetry*

Theoretical dosimetry covers analytical and numerical dosimetry. The approach for solving the equations following these two techniques is identical.

At LFs, where the dimensions of the body are small compared to the wavelength, the study of coupling between the external fields and the biological organism is considerably simplified, since the following conditions are fulfilled [AND 06]:

1) absence of propagation phenomena. The induced fields instantaneously follow the external fields. This means conduction currents dominate in comparison with displacement currents [3.4];

2) the electric and magnetic fields are decoupled, allowing them to be processed separately;

3) the penetration depth is not considered (section 3.2.3).

This means electrostatic and magnetostatic methods apply. This regime is called the quasi-static regime. The formulation of Maxwell's equations is greatly simplified. Furthermore, the magnetic field flowing through the exposed body is not significantly disturbed. The field in the body is equal to the field which would be there in its absence. However, an external electric field is disturbed by the exposed element, requiring more calculations and the consideration of the volume outside the body.

The reciprocal independence of the electric and magnetic fields makes it possible to deal with the problems of electric and magnetic dosimetry completely separately. There is no link between the two radiation sources or between the electric fields and the currents induced by each source.

The magnetic field problem is solved in two main steps:

– external: determining the distribution of the magnetic field in the place of the body, either from a modeled current source or from extrapolated field measurements;

– internal: determining the distribution of the electric field induced in the body by the magnetic field, then deducing the current density from it, along with the conductivity.

The electric field problem is also solved in two main steps:

– external: determining the distribution of surface electric charges, supposing the body to be homogeneous and perfectly conductive;

– internal: spreading the currents and current densities depending on charge distributions, then deduce from this the induced electric fields, with the conductivity.

Here is a representation of these two resolutions (Figure 8.22).

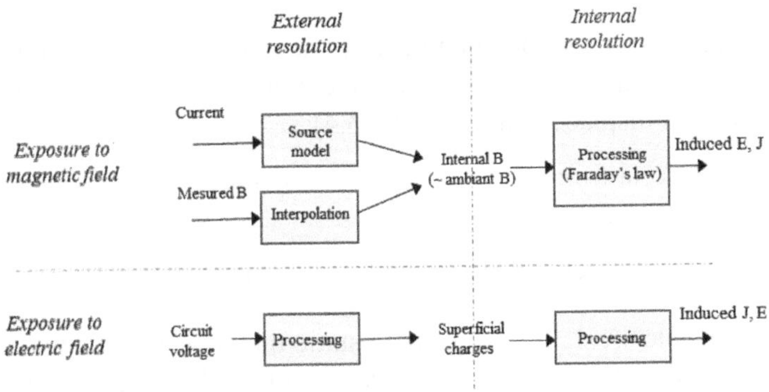


**Figure 8.22.** *Breakdown of dosimetry calculations at low frequencies*

The problems may be solved at first, in the frequency domain (at an arbitrary frequency) for determining the coupling coefficients (output signal/input signal), and then, second, in the time domain, considering the waveform of the signal. The coupling coefficients thus take into account the geometric aspects. Temporal processing is used to calculate the *exposure index* using weighting filters. This approach considerably simplifies the assessment. The processing is vectorial (taking into account the direction of the external and internal fields).

For assessing simultaneous exposure to a low-frequency electric and magnetic field (section 5.2.2.3), the superposition principle applies, since the set of equations is linear. Exposure to each field type is dealt with separately, and the results are then added together [ZOP 09].

At HFs, the wavelength is close to or smaller than the dimensions of the body. Phenomena of propagation and field attenuation with depth and electric and magnetic field coupling can no longer be disregarded. Maxwell's equations must be solved numerically.

These problems can be solved by means of analytical and numerical approaches.

### 8.6.3.1. *Analytical dosimetry*

Analytical dosimetry consists of directly solving the equations describing the problem, taking care to reduce them by using simplified forms of the human body. The usual forms are discs, cubes, spheres, ellipsoids and even simplified and homogeneous forms of the human body [STD 03]. It is easiest to solve the equations when the field is uniform. The result produces a good approximation.

The equations are solved directly in the simplest cases. Iterative resolution techniques may be implemented in slightly more complex cases, without using sophisticated simulation methods based on numerical human models.

This approach can be illustrated for LFs by a homogeneous conductivity sphere exposed to a magnetic field. According to its diameter, it may represent a head or a hand, for example.

The magnetic field present in the volume is considered equal to the ambient field ($\mu_r = 1$). It is obtained without difficulty through direct calculation (for example Biot–Savart's law [1.4]).

By applying Faraday's law, the distribution of electric fields and induced currents for a homogeneous field can then be determined in the time domain $B$:

$$E = \frac{r}{2} \cdot \frac{dB}{dt} \ (\text{V}\cdot\text{m}^{-1}) \qquad J = \sigma \cdot \frac{r}{2} \cdot \frac{dB}{dt} \ (\text{A}\cdot\text{m}^{-2}) \qquad [8.11]$$

where $r$ corresponds to the distance to the sphere center, σ is the conductivity of average value 0.2 S·m$^{-1}$ (section 3.2.1), $E$ and $J$ are maximum in a uniform field when $r = R$, radius of the sphere. In the frequency domain and for a sinusoidal field with a magnitude of $B$, the maximum current density $J_{max}$ is therefore:

$$J_{max} = \sigma . \pi . f . R . B \qquad [8.12]$$

These relationships illustrate the coupling phenomena presented in Chapter 4:

– the induced electric fields are proportional to both the strength and the frequency of the magnetic field;

– the induced current density is proportional to the induced electric field and the tissue conductivity;

– the induced electric field and the current density are more significant on the surface. They are zero at the center.

A disc may be used instead of a sphere. It produces the same result, subject to certain precautions of position and orientation. The 3D problem is transformed advantageously into a bidimensional problem, simplifying the resolution again.

An iterative method may be applied to the disc when the field is not uniform. To avoid this processing, standards [STD 19, STD 20, STD 21] propose correction factors in table format (these correction factors have been defined by iteration). They have been adapted to the assessment of exposure in arc welding [STD 07] and resistive welding [STD 08].

Although the models are simple, the results of analytical calculation are precise, making it possible to validate an experimental or numerical approach [HIR 11].

#### 8.6.3.2. *Numerical dosimetry*

Numerical dosimetry, like all simulation techniques, involves theoretical models. The models used in dosimetry are realistic, from both a morphological and an anatomical point of view. They are composed of heterogeneous human tissues, each with their own electrical properties. Animated by increasingly powerful simulation methods, they provide solid knowledge of the distribution of internal electric fields, current density and SAR. The mean limitations of this technique are the implementation costs and the time needed for processing and analysis. A high level of expertise is also required.

The method consists of immersing the model in an electromagnetic field and solving Maxwell's equations. These equations connect the electromagnetic quantities ($E$, $B$, $J$, etc.). This resolution, which uses iterative techniques, converges toward an approximate solution, which cannot be obtained directly in the majority of real cases. The most complex exposure conditions can be processed (specific postures, multiple exposure, etc.).

The electric field, current density and induced magnetic field are calculated for each elementary volume (voxel) of all the space occupied. The increased power of calculation systems makes it possible to use models with millimetric resolution for better precision.

The general approach of this method involves the following steps:

– calculation of the unperturbed electric, magnetic or electromagnetic incident field;

– choice of a voxel-based subject (or part of it) from a library of numerical models;

– attribution of dielectric properties to each voxel;

– formulation of Maxwell's equations to mathematically represent the situation considered;

– discretization of these equations for numerical resolution;

– choice of numerical resolution algorithm;

– convergence toward the expected result;

– postprocessing of results and, afterward, checking of results obtained.

#### 8.6.3.2.1. Calculating the incident field

The electromagnetic field may be determined from a source model, by means of prediction formulae (see Chapter 1), using the physic laws (Biot–Savart law, for example). A precise source model is preferable in high-frequency near field. This model may be produced by meshing a CAD file produced by the engineering design department. It must then be powered (virtually) with the appropriate signal.

The incident field may also be obtained using a 3D map of the exposure area. Field measurements must be interpolated with the spatial resolution of the human model.

#### 8.6.3.2.2. Choice of human model

The models are heterogeneous, with tissues and internal organs (heart, brain, muscles, lungs, bones, etc.) differentiated by their electrical properties (Chapter 3; [STD 03]). They are discretized into elementary cubic cells (voxels).

Many 3D digital human models are available in order to represent exposure situations precisely. Models of men, women and animals are also available, as well as models of children, pregnant women and fetuses. These models are distinguished by resolution (voxel dimensions), tissues, size, corpulence and posture.

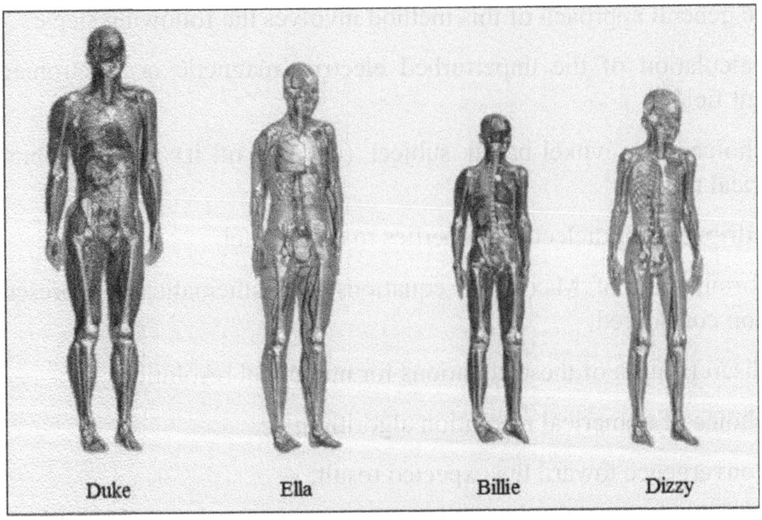


**Figure 8.23.** *Some members of the virtual family, distinguished by size and morphology. Some tissues are not shown, allowing the internal organs to be seen (source: IT'IS Foundation [ITI 10, ITI 14])*

There are models for various ethnic origins, taking into account specificities of size, weight and morphology. Each model is unique. Today, the models are developed from MRI images.

Among the available models are as follows:

– the virtual family that includes Duke and Ella (IT'IS Foundation, Switzerland);

– the virtual human project (National Library of Medicine, United States);

– Norman and Naomi (Health Protection Agency, United Kingdom);

– Taro and Hanako (NICT Japan, Japanese man and woman with average corpulence).

The modeled body is meshed in cubic voxels of millimetric dimensions. They have conductivity and permittivity, depending on the nature of the tissue they represent. Nearly 100 tissues are thus differentiated. The simple models use homogeneous cubic voxels of average conductivity, between 0.5 mm and 5 mm in size. The resolution and model choices are dictated by the compromise between the desired precision, the calculation time and the memory capacity of the computing tools.

**Figure 8.24.** *Example of modeled posture (source: IT'IS Foundation [ITI 10, ITI 14])*

The modeling of an adult of 1.76 m and 73 kg requires 36 million voxels of 2 mm. On top of this, there are 8 million elements, taking into account the surrounding free space.

To reduce the calculation time, the models with the lowest resolution (bigger voxels) are used, in order to obtain coarse results, which will be used to initialize calculation at the desired resolution.

#### 8.6.3.2.3. Formulation of equations

Maxwell's equations must be adapted to the exposure situation. For example, the approximation of quasi-static regimes, applicable at low frequencies, prevents long loops of calculation, since the internal and external problems are decoupled. Average conductivity is often used at high frequencies. Magnetic permeability is equal to that of the vacuum in all space, including inside the exposed body.

Conditions must be imposed on the limits for the purpose of solving the equations. They describe the behavior of fields between two media with different conductivities and permittivities. These equations must then be discretized in space and in the time or frequency domain. The spatial calculation interval must be adapted to the segmentation of the human model.

#### 8.6.3.2.4. Solving the equations

Solving these equations produces the desired dosimetric quantities: strengths and distribution of electric fields and induced currents.

Maxwell's equations are solved numerically. The simplest approach is to seek approximate solutions through iterations. The iteration algorithm is followed by a convergence test and a stability test. The convergence test consists of measuring the distance between two successive solutions. To increase the speed, the convergence begins with coarse meshing, before switching toward the meshing necessary for the required precisions (generally 2 mm).

The most suitable method depends on the following parameters:

- exposure conditions: quasi-static regime or complete equations;
- time domain or frequency domain;
- conditions for the limits of the studied volume;
- memory capacities and processing power;
- computing time;
- etc.

The precision of the results depends primarily on the resolution, the quality of source modeling and the quality of the human model.

A description of these methods does not lie within the scope of this book. A few examples will be cited, however [STD 03]:

– finite element method (FEM);

– method of moments (MoM);

– finite-difference time-domain or frequency-domain methods (FDTD or FDFD);

– impedance method (IP): the distribution of the current induced in the body is determined, supposing the body to be equivalent to an impedance network.

#### 8.6.3.2.5. Postprocessing

After the convergance of the results, processing consists of:

– verifying the consistency of the results (absence of divergence in the results, etc.);

– calculating the induced electric field value at the 99th percentile (thus filtering out aberrations in the results at the tissue edges);

– determining the SAR or current density and their distribution at HFs from the electric field levels;

– calculating the average SAR over the whole body or a small mass (1 or 10 g) of adjacent tissues or the average current density over a small surface (1 cm$^2$) or even the internal electric field over a short length (2 mm);

– selecting the organs and tissues affected by the basic restrictions at LFs to distinguish between sensory and health effects;

– calculating the exposure indices.

### 8.6.4. *Summary of dosimetry*

Dosimetry covers two broad domains: experimental dosimetry and theoretical dosimetry, which may be analytical or numerical.

Experimental dosimetry is difficult to practice on living bodies for technical reasons and, evidently, for ethical reasons. Today, experimental dosimetry on phantoms makes it possible to validate numerical dosimetry or to study particular configurations (*in vitro* measurements, for example).

Analytical dosimetry is always practiced because of its relative simplicity. It is limited to simple situations, when the propagation area is free and at low frequencies. It does not take into account complex environments (presence of elements altering field distribution) or specific postures. It is always used, although the results are deemed conservative. Its (relative) simplicity allows for broader usage. Furthermore, it provides an overall picture of the problem, explaining the relationships between various parameters and their physical significance. It is also used as a reference for validating a numerical approach.

Numerical dosimetry, although more complex to perform, is overtaking the other techniques. It requires powerful hardware and software computing tools. Expertise in numerical calculation, for extracting the results, and expertise in exposure, for interpreting them, are also needed. Furthermore, it is time consuming.

In return, this approach allows all exposure configurations to be studied. It is reserved for complex situations, when alternative solutions have been shown to be limited. Dosimetry is particularly recommended for specific situations when the reference levels or action levels are exceeded. It is not appropriate for large-scale assessment for reasons. It should be recalled that this technique is not yet standardized at low frequencies.

# 9

# Implementation of Directive 2013/35/EU

This chapter concerns the exposure of workers in an occupational environment. Furthermore, it will help readers to understand the practical relevance of the exposure limits system presented in Chapter 5.

NOTE.– Based on the electromagnetic field (EMF) Directive (Directive 2013/35/EU), this chapter takes into account neither the legislation introduced during its transposition by the European Union member states nor the existing and non-repealed national specificities.

Although the exposure limit values (ELVs) and action levels (ALs) have been retained by all states, definitions have been clarified by some and points have been added or removed by others, because they seemed lacking or superfluous. Furthermore, some states have aimed to harmonize it with other national texts.

This means attentive reading of the transposed text is sometimes required. However, understanding the directive provides a strong knowledge baseline for the implementation of national regulations.

In order to understand this chapter in detail, reading of Chapter 5 is recommended.

## 9.1. Context of the directive

### 9.1.1. *Introduction*

Electromagnetic radiation is divided into three domains according to its fundamental physical properties, which themselves depend on the frequency (Chapter 1):

– from 0 Hz to 300 GHz (1 GHz = $10^9$ Hz): radiation domain commonly called electromagnetic fields. The wavelengths ($\lambda$) are greater than 1 mm;

– from 300 GHz ($\lambda$ = 1 mm) to 300 PHz (1 petahertz = $10^{15}$ Hz; $\lambda$ = 1 nm): domain corresponding to optical radiations, including infrared radiations (heat), visible radiations (light) and ultraviolet radiations (tanning, bactericidal effects);

– ionizing radiations in the highest frequency domain, i.e. beyond optical radiations: these radiations have enough energy to be able to destroy atomic bonds. These rays are generally characterized by their energy (eV).

A specific European directive covers each domain [EC 01, EC 10, EC 08]. Workers are, therefore, normally protected from the risk of exposure from static fields to electromagnetic radiation of infinite frequencies. The Optical Radiation Directive takes into account only artificial optical radiation.

These directives have been edited in order to help employers comply with their obligations. The EMF Directive is one of these. It defines the minimum requirements for the protection of workers from risks due to electromagnetic fields of frequencies between 0 Hz and 300 GHz. The member states had the opportunity to define stricter exposure limits during transposition. National legislative rules take precedence, especially if they include stricter limits, fixed previously, which have not been repealed. The EMF Directive does not define limits for the general public.

Generally, employers are obliged to protect their employees from risks due to physical agents, as cited in Framework Directive 89/391/CEE [EC 02]. This directive establishes the basic rules for the health and safety of workers. It aims to eliminate the risk factors of disease and occupational accidents.

#### 9.1.1.1. *Effects*

The EMF Directive covers known direct and indirect effects that may be caused by electromagnetic fields from 0 Hz to 300 GHz and which may occur in the short term. The ELVs have been established on the basis of scientifically validated studies. The directive does not cover long-term effects for which there is currently no convincing scientific evidence.

Direct biological effects are effects on the human organism, directly caused by its presence in an electromagnetic field. Direct effects in the high-frequency spectrum (telecommunication, HF industrial applications, etc.) are thermal effects, which may in extreme cases cause tissue necrosis. Direct effects at low frequencies are the electrostimulation of sensory functions and excitable tissues. Perceptions and transient disorders may follow. Under stronger fields, uncontrolled muscle contraction is possible. In this case, there is a health and safety risk to the exposed person.

Indirect effects are effects caused by the presence of an object in a strong electromagnetic field, which may lead to a health or safety hazard. An example of these effects is contact currents, which may flow through the person. These may be disturbing or even painful and may perturb active implanted or body-worn medical devices. The projection of ferromagnetic objects in a static magnetic field and the combustion of materials through spark discharges are also considered indirect effects that may affect safety. These effects were introduced at the end of Chapters 4 and 5. Risks related to contact with a live conductor are not covered by the EMF Directive.

### 9.1.2. *EMF directive development*

An initial directive (2004/40/EC) was published in the Official Journal of the European Union in 2004. It was originally due to be applied in 2008. However, it met with implementation difficulties in the medical sector, in some industrial activities and in the armed forces. Its application date was therefore postponed for the first time in 2012 and again in 2013, before it was definitively repealed.

The European Commission had used these years to study the possible impacts of such a measure more thoroughly. It closely examined the arguments put forward by the parties involved. The environmental, economic and social impacts were also studied in a European context [EC 06].

It initiated preliminary work, with the aim of using an MRI for medical purposes. For this, working groups made up of practitioners, patients and manufacturers were established. Furthermore, the Commission consulted social partners and government representatives. It gathered additional information and the expertise of representatives from industry, national federations, standardization committees and scientists, including those of the International Commission on Non-Ionizing Radiation Protection (ICNIRP), World Health Organization (WHO), North Atlantic Treaty Organization (NATO), Scientific Committee on Emerging and Newly Identified Health Risks (SCENIHR) and other international organizations.

The development of the directive was laborious, since the subject is sensitive and electromagnetic fields are complex. They are abstract, imperceptible and difficult to measure, and the biophysical and biological mechanisms are little known.

Searching for a compromise delayed the publication of a new text. It inevitably became technical to read, but ultimately, it provides all the safeguards for prevention in the context of the health and safety of workers. It does not hinder industrial development or the use of advanced medical techniques and does not impose exorbitant costs on companies, and in particular on SMEs.

As presented in Chapter 3, the biophysical effect reported following exposure to high-frequency electromagnetic fields is the heating of tissues through energy absorption. There is a real health risk for the person if exposure is intense (i.e. well above the ELVs). This effect has been studied in detail for more than half a century. To protect their employees, telecommunication and radio broadcasting operators had anticipated Directive 2004/40/EC, implementing it without much difficulty. Also, the technical section relating to thermal effects was retained in the new directive, especially since the limit thresholds at high frequencies had not evolved since the publication of the 1998 ICNIRP Guidelines [ICN 98].

Discussions around the new text focused mainly on low frequencies associated with non-thermal effects (sensory and neuromuscular stimulation).

In this domain, the most recent knowledge (2010 ICNIRP Guidelines [ICN 10]) allows new limits to be fixed.

Maximum thresholds were defined in order to ensure the protection of exposed workers, while also facilitating the pursuit and development of activities. To avoid any ambiguity, the effects, ELVs and ALs were defined precisely and, if necessary, supplemented by the Practical Guide [EC 04]. The exposure level (exposure indices) calculation methods were clarified for low frequencies (weighted peak method in the time domain, see Chapter 6).

The EMF Directive ensured that all workers, whatever their categories and whichever European Union member state they lived in, were protected at a minimum level in order to avoid potential distortions of competition. Its application guarantees overall homogeneity of the European legal framework. Clear and well-established limits facilitate the demonstration of compliance to reassure workers and legally protect employers. More conservative ELVs could have put economic activity at a disadvantage by reducing competitiveness of export sectors and causing a loss of attractiveness in the settlement of companies.

The social aspect was not omitted during the development of the directive. The employability of people at particular risk, access to healthcare (MRI) and the concerns of workers about electromagnetic fields were considered.

Furthermore, if convincing scientific evidence of long-term effects became available, the Commission would be duty-bound to present a legislative proposal to protect against them.

All these elements ensure that many countries outside the European Union might be inspired by this directive.

## 9.2. Implementation of the directive

### 9.2.1. *Introduction*

The implementation of the directive within a company may be divided into various phases, introduced then detailed in the following sections. They comply with the general provisions of the Framework Directive [EC 02].

#### 9.2.1.1. *Preliminary phase*

The preliminary phase allows the requirements of the EMF Directive to be understood. It includes analysis of the situation through more or less thorough study of legislative and standardized documents, depending on the activity domain, the diversity of equipment, etc. The aim is to verify whether this directive applies to the company. If necessary, this first phase must lead to setting up the organization and an initial action plan. Internal communication is prepared.

#### 9.2.1.2. *Risk assessment phase*

This phase begins with an inventory of the equipment whose radiation may expose workers beyond the regulatory limits (identification of sources and their use). The best way of assessing this equipment (existing written evidence, calculation, measuring, etc.) must then be determined and applied.

#### 9.2.1.3. *Risk prevention phase*

This phase consists mainly of implementing the protective or preventive measures (technical or organizational measures).

#### 9.2.1.4. *Miscellaneous phases*

Employers are obliged to train personnel or provide them with information on the risk associated with electromagnetic fields when it is present. Furthermore, this phase includes backing up the results of the risk assessments and the preventive measures that have been implemented. Health monitoring is introduced in this phase.

For most of the companies involved, these activities have a limited period, unless the workplaces regularly undergo technical or organizational changes.

### **9.2.2. *Preliminary phase***

#### 9.2.2.1. *Is the company involved?*

A large majority of workers are exposed only to radiation strengths much lower than the limits recommended for the general public, and consequently much lower than those fixed by the EMF Directive. Some workplaces in the industrial, transport, telecommunication, electrical energy management and medical sectors may expose workers to high field levels. Research, production and maintenance activities are also involved.

Hence, the first step consists of checking whether the company may be involved in the EMF hazard. The tables in Appendix 1 may be used to identify whether the presence of equipment or working situations is likely to generate an electromagnetic field risk [EC 04]. Additional information on radiation and other data relating to safety may be provided by manufacturers, equipment installers or documents published by a specific professional branch. These tables are as follows:

– Table A1.1: list of equipment and workplaces where it is unlikely that electromagnetic fields would be a risk for most workers. No *a priori* assessment is necessary;

– Table A1.2: list of equipment where electromagnetic fields may pose a risk to all workers. The applicable ALs or ELVs may be exceeded. Specific assessment is necessary;

– Table A1.3: a complement to Table A1.2, list of equipment where electromagnetic fields may pose a risk to workers at particular risk, i.e. pregnant workers or workers with passive implanted medical devices (excluding workers with active implanted medical devices). This equipment must be assessed in relation to the limits in the EC Recommendation (general public);

– Table A1.4: a complement to Tables A1.2 and A1.3, this lists the equipment and situations where electromagnetic fields may pose a risk to workers with active implanted or body-worn medical devices (electronic). Localized individual exposure may exceed the reference levels in the 1999/519/EC recommendation or cause interference problems with these devices (Chapter 5). The occupational doctor must take account in the workplace assessment of the information supplied by the healthcare team responsible for implanting the device or for subsequent care.

The four tables are not exhaustive.

The distribution of equipment or working situations may be represented in Figure 9.1.

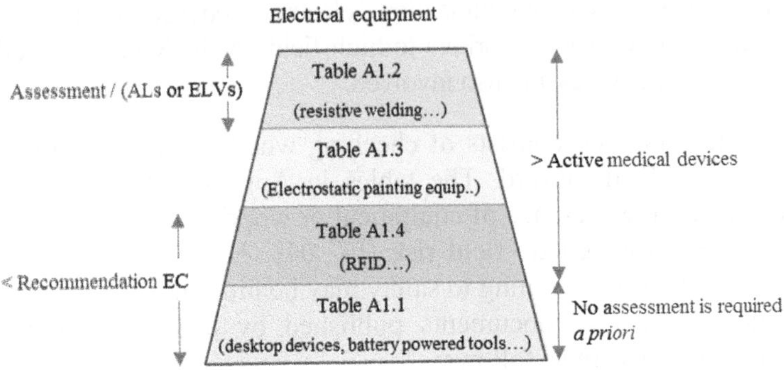


**Figure 9.1.** *Distribution of electrical equipment*

Table A1.2 summarizes four equipment categories:

– those whose ALs may be exceeded (the EC recommendation is exceeded);

– those whose ALs are exceeded, but whose sensory effect ELVs are complied with;

– those whose ALs are exceeded, but whose health effect ELVs are complied with (the sensory effect ELVs are exceeded);

– those whose health effect ELVs are likely exceeded.

Without exemption, the situations of the fourth category must occur only by accident, in particular when the preventive and protective measures are not respected.

An overall understanding of the EMF Directive is sufficient if the company does not have equipment likely to radiate electric, magnetic or electromagnetic fields of levels greater than the limits recommended for the general public. This is generally the case for tertiary companies, where the majority of the equipment is available on the market open to the general public, and is therefore normally compliant with the EC Recommendation.

However, if the company has equipment that is professional or specific to its activity, it must appropriate the EMF Directive and EC Recommendation (general public) more strictly the more significant and diversified the electromagnetic radiation. Analysis of the EC Recommendation (general public) remains necessary, since it allows people at particular risk and non-employees of the company, such as visitors, to be considered.

Knowledge of Framework Directive 89/391/CEE (whose aim is to implement measures to promote the improvement of the health and safety of workers [EC 02]), and implementation of specific directives (risks relating to physical or chemical agents) make it possible to put the EMF Directive into practice more quickly. Implementation of this directive does not stand out from other preventive processes: identification of sources (or hazards), assessment of workplaces and, if necessary, implementation of technical or organizational measures, training/informing of employees and documentation.

#### 9.2.2.2. *Organization*

If the company is involved in the electromagnetic field risk as defined by the EMF Directive, it must identify the team or person in charge of organizing the process, guiding the risk prevention and ensuring internal and external communication. It must rely on people who are competent in risk prevention, occupational healthcare, "human exposure to electromagnetic fields" and human management.

Very few companies have the internal capacity required to carry out assessment properly, so external qualified resources may be used for a relatively limited work burden over time, guaranteeing a quick increase in internal competence.

#### 9.2.2.3. *Locating the electromagnetic field risk*

Electromagnetic fields are scarcely understood. There is a high degree of confusion between electric, magnetic and electromagnetic fields. Radiation sources are themselves poorly understood. The largest equipment does not necessarily cause the strongest exposure. People have preconceptions of how sources radiate in terms of both strength and propagation direction.

The risk relating to electromagnetic fields is not "intuitive", unlike other occupational risks. There is also confusion between biophysical, biological and health effects (Chapters 3 and 4).

During presentations, the electromagnetic field risk should be put into perspective in relation to other risks. It must not be dealt with in a specific way, since no direct health effect caused by low-frequency electric and magnetic fields ever seems to have been reported in industry (a few incidents have been reported following exposure to an intensive high-frequency field).

### *Notices*

As presented in Chapter 5, the exposure limits to electromagnetic fields are expressed in a system composed of a double set of values. Their definitions are different from those relating to the risks due to other physical or chemical agents, although their terminology is identical, which adds to the confusion. As a reminder:

– *ELVs*: these values are limits established from critical effects (lowest level effects which may disturb the person or harm his health), taking into account a safety margin. There is no risk when exposure is lower than the ELVs. The directive distinguishes ELVs relating to **sensory effects**, which may be exceeded in certain conditions (level, exposure duration and information) and ELVs relating to **health effects** (health and safety). In the potential event of the ELVs being exceeded in a workplace, preventive and protective measures must necessarily be implemented. The physical quantities used for ELVs between 1 Hz and 6 GHz are physical quantities within the human body; the most significant are expressed in terms of induced electric field and specific absorption rate (SAR);

– *ALs*: these are operational levels that are not directly related to risks, defined to simplify the compliance assessment process due to the complexity of assessment in relation to the ELVs. When the ALs are respected, the ELVs are deemed to be respected. However, the non-respect of the ALs does not necessarily imply that the ELVs have been exceeded. The ALs should be compared with the measurements or calculations for electric, magnetic and electromagnetic fields in the workplace. As an example, it will be remembered that there are three ALs depending on the exposed body part for exposure to magnetic fields (low ALs, high ALs and limb ALs – section 5.2.2.1).

The EMF Directive does not propose limit values applicable to the general public, for which Recommendation 1999/519/EC exists. The basic restrictions and reference levels of the latter are lower than the ELVs or ALs applicable to workers (Chapter 5).

The links between these values are illustrated by Figure 9.2 in section 9.2.3.4.

#### 9.2.2.4. *Implementation of the prevention process*

It will be remembered that the employer is responsible for assessing the risks to workers due to electromagnetic fields in the workplace. It is his responsibility to demonstrate the effectiveness of the preventive measures he has implemented to prevent these risks. If necessary, he measures or calculates the electromagnetic field levels to which workers are exposed.

One option is to integrate electromagnetic risk management into the company's overall risk management program. In this way, this risk may be managed like other occupational risks present in the company (noise, vibration, chemicals, etc.). This means it can be treated consistently. Thus, it is not differentiated from other risks and its management and monitoring are guaranteed.

A more flexible and reactive alternative is to manage the risk when it occurs. This assumes that the EMF topic is well understood. Management is ensured by regular and scheduled monitoring (for example, a monthly progress report). In all cases, the risk may be the subject of an improvement plan.

The process may be formalized according to national requirements (for example, by the "single risk assessment document" (document mandatory in France, which consolidates all risks present on a facility) or preventive action plan.

The process and the risk analysis results may be presented to the employees' representatives on the health, safety and environment (HSE) committee. This is the first step in the process visible to the entire company. This first presentation is important and is, above all, informative. It allows the company to show their motivation and control of the topic, and the actions are organized. It makes it possible to identify people able to respond to the questioning of employees on the risks and possible health effects. The risk is poorly understood by employees, so the questions asked often lie outside the occupational dimension (mobile telephone, base station antenna, etc.).

The information must be adapted to the context of the company or facility (differences in culture, education, knowledge, EMF news, etc.). It may be based on an extensive body of literature. The availability of authoritative information to employees should not pose a problem. The websites of

ICNIRP, WHO, etc. provide explanations. The subject is complex, and scientific, biological and medical knowledge is required to analyze it in detail.

Workers on a given facility should be informed each time an assessment is to be carried out there. Indeed, it is difficult to perform a measuring campaign in workplaces with the company's employees, and particularly with external people, without informing workers using the equipment to be assessed. They must be observed before the actual measurements are taken. The workers' activities may be disrupted by this operation and their cooperation may be necessary.

### 9.2.3. *Risk assessment*

Risk assessment depends on the company's activity sector. Its purpose is to reduce occupational accidents and illnesses to improve employees' working conditions and to meet the regulatory obligation.

Before electromagnetic field risks can be prevented, the following steps must be taken:

- identification of equipment radiating electromagnetic fields;
- identification of workplaces involved;
- risk assessment.

The EMF Directive states that it is not necessary to carry out a risk assessment if the equipment is designed for public use and is used as intended. These pieces of equipment are deemed compliant with the EC Recommendation (1999/519/CE) on the limitation of the exposure of the general public to electromagnetic fields or with other stricter directives. However, this provision is not always sufficient for protecting people at particular risk, such as those holding an active implantable or body-worn medical device.

#### 9.2.3.1. *Identification*

Before assessing the risks, the equipment and workplaces potentially at risk must be identified. This identification takes into account the use of the equipment. An area of the company may be at risk, without it necessarily being a workplace (e.g. passage area).

Identification does not mean listing all electrical equipment with the argument that it all radiates. Tables A1.1–A1.4 in Appendix 1 may be used to restrict the number. Most equipment produces only very low-strength electromagnetic fields. These tables are insufficient for establishing an exhaustive inventory, especially in an industrial company or in some specific sectors, such as research or activities relating to defense.

Generally, strong fields may be encountered when the voltage is high (>100 kV at 50 Hz), when the current is strong (>100 A at 50 Hz), or when the equipment is designed to radiate. Equipment meeting these criteria is included in the inventory, independent of all physical consideration. It is important to remember that in general, the higher the frequency, the lower the thresholds to be respected.

The following elements must be taken into account to complete the list:

– warning or instructions in the manufacturer or distributor's documentation;

– industrial handbooks;

– presence of specific signage;

– restrictions contained in the safety sheet;

– reports from employees of anomalies due to electromagnetic fields (e.g. attraction of objects or heating);

– information provided by the occupational doctor.

"CE" marking is not a guarantee that the product complies with the EC Recommendation relating to the general public [REC 99]. It is often limited to the regulation regarding frequencies and electromagnetic compatibility, while it ought to include the safety of people.

The inventory must contain the general technical characteristics of the source, its use and the workplaces. These elements will make it possible to prioritize the assessments (e.g. depending on the probability gravity product) and to bring together the equipment for the purpose of communalizing the assessments. Here is a list of some pieces of information that should be included in an inventory list:

– type of equipment;

– type of operation;

– nature of radiation;

– estimation of frequency;

– estimation of the working load (current used by the industrial process, power radiated, etc.);

– access conditions;

– location within the facility.

The frequency and working load are probably the most important pieces of data, since they make it possible to quickly rule out a piece of equipment. The inventory must take into account current and occasional operations, such as maintenance, repair and cleaning. It must not omit mobile sources.

Identification allows the resources and time required to implement the directive to be assessed. The implementation of preventive measures and material investments may be planned.

#### 9.2.3.2. *Workplace compliance*

##### 9.2.3.2.1. Implementation of assessment

Once the inventory has been consolidated, the risk assessment process begins. The equipment and its use must be taken into consideration.

This process aims to display the assessment results and, if necessary, to propose preventive or protective measures, including updating the safety sheet at work.

The following information must be collected on the activity:

– localization of the workplace;

– description of the workplace;

– operation time or effective exposure duration;

– detailed characterization of the equipment used that generates electromagnetic fields;

– use of the equipment with estimation of minimum distances and justification of the difference between the practical use and that recommended by the safety sheet. Observation of how the worker does his job is necessary.

This information can be used to assess the compliance level. This may be determined in several ways:

– by design, whatever the use intended by the manufacturer. For example, analysis of these characteristics may be sufficient to categorize it: power consumed, radiation power, operational current, etc.;

– by compliance with the EC Recommendation (general public);

– by the manufacturer's certificates and notices;

– by respect of the "product" directives, which are more restrictive than the EMF Directive;

– by respect of the "product" standards, considering its use;

– by the preventive measurements already implemented;

– by comparison with similar products or those radiating more strongly;

– by theoretical analysis;

– by measurement (assessment in relation to the ALs in most cases);

– by numerical simulation (assessment in relation to ALs or ELVs).

The choice of option depends mainly on the number of pieces of equipment and the company expertise relating to electromagnetic fields.

The following parameters shall be used when calculations are necessary:

– the type of exposure (nature of the field, frequency, level, etc.);

– the frequency for high-frequency radiation and the waveform for low-frequency radiation;

– the exposure conditions (duration, etc.);

– the distance between the source and the exposed person;

– the distribution of the field over the body of the employee;

– the spatial variations of the field at the workplace.

Radiation measurements must be taken in the places where the worker is located. Measuring is introduced in sections 8.3.2 (magnetic fields), 8.3.4 (electric fields) and 8.4 (electromagnetic fields).

Thus, the assessment possibilities range from simple estimation to complex quantitative analysis. Measuring is sometimes the most practical method, although calculations can be very precise. Numerical simulation in relation to the ELVs is a heavy method in the vast majority of cases.

Assessment does not cover all the situations at a workplace, but it is necessary to ensure that it takes into account the most negative configuration/s (within the limits of intended usage). It is important not to forget to take into consideration indirect effects and employees at particular risk.

In theory, in a company, all field types and the whole frequency spectrum from 0 Hz to 300 GHz should be analyzed. In practice, this is reduced in the very large majority of cases, either to low-frequency magnetic fields (50 Hz or complex waveforms, see Chapter 6) or to one or even two narrow bands of high-frequency electromagnetic fields. These fields may be very different from one piece of equipment to another. In some cases, particular attention must be paid to workplaces subjected to multiple exposure sources. However, in general, the radiation level of the source at a station often becomes negligible at the neighboring station located only a few meters away. It suffices to determine the distance at which it is possible to disregard it in comparison with other contributions.

#### 9.2.3.2.2. Assessment in relation to the ALs

Low-frequency electric and magnetic fields must be dealt with separately, since each has its own ALs. Except in the presence of high-voltage equipment, electric fields may generally be considered negligible [EC 04].

Compliance with low ALs relating to low-frequency electric fields enables annoying spark discharges to be avoided in the workplace. Respect of high ALs prevents sensory and health effects and guarantees that the ELVs are not exceeded (see Chapter 5).

The EMF Directive defines three categories of ALs for low-frequency magnetic fields: low ALs, high ALs and limb ALs. They have been deduced from the ELVs, as for the majority of ALs. Compliance with them guarantees compliance with the ELVs. Low ALs relate to sensory effects. They may be exceeded, on the condition that the sensory effect ELVs are not exceeded or that the following conditions are met:

– the excess is temporary;

– the workers have been informed of the possible transient symptoms or sensations;

– when such effects are reported, the risk assessment must be updated if necessary;

– the health effect ELVs, in the absence of high ALs, are not exceeded.

NOTE.– The health effect ELVs must never be exceeded, regardless of the situation.

The first condition is often fulfilled, since many industrial processes work only intermittently.

The EMF Directive defines ALs for high-frequency electromagnetic fields (100 kHz–6 GHz). They are quantified in terms of electric and magnetic fields. They are derived from the health effect ELVs (SAR). Knowledge of a field defines the level of the other beyond a distance equal to around one wavelength [1.21]. Below this, the electric and magnetic fields must be determined independently. In an occupational environment, strong fields are generally present only in close proximity to the source, i.e. in near field. The induced current may be measured in the band between 100 kHz and 110 MHz (section 8.5.2). This avoids performing the assessment in relation to the ELVs.

The directive defines ALs for static magnetic fields. The aim is to prevent a malfunction of active implantable medical devices and to limit the risk of the projection of ferromagnetic objects near sources of strong radiation (>100 mT). The risk is negligible below 3 mT.

Since electromagnetic fields may cause direct and indirect effects, the resulting risks must be dealt with separately.

#### 9.2.3.2.3. Assessment in relation to the ELVs

The EMF Directive established separate ELVs for sensory and health effects.

At low frequencies, the sensory effect ELVs are defined from 1 Hz to 400 Hz. They apply to exposure of the central nervous system of the head. Health effect ELVs are defined up to 10 MHz for the whole body, with the aim of protecting people against stimulation of the peripheral nervous system, an effect which can be damaging to health and safety (see Chapter 4)[1].

The sensory effect ELVs, defined in terms of induced electric fields, may be exceeded, provided that this is justified by the practice or procedure under the conditions cited for the exceeding of low ALs associated with magnetic fields:

– the excess is temporary;

– the workers have been informed of the possible transient symptoms or sensations;

– when such effects are reported, the risk assessment must be updated if necessary;

– the health effect ELVs are respected.

Such conditions also apply to sensory effect ELVs between 0 and 1 Hz, although they are expressed in terms of magnetic induction.

Health effect ELVs are defined from 100 kHz to 300 GHz in order to prevent heating, which might cause thermal stress, heatstroke or damaging effects on the tissues. The quantities used are the SAR between 100 kHz and 6 GHz and the power density beyond this range. These values are averaged over time. They are an average whole body SAR, a localized SAR for the head and trunk and a localized SAR for the limbs.

Between 300 MHz and 6 GHz, there is also a sensory effect ELV. Its aim is to prevent microthermal auditory effects resulting from exposure to pulsed fields. This ELV is defined in terms of specific absorption (SA), the energy quantity of a pulse absorbed by a unit of mass. They may also be exceeded temporarily under the conditions cited previously.

---

1 In practice, the health effect ELVs supplement the sensory effect ELVs above 400 Hz. Therefore, it is much simpler to use the weighted peak method in the time domain. This little tip does not skew the principle required by the EMF Directive. This is the principle adopted for low ALs (magnetic fields) and the basic restrictions from the 2010 ICNIRP Guidelines [ICN 10].

The use of ELVs requires some numerical dosimetry skills (see section 8.6). Therefore, they must be reserved for specific exposures or to manufacturers wishing to certify their equipment when exposure exceeds the ALs or when the measurements of external fields are not adapted.

ALs have been fixed for the harshest exposure conditions, especially when there is a uniform field. In most cases, e.g. in the industrial environment, the source is very close and exposure is non-uniform. The established limits produce a very conservative exposure assessment. Therefore, to avoid using dosimetric simulation methods, ICNIRP allows standardization committees to establish limits peculiar to specific exposures. Standards [STD 18, STD 07] and [STD 08] are examples applying to exposure due to "power lines", arc welding and resistive welding, respectively.

#### 9.2.3.3. *Workers at particular risk*

The exposure assessment of workers at particular risk, mainly pregnant women and those wearing active medical devices (implanted or worn), adds a certain degree of complexity. The ALs are not sufficient to guarantee them adequate protection. An assessment separate to that for workers *without* particular risk is necessary.

The practical guide [EC 04] suggests that the EC recommendation be applied to pregnant workers. These levels cannot guarantee adequate protection in a minority of cases for those wearing active devices, since some equipment is particularly susceptible to interference (old devices or those adapted to certain pathologies). Advice for facilitating this assessment may have been provided by the medical team responsible for the health of these people. *In fine*, the occupational doctor must declare their ability at their workstation.

In this context, it is useful to note that, for a number of work situations listed in the tables A1.1–A1.4, a distinction is drawn between the presence of a piece of equipment and its use, since some of them may radiate strong, but very localized fields. Their presence is unlikely to damage active medical devices. However, their use requires a specific assessment that may result in a reminder of several preventive measures that are already likely followed by the holder, such as warnings about the use of a mobile telephone or induction cooking hob.

### 9.2.3.4. *Limitations system: flow chart*

The values system (ELVs and ALs) proposed by the EMF Directive may appear complex, but its aim is to minimize the assessment work, in particular for SMEs. It proposes many options. Figure 9.2 presents a simplified flow chart describing this system.

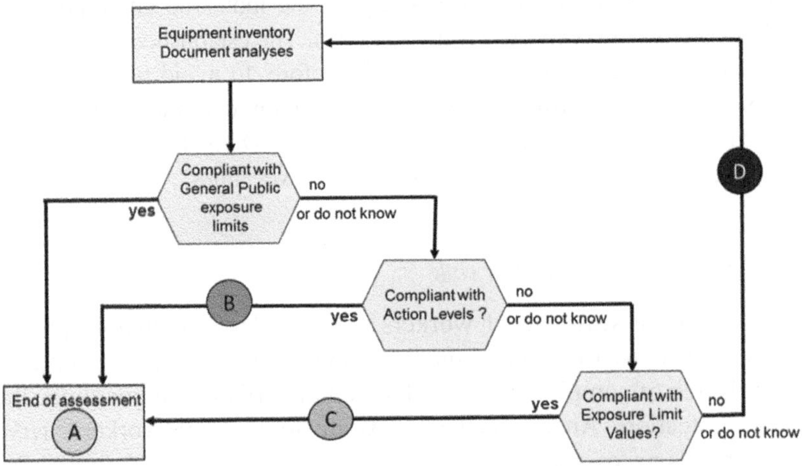


| | |
|---|---|
| A | Completed assessment.<br>Indirect effects to be considered if required.<br>Safe backup of assessment results.<br>Informing workers of the risk assessment results.<br>Additional requirements may be applied to the employer. |
| B | The risk analysis must be supplemented for people at particular risk. |
| C | The health effect ELVs are respected over the whole body and:<br>– the sensory effect ELVs are respected in the head or;<br>– the sensory effect ELVs are exceeded, but the following conditions are fulfilled:<br>+ temporary exceeding;<br>+ involved workers are informed;<br>+ preventive and protective measures if there are transient symptoms.<br>The risk analysis must be supplemented for people at particular risk. |
| D | Non-compliance: impossible to demonstrate that the ALs or ELVs have not been exceeded; implementation of organizational or technical measures to reduce exposure. |

**Figure 9.2.** *General flow chart showing the implementation of Directive 2013/35/EU*

#### 9.2.3.4.1. Description of the flow chart

At each step it is always possible to implement preventive or protective measures to reduce exposure, instead of continuing with the assessment. Sometimes, it is better to introduce preventive measures than to perform costly and complex assessments (vs. ELV) whose outcome is uncertain and to ultimately conclude that they must be implemented. However, thorough investigations may prevent ergonomic modifications of a workplace, which may lead to very real afflictions, such as musculoskeletal disorders.

*Input*

At an input are the equipment from tables A1.2–A1.4 in Appendix 1 and those whose compliance with the EC Recommendation (general public) is still in question. The technical documents available should also be an important source of information.

*Output*

Administrative activities (safe backup of assessment results, information, etc.) must be undertaken when compliance with the "general public" Recommendation, ALs, or ELVs has been demonstrated. The occupational doctor must consider people at particular risk. Indirect effects must also be taken into account.

*Step 1*

Does the equipment (source of electromagnetic fields) used in the company comply with the EC Recommendation? The pieces of equipment and workplaces in table A1.1 in Appendix 1 (non-exhaustive list) are assumed to not exceed the basic restrictions or reference levels of the Recommendation and it is unlikely that electromagnetic fields are a risk for most workers.

*Step 2*

Does the equipment used by the company comply with the ALs? The ALs are defined in terms of field quantities outside the exposed body. They may be measured or calculated relatively simply They include:[2]

– for direct effects:

2 The quantities and their units associated with the ALs are summarized in Table 5.22 (Chapter 5).

- low ALs for exposure to LF magnetic fields in the head (sensory effects),

- high ALs for exposure to LF magnetic fields in the head and trunk (health effects),

- limb ALs for exposure to LF magnetic fields in the limbs (health effects),

- high ALs for exposure to LF electric fields,

- AL for exposure to HF electromagnetic fields,

- AL for currents induced in the limbs;

– for indirect effects:

- low ALs for exposure to LF electric fields,

- ALs for contact currents,

- ALs for static magnetic fields.

If the ALs are respected, no additional measure is necessary. Compliance has been demonstrated. Low and high ALs do not cover the same risk; they must be considered independent.

In some situations, it is impossible to demonstrate compliance with the ALs (e.g. mobile telephone). If the ALs are exceeded (or if this is unknown, for whatever reason), step 3 is begun, in the knowledge that protective and preventive measures can always be implemented.

Non-compliance with the ALs does not necessarily imply non-compliance with the corresponding ELVs. It is possible to exceed the ALs and respect the ELVs (the ALs were derived from the ELVs, considering the harshest exposure conditions). On the other hand, compliance with the ALs does imply compliance with the corresponding ELVs.

### *Step 3*

In this step, the investigations are thorough, since compliance must be demonstrated in relation to the ELVs, which are physical quantities within the body[3] between 1 Hz and 6 GHz. Therefore, they cannot be easily

3 The physical quantities, the units and the frequency domains associated with various ELVs are summarized in Table 5.21 (Chapter 5).

measured or calculated. This step can be divided into the following two phases.

### Phase 1

Does the equipment used by the company comply with the sensory effect ELVs? The sensory effect ELVs are as follows:

– ELVs for exposure of the central nervous system of the head to a low-frequency magnetic or electric field;

– ELV for exposure to a static or quasi-static magnetic field;

– ELV for localized exposure (head) to high-frequency electromagnetic fields.

### Phase 2

Does the equipment used by the company comply with the health effect ELVs? The health effect ELVs apply to the whole body. They include the following:

– ELVs for exposure to a low-frequency magnetic or electric field;

– ELV for exposure to a magnetic field of frequency lower than 1 Hz;

– ELVs for exposure to high-frequency electromagnetic fields.

If the health effect ELVs are *respected* over the entire body and the sensory ELVs are respected, the use of the equipment is compliant.

If the health effect ELVs are *respected* over the entire body, if the sensory effect ELVs are exceeded and if the following conditions are met:

– exposure is temporary;

– worker has been informed of a possible risk;

– preventive measures have been taken in the event of reported transient symptoms.

The use of the equipment is then deemed compliant. If not, i.e. if the previous conditions are not fulfilled, although the health effect ELVs are respected, preventive or protective measures (technical or organizational) must be implemented to protect people from sensory effects (in the head).

If the health effect ELVs are *exceeded* or if it cannot be proven that they are respected, measures must be implemented to protect body part/s that may be subject to a health risk (unless a specific exemption has been accorded nationally).

The directive therefore offers the possibility of an incremental approach to risk assessment (simple, right?).

### 9.2.4. *Risk prevention*

#### 9.2.4.1. *General principles*

The measures must be based on the general risk prevention principles outlined in the Framework Directive [EC 02]. These technical and organizational measures aim to guarantee compliance with the ELVs. Measures must be taken, as far as is reasonably possible, to eliminate the identified hazards or reduce them to a minimum. These measures must take into account the technical progress and availability of measures for monitoring the electromagnetic field level at the source. These measures may be:

– other work methods resulting in lesser exposure to electromagnetic fields;

– substitution by equipment radiating weaker electromagnetic fields, taking into account the work to be done;

– implementation of technical measures to reduce exposure to electromagnetic fields, such as the use of interlocking and shielding devices and protection systems;

– appropriate delimitation and access control measures;

– in the event of exposure to electric fields, measures and procedures for managing spark discharges and contact currents through technical means and employee training;

– appropriate maintenance programs for equipment, areas and workplaces;

– design and layout of workplaces and stations;

– limitations of the duration and strength of exposure;

– availability of appropriate personal protective equipment.

Technical protection measures must be prioritized over organizational measures. Personal protective measures are the last resort.

#### 9.2.4.2. *Elimination of hazard*[4]

Priority must be given, as far as reasonably possible, to eliminating or reducing to a minimum the hazards identified by the most recent assessment. One possible solution is to purchase equipment whose use complies with the EMF Directive or with other more restrictive documents.

NOTE.– Taking into account the human exposure from the design phase of a piece of equipment results in an additional cost that may be marginal. This restriction supplements electromagnetic compatibility.

One quick and relatively simple solution may be a change in the way equipment is used. Some "beliefs" mean people needlessly expose themselves to strong fields. Therefore, their exposure may be reduced when they are given information or when the station is rearranged, which does not need to be complex.

Examples of recommendations to limit exposure during arc welding operations [STD 14]:

– do not place the wire over the shoulders (near the head);

– never wind the wire around a body part;

– bring the active and return wires close together and tie them to one another over the greatest possible length;

– keep the wires on the same side of the body;

– connect the return wire as close as possible to the welding area;

– position the head and trunk as far as possible from the welding circuit;

– do not work close to, sitting on or leaning on the welding current generator.

4 Hazard is the intrinsic property or ability of something to cause harm effects. Risk is the likelihood that any hazard will actually cause somebody harm under conditions of use or exposure.

Substituting one technology for another may introduce a much higher risk, for example, replacing resistive welding (section 2.2.2.5.2) with laser welding.

#### 9.2.4.3. *Technical measures*

Technical measures include:

– shielding for small pieces of equipment;

– use of an absorbent screen to protect against high frequencies;

– remote activation command outside the radiation area;

– impossibility of producing fields if a human body is detected;

– access limitation in the close vicinity of sources, with warning signs;

– automatic interruption of radiation before operation;

– manual interruption of radiation before access is authorized (operation to be planned in general).

The shielding of low-frequency magnetic fields (<10 kHz) transmitted during welding or induction operations is not effective in the vast majority of cases. The best form of protection is still further distance from the source.

Technical measures against the indirect effects that may be caused by contact currents or spark discharges are generally earthing, to prevent insulated conductive objects from charging, and the removal of large metal objects, which are not currently being used.

*Example of technical measures applicable to UHF RFID installations*

Generally, UHF RFID systems (section 2.2.3.1.6) installed according to the manufacturer's recommendations comply with the EMF Directive and the "general public" recommendation. Nonetheless, exposure can be reduced by implementing a few technical measures [CNR 12]. A non-exhaustive list of good practices might be to:

– reduce the antenna aperture (width of the radiation pattern);

– limit the transmission power to only what is needed;

– point reader antennae downwards (to limit the extent of the readable area);

– point the antennae toward the tags;

– increase the number of antennae per reader (so they are as close as possible to the tags in order to distribute a reduced transmission power);

– reduce the number of readers (a reader transmits only on one antenna at a given moment and these may be distributed over several stations);

– transmit only in the presence of tags (requires mechanical, optical or other detectors);

– stop transmission when the tags have been identified;

– install screens or absorbent materials between the RFID antennae and the people present on the site;

– install a visual or acoustic operating indicator;

– ensure respect of the regulations relating to radiofrequency transmission.

When technical measures are not sufficient, they must be accompanied by organizational measures.

#### 9.2.4.4. *Organizational measures*

The main organizational measures are signage of areas where the ALs are exceeded, and updating of the safety sheets at the work station. In terms of signage, item 5.5 of the directive states:

> "Workplaces where workers are likely to be exposed to electromagnetic fields that exceed the ALs shall be indicated by appropriate signs ... where access to these areas is suitably restricted for other reasons and workers are informed of the risks arising from electromagnetic fields, signs and access restrictions specific to electromagnetic fields shall not be required".

There are at least two standardized sign categories in relation to EMF for this:

– warning signs indicating a potentially hazardous situation and its nature (Figure 9.3);

– prohibition signs indicating an imminently hazardous situation (Figure 9.4).

a) Magnetic field

b) Non-ionizing radiation

**Figure 9.3.** *Warning sign relating to electromagnetic fields*

a) No access to wearers of active medical devices

b) No access to wearers of metal implants

**Figure 9.4.** *Restriction sign relating to electromagnetic fields*

These symbols may be accompanied by an explanatory note and safety instructions. Mandatory information and notice signs might also be displayed.

NOTE.– The pacemaker symbol applies to any wearer of an active medical device, whether implanted or body-worn.

As well as signage, the organizational measures are as follows:

– issuing or updating of safety sheets at the work station;

– installation of visual or acoustic radiation indicators;

– delimitation and restriction of areas (horizontal and vertical markings, map display, etc.);

– implementation of access procedures for safety interlocked areas (these measures also aim to limit the duration and therefore the average strength of exposure over time);

– availability of documents about the organizational measures applied;

– provision of information relating to site safety (prevention plan);

– design and layout of workplaces and stations, for example: exposure may be reduced without special restrictions through proper positioning of wires, an operation that can be simply performed;

– adaptation of maintenance programs; specific exposure situations may be encountered, since personnel may have to access an area that is usually inaccessible;

– implementation of regular maintenance programs to ensure the proper functioning and state of equipment (including inspection of wires, shielding, protective measures, etc.).

The training or informing of workers is discussed later on.

#### 9.2.4.5. *Personal protective equipment*

Personal protective equipment is used when risk reduction measures are insufficient in some specific exposure situations. This may involve a transitory period, when work is ongoing while replacements of equipment or the implementation of new work methods are awaited. It may also involve maintenance situations during which radiation cannot be interrupted. Is the broadcast carrier permitted to interrupt a television show?

Conductive clothing exists to protect people from low-frequency electric fields. Some models are adapted to work under very high voltages at 50 or 60 Hz. They protect against the indirect effects of contact currents and spark discharges. However, this clothing does not protect against magnetic fields.

There are also conductive suits that limit exposure to high-frequency electric fields (electric fields are mitigated by a minimum factor of 5). Made of stainless steel microfibers, they are used for work on radio and television broadcasting pylons (from 80 MHz to 1 GHz). An insulating oversuit is required to protect people against indirect effects.

#### 9.2.4.6. *Reexamination of risk assessment*

The employer must reexamine the risk assessment each time there are reasons to suspect it is no longer valid. Periodic reexamination of the risk assessment is recommended in order to verify the effectiveness of the preventive and protective measures implemented and to take into account site alterations. Industrial configurations are fairly stable, so the elements that may trigger a new assessment are as follows:

– change in the manufacturing process and implementation of new work equipment;

– rearrangement of a work area or new definition of a workplace;

– significant transformation of a station following a new work organization;

– reports of transient symptoms, such as vertigo, nausea or disturbing sensory perceptions, etc.

The periodicity of the assessment must depend on the margin in relation to the limits and the quality of the preventive maintenance program. Updating the assessment is quicker than the initial assessment, particularly if a procedure has been identified.

The assessment must be updated when people's status changes. These people may be equipped with a medical device or be pregnant. Such a change should trigger a specific risk analysis.

### 9.2.5. *Other activities*

#### 9.2.5.1. *Training/information*

Information or training should be provided when the sensory effect ELVs, in the absence of the low ALs in the head, are exceeded or if the absence of risk is not demonstrated.

It would be logical to provide training when there is a risk that the health effect ELVs or, by default, the high ALs, will be exceeded. This risk should not exist if the assessment is complete and the rules for use or operation are respected. Training is stricter than providing information. It involves formal convocation of people, verification of their presence, acquired knowledge testing and availability of support.

When the ELVs, and therefore the ALs, are not exceeded, there is no risk and no obligation to provide information. Information must be provided for people who may be exposed above the limits. This may involve, for example:

– in the radiofrequency domain (telecommunication, radio broadcasting, radar, electromagnetic compatibility, industrial processes, etc.): engineers, technicians, production agents;

– in the construction domain (presence of very powerful transmission antennae on roofs and towers: radio, television, etc.): roofers, house painters, window-cleaners, construction professionals, air conditioning technicians, firefighters and public health personnel;

– in the production domain: production employees, maintenance employees, etc.

Occupational doctors and risk prevention engineers (HSE) must be kept informed in order to respond to the concerns of the workers. The management should also be aware.

The information provided depends on the context of the company. It must be aimed at the activity of people (possible risks) in order not to drown the relevant content, impoverishing the information actually provided. It does not mean hiding information, since most of it is public, but it must be analyzed with a critical eye.

Information includes:

– physical quantities and units used;

– possible effects caused by electromagnetic fields and the associated risks;

– concept of ALs and ELVs;

– means of detecting and reporting sensory and health effects;

– results and assessment method used;

– indirect effects of radiation if they appear;

– measures implemented (including signage, instructions, organization, etc.);

– safe work practices that minimize risks;

– work practices to avoid;

– additional measures required with regard to employees at particular risk;

– conditions required for health monitoring.

More technical information may include, in addition to the points cited above:

– preventive and protective alternatives and their implementation;

– how to perform basic measurements (direct reading on a broadband device);

– how to perform complex measurements (selective measurements at low and high frequencies);

– concept of exposure indices;

– spatial and temporal averaging techniques;

– influence of the body and the environment on measurements;

– sources of uncertainty and correction factors;

– verification of the proper functioning of measuring devices;

– identification of areas at risk, areas where the health effect ELVs may be exceeded.

Beyond this, technically advanced training is possible:

– analysis of generic or product standards;

– measurements for dosimetry assessment;

– analytical dosimetry;

– numerical dosimetry.

#### 9.2.5.2. *Health monitoring*

When an employee is exposed to electromagnetic field levels exceeding the ELVs and observes an effect on his health, their employer must ensure that he can benefit from health monitoring and, if necessary, medical examinations.

Examinations or monitoring must be arranged at the times chosen by the employee. Medical files are stored in an appropriate format, in view of further consultation, in compliance with the confidentiality rules.

If exceeding the ELVs causes the employee's health to deteriorate, the risk assessment must be repeated. The occupational doctor is aware of this assessment, as well as that performed for other risks.

It is difficult to diagnose the effects of exposure to electromagnetic fields – with the usual practices, at least. Knowledge is insufficient [LAM 13], particularly for frequencies lower than 100 kHz, and there is no recognized method for health examinations.

Two completely separate situations must be taken into account: exposure to low frequencies and exposure to high frequencies, whose possible acute effects are, respectively, stimulation of the central nervous system or neuromuscular system and thermal effects.

During overexposure to low frequencies, the clinical symptoms usually reported are, for electric fields, cutaneous reactions (oscillation of the small hairs on the body or electric shocks) and, for magnetic fields, magnetophosphenes. At high frequencies, the effects may be localized heating on the skin, in the wrists and ankles, and burns.

To avoid health effects, sufficient information on the possible risks relating to work in the direct vicinity of the sources of strong electromagnetic fields must be provided. This information includes safe work practices.

#### 9.2.5.3. *Archiving*

The directive does not specify how data relating to the electromagnetic field risk should be archived: assessment results are kept in a format that enables them to be traced and consulted at a later date. If applicable, these results include the risks identified for workers at particular risk.

This point is supplemented by the national texts.

# Appendix 1

# Equipment and Workplaces Classification

## A1.1. Non-exhaustive list of equipment and workplaces where it is unlikely that electromagnetic fields would be a risk for most workers

| Wireless communications |
|---|
| Workplaces containing phones, mobile phone and cordless (including base station for DECT cordless phone). |
| Workplaces containing wireless communication devices (e.g. Wi-Fi or Bluetooth) including access points for WLAN. |
| **Office** |
| Audio-visual equipment (e.g. televisions, DVD players). |
| Wired communication equipment and networks. |
| Computer and IT equipment. |
| Electric fans and fan heaters. |
| Office equipment (e.g. printers, photocopiers, paper shredders, electrical staplers). |
| Landline phones and fax machines. |
| **Infrastructure (buildings and grounds)** |
| Workplaces accessible to the general public that meet the reference levels specified in Council Recommendation 1999/519/EC. |
| Alarm systems. |
| Base station antennas outside an operator's designated exclusion zone. |
| Workplaces containing electric garden appliances. |
| Household and professional appliances as long as they do not contain transmission |

| |
|---|
| equipment such as WLAN, Bluetooth and mobile phones. |
| Lighting equipment including desk lamps. |
| **Electrical supply** |
| Electrical circuit where the conductors are close together and have a net current of 100 A or less, includes wiring, switchgear, transformers, etc. (exposure to magnetic fields). |
| Electrical circuits within an installation, with a phase current rating of 100 A or less for the individual circuit, includes wiring, switchgear and transformers (exposure to magnetic fields). |
| Electrical installations with a phase current rating of 100 A or less, includes wiring, switchgear and transformers (exposure to magnetic fields). |
| Overhead bare conductor up to 100 kV or overhead line up to 150 kV above the workplace (exposure to electric fields). |
| Overhead bare conductors at any voltage (exposure to magnetic fields). |
| Underground or insulated cable circuit at any voltage (exposure to electric fields). |
| **Light Industry** |
| Coating and painting equipment. |
| Control equipment not containing radio transmitters. |
| Workplaces containing glue guns (portable). |
| Workplaces containing heat guns (portable). |
| Hydraulic ramps. |
| Measuring equipment and instrumentation not containing radio transmitters. |
| Workplaces containing tools (electric handheld and transportable). |
| **Medical** |
| Medical equipment not employing EMF for diagnosis or treatment. |
| **Miscellaneous** |
| Equipment placed on the European market as compliant with Council Recommendation 1999/519/EC or harmonized EMF standards. |
| Battery chargers, non-inductive coupling designed for household use. |
| Non-electrical equipment of all types except those containing permanent magnets. |
| Portable equipment (battery powered) not containing radiofrequency transmitters. |

## A1.2. Non-exhaustive list of equipment where electromagnetic fields may pose a risk to ALL workers

Assessment required against applicable action levels or exposure limit values as specified in the EMF Directive (see Chapters 5 and 9).

| Infrastructure (buildings and grounds) |
|---|
| Base station antennas, inside an operator's designated exclusion zone. |
| Lighting equipment, radiofrequency or microwave energized. |
| **Electrical supply** |
| Electrical circuit where the conductors are close together and have a net current greater than 100 A (exposure to magnetic fields). |
| Electrical circuits within an installation, with a phase current rating greater than 100 A for the individual circuit (exposure to magnetic fields). |
| Electrical installations with a phase current rating greater than 100A (exposure to magnetic fields) |
| Overhead bare conductor over 100 kV or overhead line over 150 kV above the workplace (exposure to electric fields). |
| **Light industry** |
| Dielectric heating and welding. |
| Induction heating and soldering. |
| Magnetic particle inspection (crack detection). |
| Industrial magnetizer/demagnetizers (including tape erasers). |
| Microwave heating and drying. |
| Radiofrequency plasma devices including vacuum deposition and sputtering. |
| Welding, manual resistance (spot welding, seam welding). |
| **Heavy industry** |
| Industrial electrolysis. |
| Furnaces, arc and induction melting. |
| **Construction** |
| Microwave drying in the construction industry. |
| **Medical** |
| MRI equipment. |
| Medical equipment using EMF for diagnosis and treatment (e.g. short wave diathermy, transcranial magnetic stimulation). |
| **Transport** |
| Radar for air traffic control, military radar, weather and long-range radar. |
| Electrically powered trains and trams. |
| **Miscellaneous** |
| Broadcasting systems and devices (radio and TV: LF, MF, HF, VHF, UHF). |

## A1.3. Non-exhaustive list of equipment in addition to those in table A1.2 where electromagnetic fields may pose a risk to workers at particular risk, i.e. pregnant workers or workers with passive implanted medical devices (excluding workers with active implanted medical devices)

Assessment required against Council Recommendation (1999/519/EC) reference levels (see section 5.7.4).

| Electrical Supply |
|---|
| Work on wind turbines. |
| **Light industry** |
| Electrostatic painting equipment. |
| Automated induction heating systems, fault-finding and repair involving close proximity to the EMF source. |
| Automated welding systems, fault-finding, repair and teaching involving close proximity to the EMF source. |

## A1.4. Non-exhaustive list of equipment in addition to those in tables A1.2 and A1.3 where electromagnetic fields may pose a risk to workers at particular risk, i.e. workers with active implanted and active body worn medical devices

Localized personal exposure may exceed reference levels in Council Recommendation 1999/519/EC – this will need to be considered in the risk assessment, which should be informed by information supplied by the healthcare team responsible for implanting device and/or subsequent care (see section 5.7.4).

| Wireless communications |
|---|
| Use of phones and cordless (including base stations for DECT cordless phones). |
| Use of mobile phones. |
| Use of wireless communication devices (e.g. Wi-Fi or Bluetooth) including access points for WLAN. |
| **Office** |
| Audio-visual equipment containing radiofrequency transmitters. |
| **Infrastructure (buildings and grounds)** |
| Use of electric garden appliances. |

| Security |
|---|
| Article surveillance systems and radiofrequency identification (RFID). |
| Tape or hard drive erasers. |
| Metal detectors. |
| **Electrical supply** |
| Work on generators and emergency generators. |
| Inverters, including photovoltaic systems. |
| **Light industry** |
| Manual arc welding processes including MIG, MAG and TIG. |
| Industrial and large professional battery chargers. |
| Corona surface treatment equipment. |
| Furnaces resistively heated. |
| Use of glue guns. |
| Use of heat guns. |
| Induction sealing equipment. |
| Machine tools (e.g. pedestal drills, grinders, lathes, milling machines, saws). |
| Use of tools (electric handheld and transportable, e.g. drills, sanders, circular saws, and angle grinders). |
| **Construction** |
| Work in close proximity to construction equipment (e.g. concrete mixers, vibrators, cranes, etc.). |
| **Transport** |
| Work in close proximity to starter, alternator, ignition systems in motor vehicles and plant. |
| **Miscellaneous** |
| Inductive or proximity coupling battery chargers. |
| Equipment generating static magnetic fields > 0.5 millitesla, whether generated electrically or from permanent magnets (e.g. magnetic chucks, tables and conveyors, lifting magnets, magnetic brackets, nameplates, badges). |
| Headphones producing strong magnetic fields. |
| Professional inductive cooking equipment. |
| Two-way radios (e.g. walkie-talkies, vehicle radios). |
| Battery powered transmitters. |

# Appendix 2

# Weighting Filters

Chapter 6 introduced the exposure evaluation using the weighted peak method applicable to low-frequency fields. It applies weighting functions whose design is not necessarily immediate for non-specialists. This appendix presents the main functions regarding the EMF Directive.

These functions are either analog filters or digital filters. In theory, they should cover the band from 1 Hz to 10 MHz. In practice, they are often limited to 400 kHz in measurement devices, due to design constraints. This limitation generally does not pose a difficulty, since the equipment rarely radiates beyond 100 kHz if low frequencies are considered.

The weighted peak method is applicable to the external electric field, to magnetic induction and to internal electric fields or current density. Thus, the weighting functions apply to action levels (reference levels) as well as to exposure limit values (basic restrictions). These functions, and therefore the structure of the filters, depend on the limit selected.

## A2.1. Weighting filter: introduction

The general public limits (reference levels) of the 1998 ICNIRP Guidelines in the frequency band from 8 Hz to 100 kHz are used to introduce the design of a weighting filter.

The design of the filter is simple when the weighting function is approximate. Its transfer function follows the weighting function (Chapter 6) as closely as possible, as shown in Figure A2.1. The weighting function is represented by a solid line up to 100 kHz, in comparison with the levels fixed by the Guidelines represented by the piecewise linear curve (dotted line).

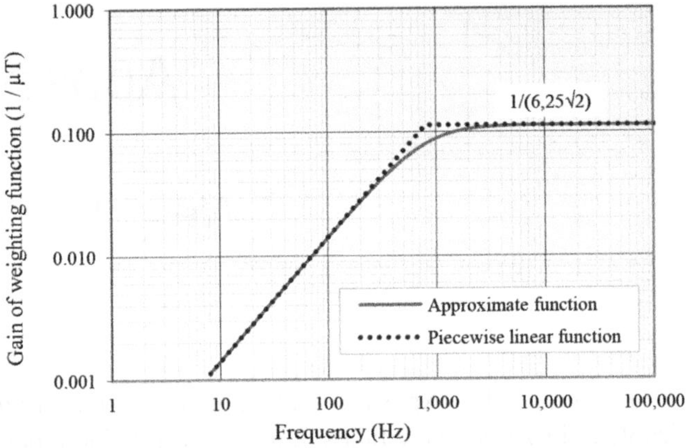


**Figure A2.1.** *Piecewise linear and approximate weighting (1998 ICNIRP – general public) from 8 Hz to 100 kHz (peak values)*

The weighting function is constant beyond 800 Hz (Figure A2.1). It presents a slope proportional to $f$ below 800 Hz (slope of 20 dB per decade or 6 dB per octave). This is the response curve of a **first-order high-pass filter** associated with a gain amplifier $\frac{1}{\sqrt{2}.6.25}$, which corresponds to the inverse of the peak reference level. The filter may be designed with a resistor and a capacitor (RC filter) [ICN 10], as shown in Figure A2.2.

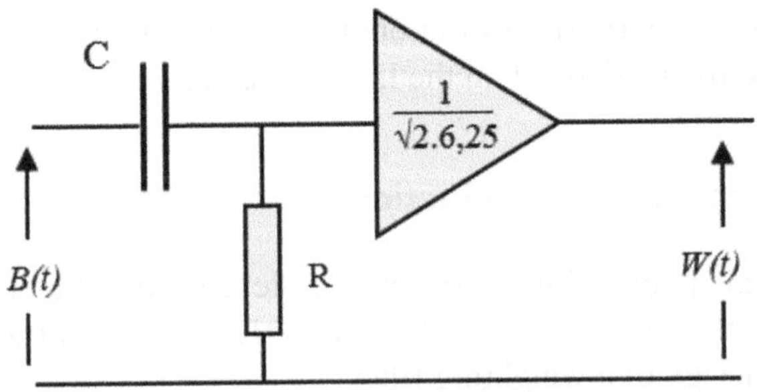


**Figure A2.2.** *Weighting filter principle*

$B(t)$ is in µT. The RC product has a value of $1/(2\cdot\pi\cdot f_c)$. $f_c$ corresponds to the break frequency usually called the cutoff frequency ($f_c = 800$ Hz).

The capacitor behaves like a short circuit at high frequencies. The circuit gain is therefore that of the amplifier. In these conditions, a high-frequency sinusoidal signal with peak amplitude lower than $6.25\sqrt{2}$ will give a peak output level lower than 1, $\max|W(t\ | \leq 1$. The limit exposure level is not exceeded in this case. The general public exposure is compliant with the 1998 ICNIRP Guidelines and the EC Recommendation.

At frequencies lower than 800 Hz, the impedance of the capacitor becomes greater than the resistance, causing a decrease in the gain of the function. Ultimately, a direct input signal (with zero frequency) is completely stopped by the capacitor; the output level is zero. This signal is thus canceled, as can be seen (or guessed) on the weighting curves (Figure A2.1).

The RC filter attenuates a sinusoidal signal at 50 Hz of a ratio of 16 in relation to a high-frequency signal. A signal with an rms value of 100 µT ($100\sqrt{2}$ µT peak) will therefore give a maximum output level equal to 1.

The gain (module) of the transfer function of this circuit is written according to the frequency $f$ (Hz):

$$|WL(f)| = \frac{1}{\sqrt{2}.Blh} \cdot \sqrt{\frac{f^2}{f^2 + f_c^2}} \quad [A2.1]$$

where $f_c$ is the cutoff frequency of the filter (Hz), and $Blh$ is the high-frequency reference level (6.25 µT if the intention is to assess exposure with regard to the magnetic field reference levels for the general public).

It can be verified that the gain of this function is zero for a frequency of zero ($f = 0$). When $f$ tends toward infinity, the term under the root tends toward 1.

If the asymptotes of the filter gain overlap the reference levels at the lowest and highest frequencies perfectly, there is a difference of 30 % at the cutoff frequency between the two functions. This difference, which corresponds to 3 dB, is accepted by ICNIRP, because, on the one hand, the reference levels have been derived cautiously from the basic restrictions, and, on the other hand, it describes more accurately the biophysical mechanisms and the electrophysiological model of a nerve (the peripheral nervous system consists of myelinated nerves, which are modeled by a cascade of RC filters under the excitability thresholds [JOK 00]).

Furthermore, a physical filter obligatorily introduces a phase difference dependent on the gain between the output signal and the input signal. The phase of the transfer function of the RC filter is written in radians:

$$\varphi(f) = \frac{\pi}{2} - \arctan\left(f / f_c\right) \quad \text{[A2.2]}$$

where $f$ is the frequency (Hz) and $f_c = 800$ Hz. It is represented in degrees by the graph in Figure A2.3.

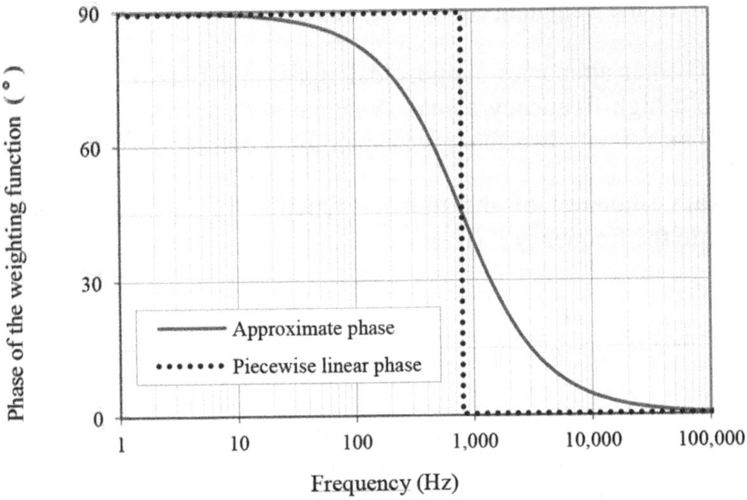


**Figure A2.3.** *Phase curve of a first-order high-pass filter*

This phase corresponds to phase $\varphi i$ introduced in [6.27].

The representation of the filter shows that the *exposure index* is proportional to the signal at frequencies higher than the cutoff frequency. It can be demonstrated to be proportional to the derivative of the signal below this frequency.

## A2.2. Weighting according to the EMF Directive (ICNIRP 2010)

The weighting functions presented in the first part of this section refer to magnetic induction (action levels). In the second part, they refer to the induced electric fields (exposure limit values).

### A2.2.1. *Exposure versus low action levels*

Low action levels (LALs) aim to protect against sensory effects. They should be considered for the exposure of the central nervous system of the head (section 5.2.2.1).

According to the principle outlined in the previous section, the weighting function to be used for calculating exposure indices is the inverse of the curve representing LALs (Figure 5.5). The filter is a little bit more advanced than that presented in the previous section, since the reference level changes slope four times.

The slope of the filter is proportional to $f^2$ below 8 Hz, to $f$ between 8 and 25 Hz, zero between 25 and 300 Hz, and then again proportional to $f$ between 300 Hz and 3 kHz, then zero above this point. The phase is 180°, 90°, 0°, 90° and 0° over these bands.

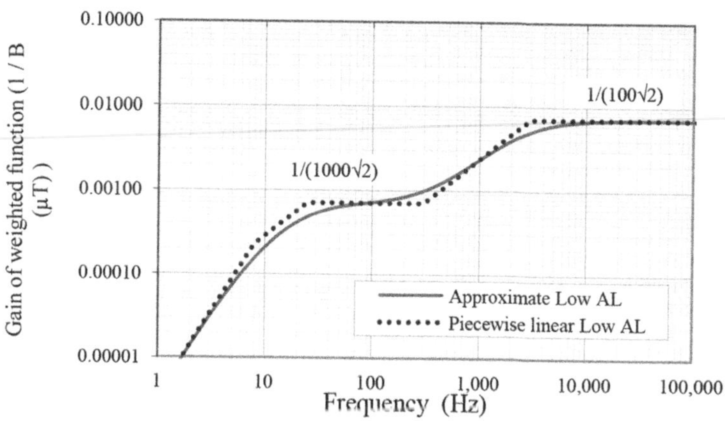


**Figure A2.4.** *Weighting function (Bode plot) according to low ALs for B as defined by the EMF Directive (peak values)*

Four elementary filters are required for design. Their cutoff frequencies are $f_1$ = 8 Hz, $f_2$ = 25 Hz, $f_3$ = 300 Hz, $f_4$ = 3 kHz. The gain at the highest frequencies is 1/(100√2). It goes down to 1/(1,000√2) around 100 Hz.

The gain according to $f$ of this function is written as follows:

$$|WL_{LAL}(f)| = \frac{1}{Blh.\sqrt{2}} \cdot \sqrt{\frac{1+\left(\frac{f_3}{f}\right)^2}{\left[1+\left(\frac{f_1}{f}\right)^2\right].\left[1+\left(\frac{f_2}{f}\right)^2\right].\left[1+\left(\frac{f_4}{f}\right)^2\right]}} \quad \text{[A2.3]}$$

*Blh* (action level at the highest frequencies) is equal to 100 μT.

As for [A2.1], when frequency ($f$) tends toward infinity, the gain tends toward 1/(*Blh*√2), and when it tends toward 0, the gain tends toward 0.

The phase of this function is a simple trigonometric function sum. It is written in radians according to the frequency:

$$\varphi(f) = \pi - \arctan(f / f_1) - \arctan(f / f_2) + \arctan(f / f_3) - \arctan(f / f_4) \quad \text{[A2.4]}$$

It is represented, thus, with the theoretical (piecewise linear) phase curve in degrees (Figure A2.5).

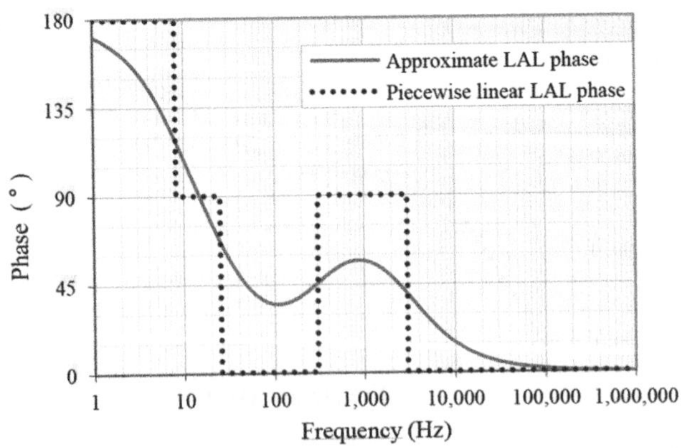


**Figure A2.5.** *Phase of the weighting function according to low ALs for B*

The constraint, to not have a gain differing by more than 3 dB from the theoretical limits and to limit the phase difference to 90°, is respected.

As indicated in section 6.5.2.3, the field must generally be processed simultaneously over three channels when it is measured over three axes. This filter should be duplicated, as shown in Figure 6.8.

### A2.2.2. *Exposure versus high action levels*

The high action levels (HALs) of the EMF Directive aim to protect against health effects, which may appear in the head and trunk of the exposed person in their workplace (section 5.2.2.1). These values were also given by the 2010 ICNIRP Guidelines. They were derived from the basic restrictions.

Assessment of the exposure level in relation to HALs follows the process applicable to LALs with a simple weighting function. This function is plotted in Figure A2.6.

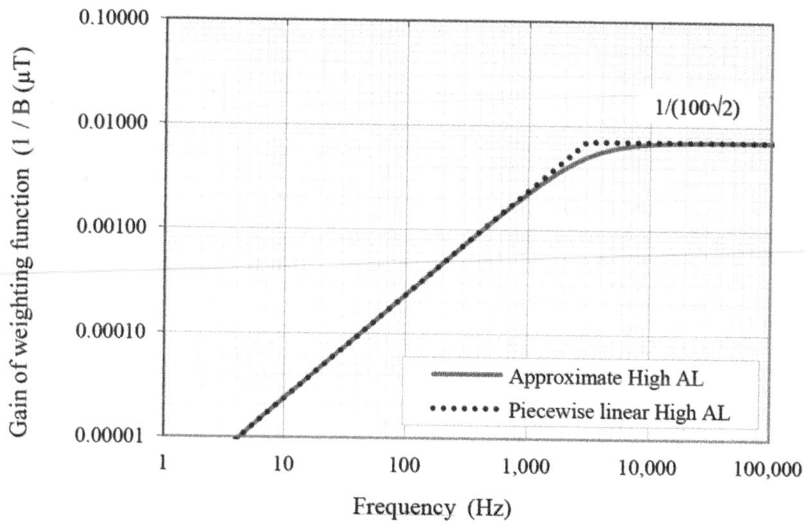


**Figure A2.6.** *Weighting function according to high ALs for B as defined by the EMF Directive (peak values)*

The module of the function represented above is written as follows:

$$|WL_{HAL}(f)| = \frac{1}{\sqrt{2}.Blh} \cdot \sqrt{\frac{1}{1+\left(\frac{f_4}{f}\right)^2}} \quad \text{[A2.5]}$$

*Blh* is equal to 100 μT. The cutoff frequency $f_4$ is equal to 3 kHz.

When the frequency tends toward infinity, the gain tends toward $1/(Blh\sqrt{2})$, and when it tends toward 0, the gain tends toward 0.

The phase of this function is also given by [A2.2]. Its curve is illustrated in Figure A2.7.

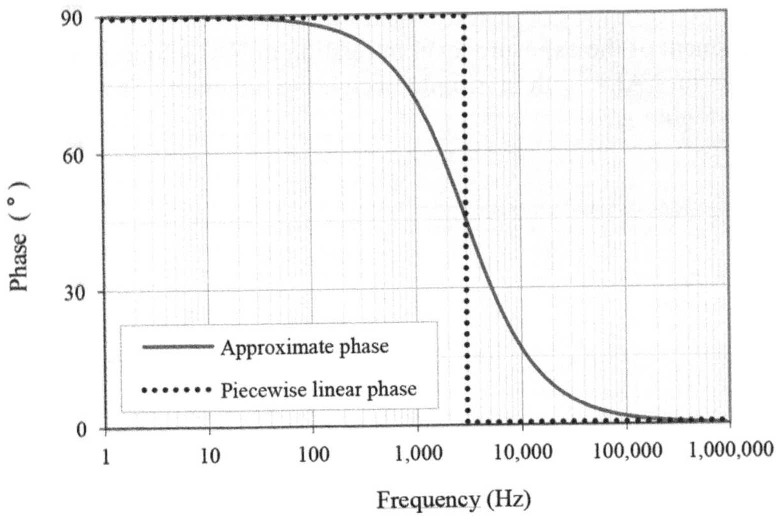


**Figure A2.7.** *Phase of the weighting function according to high ALs for B*

### A2.2.3. *Exposure versus action levels relating to limbs*

The EMF Directive defines action levels aiming to protect the limbs (lbAL) of an exposed worker from health effects (section 5.2.2.1). These values were also derived from the basic restrictions defined by the 2010 ICNIRP Guidelines.

The resulting weighting function is illustrated by Figure A2.8.

The gain of this function is written as follows:

$$|WL_{lbAL}(f)| = \frac{1}{\sqrt{2}.Blh} \cdot \sqrt{\frac{1}{1+\left(\frac{f_4}{f}\right)^2}} \quad \text{[A2.6]}$$

*Blh* is 300 μT here. The cutoff frequency $f_4$ is 3 kHz. This notation is the same as [A2.5].

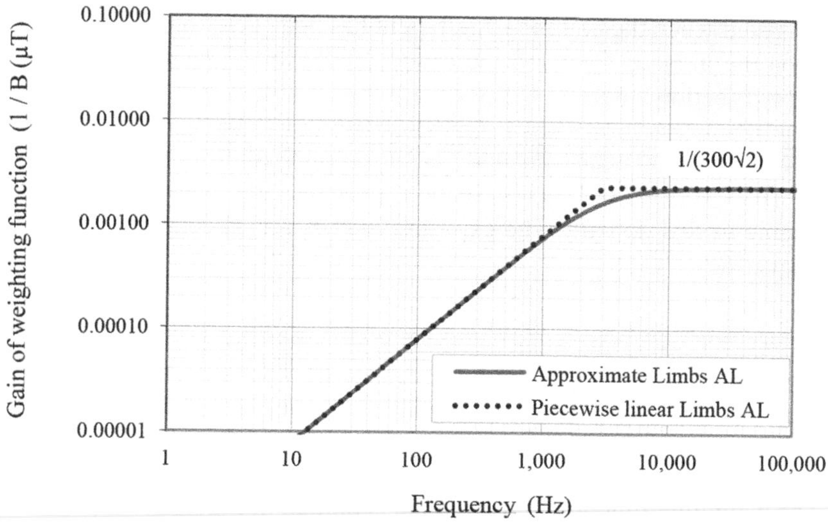


**Figure A2.8.** *Weighting function according to limb AL(B)s (peak values)*

There is a constant ratio of 3 between the gain of this function and that of the previous one [A2.5]. For this reason, the limb AL *exposure index* can be obtained directly by dividing the result from the high AL function by 3:

$$IE_{lbAL}^{WPM} = \frac{IE_{lIAL}^{WPM}}{3} \quad \text{[A2.7]}$$

The phase curve associated with [A2.6] is identical to the previous phase curve (Figure A2.7).

### A2.2.4. *Exposure indices for induced electric fields*

It is more practical to assess the exposure with the action levels (or reference levels) than with the exposure limit values (or basic restrictions). However, when the former are exceeded, an alternative is to verify the exposure in relation to the latter.

At low frequencies (1 Hz $\leq f \leq$ 10 MHz), the exposure limit values are defined in terms of the induced electric field in the EMF Directive. This field is calculated when the coupling coefficient has been determined (relationship between the induced field and the external field for a given exposure – see concept in section 8.6).

The weighted peak method is also applied to induced electric fields [ICN 10, EC 01]. This section presents the weighting functions applicable to this domain. The exposure indices are calculated as described in Chapter 6.

#### A2.2.4.1. *Exposure versus sensory effect exposure limit values*

The sensory effect exposure limit values (ELVSs) aim to protect against sensory effects. They should be considered for the exposure of the central nervous system of the exposed person's head.

The weighting function in peak value is plotted in Figure A2.9 (inverse of the CNS curve in Figure 5.3). It will be noted that the curve is not limited to 400 Hz, unlike the limits. Beyond this point, the curve represents the health effect exposure limit values. This does not alter the results at all, because compliance with these values must be verified either way, and this avoids curtailing the spectrum of the signal and simplifies the processing. This example is given in rms value in the appendix of the 2010 ICNIRP Guidelines [ICN 10].

The cutoff frequencies are $f_1$ = 8 Hz, $f_2$ = 25 Hz, $f_3$ = 400 Hz, $f_4$ = 3 kHz. The minimum limit value ($ELVS_{min\ peak}$) is 70 mV·m$^{-1}$ (50 mV·m$^{-1}$ rms).

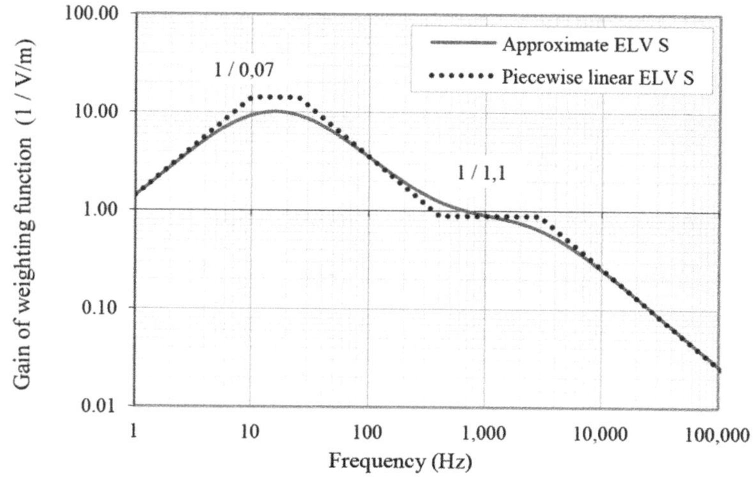


**Figure A2.9.** *Gain response of the weighting function for sensory effect ELVSs (peak values)*

The theoretical function (piecewise linear) reaches the inversed value, i.e. $1/(0.07\ \text{V}\cdot\text{m}^{-1})$ at 20 Hz. The approximate function is expressed as follows:

$$|WL(f)| = \frac{1}{ELVS_{min\ peak}} \cdot \sqrt{\frac{1+\left(\frac{f}{f_3}\right)^2}{\left[1+\left(\frac{f_1}{f}\right)^2\right]\cdot\left[1+\left(\frac{f}{f_2}\right)^2\right]\cdot\left[1+\left(\frac{f}{f_4}\right)^2\right]}} \qquad [A2.8]$$

This expression shows that when the frequency $f$ tends toward zero or toward infinity, the gain tends toward 0.

The phase of the function is a sum of trigonometric functions (arctan). It is therefore represented (Figure A2.10) together with the piecewise linear phase curve.

The *exposure index* for induced fields (exposure limit values) is calculated as for magnetic fields (action levels).

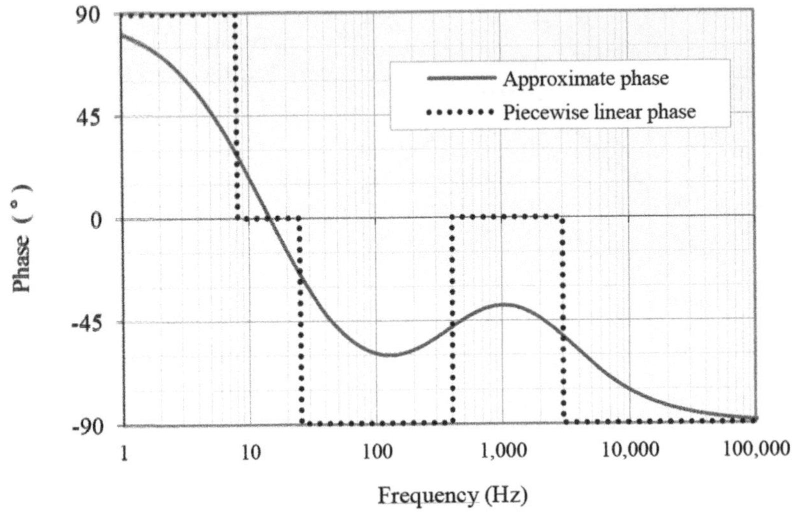


**Figure A2.10.** *Phase response of the weighting function for sensory effect ELVS*

### A2.2.4.2. *Exposure versus health effect exposure limit values*

Health effect exposure limit values (ELVHs) aim to protect against health effects. They should be considered for the exposure of the peripheral nervous system present throughout the exposed person's body (head, trunk and limbs, section 5.2.1.1).

The weighting function is plotted in Figure A2.11 (inverse of the PNS curve in Figure 5.2). The peak values are taken into account.

The cutoff frequency ($f_4$) is 3 kHz. The approximate curve is expressed as follows:

$$|WL(f)| = \frac{1}{ELVH_{\text{min crête}}} \cdot \sqrt{\frac{1+\left(\dfrac{f}{f_3}\right)^2}{\left[1+\left(\dfrac{f_1}{f}\right)^2\right]\cdot\left[1+\left(\dfrac{f}{f_2}\right)^2\right]\cdot\left[1+\left(\dfrac{f}{f_4}\right)^2\right]}} \qquad \text{[A2.9]}$$

where $ELVH_{min\ peak}$ (low-frequency limit) is equal to 1.1 V·m$^{-1}$ corresponding to 800 mV·m$^{-1}$ rms.

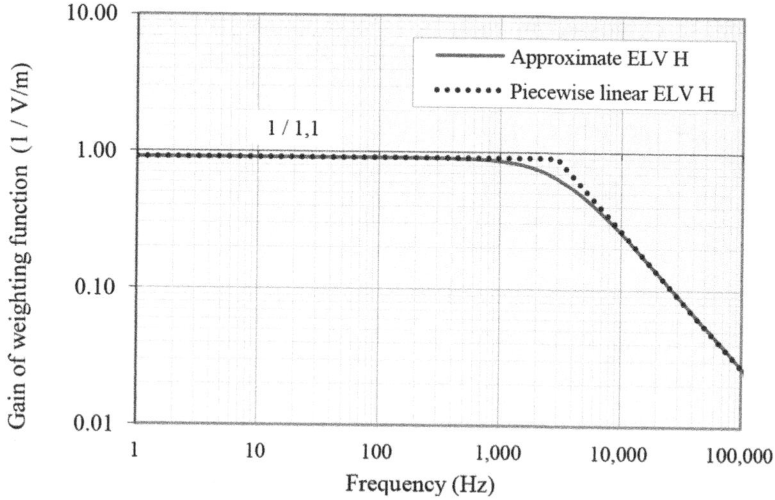


**Figure A2.11.** *Gain response of the weighting function for health effect ELVs*

This function corresponds to the frequency response of a first-order low-pass filter. Its phase response is illustrated in degrees by Figure A2.12.

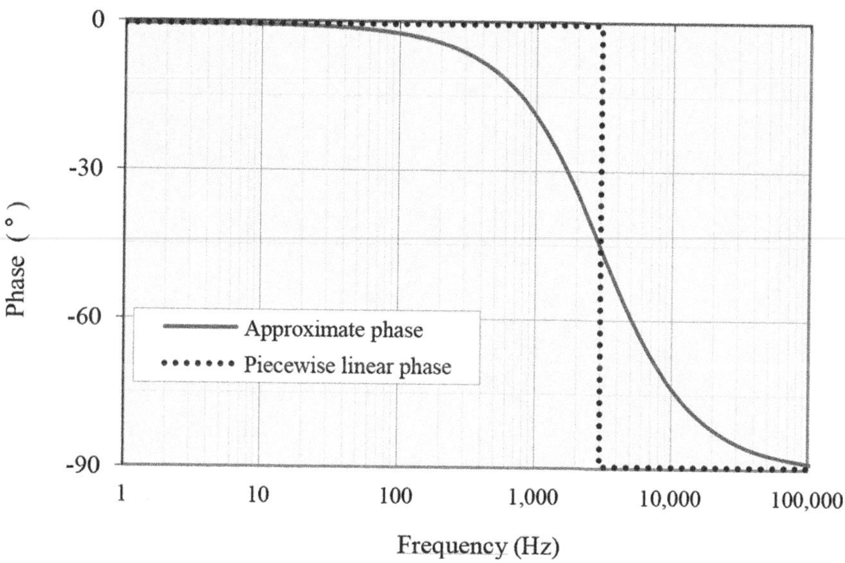


**Figure A2.12.** *Phase response of the weighting function for health effect ELVS*

The phase expressed in radians as a function of the frequency is the following:

$$\varphi(f) = -\arctan\left(f / f_4\right) \qquad \text{[A2.10]}$$

# Glossary

To facilitate the reading of this book, this glossary recalls the definition of key terms used. These terms refer to human exposure to electromagnetic fields from 0 Hz to 300 GHz. The proposed definitions are not necessarily valid in another context.

**Absorption:** Energy wave dissipation in an medium in which it propagates, for example in the form of heat. The absorption effect results in attenuation of the wave.

**Action levels:** See reference levels. Term generally used in the occupational exposure domain.

**Antenna:** Part of a radiofrequency transmitter or receiver designed to radiate or to collect electromagnetic waves.

**Attenuation:** Progressive reduction of energy suffered by radiation as it passes through a medium, especially as a result of absorption or scattering.

**Basic restrictions:** Exposure limit values expressed in terms of quantity in relation to all known mechanisms of biophysical interactions that may lead to health effects. Same as exposure limit values (ELVs).

**Contact current (Ic):** Refers to the current that appears when a person comes into contact with an object in an electromagnetic field expressed in ampere (A).

**Current density:** Ratio of the current flowing through a given cross-sectional area to the value of the cross-sectional area. The current density is expressed in ampere per square meter ($A.m^{-2}$).

**Decibel (dB):** Tenth of Bel. Dimensionless ratio between two parameters using logarithms to the base 10, commonly used in telecommunication. 20 dB corresponds to a ratio of 100.

**Dielectric:** Medium in which all the energy to establish an electric field is recoverable when you stop applying a field or voltage. A perfect dielectric has zero conductivity and all absorption phenomena are absent. The absolute vacuum is the only known perfect dielectric. The usual synonym is insulation.

**Dosimetry:** Measurement or determination by calculation of internal electric field strength, induced current density, specific absorption (SA) or specific absorption rate (SAR) in humans or animals exposed to electromagnetic fields.

**Duty cycle:** Ratio of a pulse duration to the pulse period of a periodic pulse series often expressed in percent. A duty cycle of 100% corresponds to a continuous wave (CW).

**Effects:**

**Acute effects:** Immediate adverse health effects at an exposure and disappearing when it ceases if irreversible disorders are not products. For electromagnetic fields, it is the immediate, proven and verifiable experimentally or scientifically. There are thermal effects and non-thermal effects.

**Biological effects:** Physiological, biochemical and behavioral changes induced in an organism, a tissue or a cell in response to a stimulation. These changes are reversible if they fall within the limits of homeostasis.

**Biophysical effects:** Interactions between electromagnetic fields and cells or organisms. Biophysics lies at the interface of physics and biology. This is the sector of study of physical phenomena applied to biological molecules. Understanding the biophysical mechanisms may explain the biological effects observed.

**Chronic effects:** Health disorder without noticeable effect in the short term at a low-dose prolonged exposure, generally occurring with a lag time of several months or a few decades, and usually irreversible in the absence of treatment.

**Critical effects:** First adverse effects that occur when increasing the level of exposure.

**Direct effects:** Effects in the human body directly caused by its presence in an electromagnetic field.

**Health effects:** Adverse health effects such as excessive thermal heating or stimulation of nerve and muscle tissue.

**Indirect effects:** Effects caused by the presence of an object in an electromagnetic field that may become the cause of a safety or health hazard.

**Long-term effect:** Effects due to prolonged exposure to levels several orders of magnitude below those causing short-term effects.

**Non-thermal effects:** Any effect unrelated to heating resulting from the action of electromagnetic fields on biological organism.

**Sensory effects:** Stimulation of sensory organs may cause temporary symptoms such as dizziness or transcient disturbed sensory perceptions. These effects might eventually cause temporary discomfort or affect cognition or other minor brain functions.

**Short-term effect:** Immediate effects caused by a high-intensity exposure.

**Thermal effects:** Effects such as tissue heating through energy absorption from electromagnetic fields in the tissue.

**Electric conductivity** ($\sigma$)**:** Quantity that, when multiplied by the electrical field strength, yields the conduction current density. It is the reciprocal of the resistivity and is expressed in siemens per meter ($S \cdot m^{-1}$).

**Electric field (E):** (Vector) quantity that corresponds to the force exerted on a small charged particle. It is expressed in volt per meter ($V \cdot m^{-1}$). A distinction has to be made between the environmental electric field and the electric field present in the body (*in situ*) as a result of exposure to the environmental field.

**Electromagnetic field (EMF):** Physical time-varying field that propagates in a medium or free space composed of an electric field and a magnetic field governed by Maxwell's equations. The propagation is divided into three regions as follows:

1) ***Reactive near-field*:** Region of the field immediately surrounding the source (antenna) wherein the reactive field predominates. The commonly accepted distance to the reactive near-field boundary is $\lambda/2.\pi$ m, $\lambda$ being the wavelength in meters.

2) ***Radiating near-field*:** Region of the field that extends between the reactive near-field region and the far-field region, wherein radiated fields predominate and the angular field distribution is dependent upon distance from the antenna.

3) ***Far-field*:** Region of the field of a antenna where the angular field distribution is essentially independent of the distance from the antenna. The fields have been spreading and have the character of a plane wave, i.e. they are locally uniform and decrease with distance in the direction of propagation. In practice, this area practically starts $\lambda + D^2/\lambda$ in the exposure domain, where $D$ is the largest dimension of the radiating element.

**Electromagnetic field polarization:** Orientation of the electric field when an electromagnetic field propagates. Polarization can be constant, i.e. vertical or horizontal, or it can rotate with the time (as circular polarization).

**Electromagnetic wave:** A wave associated with electric and magnetic fields which are time-varying and described by Maxwell's equations.

**EMC:** Electromagnetic compatibility.

**EMF:** Electromagnetic field.

**Equivalent isotropically radiated power (EIRP):** Power supplied to an antenna multiplied by the relative gain of the antenna in a given direction relative to an isotropic antenna (gain = 1).

**Equivalent radiated power (ERP):** Power supplied to an antenna multiplied by the relative gain of the antenna in a given direction relative to the half-wave dipole.

**Exposure:**

**High-intensity exposure:** Very strong exposure able to cause critical effects.

**Intermittent exposure:** Succession of exposure periods alternating with periods of non-exposure of a few seconds to a few hours each.

**Medium-intensity exposure:** Exposure to a level above the basic restrictions, but below levels able to cause critical effects.

**Prolonged or extended exposure:** Exposure duration from a few weeks to several years or decades.

**Repeated exposure:** Exposure during a period of several minutes to several hours a day and repeated from 1 day to another.

**Short-term exposure:** Single exposure from a fraction of a second to several hours a day or week.

**Single exposure:** Continuous exposure for a short time.

**Weak-intensity exposure:** Exposure to a level below the basic restrictions.

**Exposure index (EI):** Observed exposure divided by the exposure limit (reference levels, basic restrictions, etc.). The exposure is compliant with the limit if the index is less than 1 (100%).

**Exposure limit values (ELV):** See basic restrictions. Term generally used in the occupational exposure domain.

**Field:**

**Qualitative sense:** region of space in which certain phenomena occur;

**Quantitative sense:** physical quantity whose knowledge allows evaluating the effects of the fields.

**Field strength:** rms magnitude (module) of an electric or magnetic field vector.

**Free space:** Ideal medium, perfectly homogeneous, with a unitary dielectric permittivity and in which there is no element able to reflect, diffract or absorb energy. The absolute vacuum has these qualities.

**Hazard:**

**Common definition:** risk, threat or insecurity;

**Scientific or safety definition:** any source of potential damage, harm or adverse health effects on something or someone under certain conditions.

**Health:** Defined by World Health Organization (WHO) as a state of complete physical, mental and social well-being and not merely the absence of disease or infirmity.

**Instantaneous:** Describes parameters to be measured or evaluated on a very short time interval (100 μs or less).

**Interaction:** Means that the fields and exposed organisms act reciprocally one on the other.

**Isotropic:** Identical properties in all directions (opposite: anisotropic).

**Isotropic antenna:** Hypothetical antenna that radiates the same intensity of electromagnetic waves in all directions.

**Limb current:** The current in the limbs of a person exposed to electromagnetic field as a result of contact with an object in an electromagnetic field or the flow of capacitive currents induced in the exposed body expressed in ampere (A).

**Magnetic field (H):** (Vector) quantity that characterizes with the magnetic induction, the magnetic field at any point, expressed in amperes per meter ($A \cdot m^{-1}$).

**Magnetic flux density or magnetic induction (B):** (Vector) quantity resulting in a force that acts on moving charges, expressed in tesla (T). In free space and in biological materials, the magnetic flux density is proportional to the magnetic field (H): $B = 4\cdot\ \cdot 10$ −7 H magnetic field (H): $B = 4\cdot\ \cdot 10$

**Magnetic induction (B):** See magnetic flux density.

**Microwaves:** The term refers to electromagnetic waves with a frequency range from approximately 300 MHz to 300 GHz.

**Penetration depth (or skin depth):** For a plane electromagnetic wave incident on the surface delimiting a medium, the distance from the surface of the medium to the point at which the field strengths or induced current densities have been reduced to $1/e$ (≈37%) of their initial value in the medium.

**Permeability (μ):** Ratio of the magnetic flux density (or magnetic induction) to the magnetic field strength at a point, expressed in henry per meter ($H\cdot m^{-1}$).

**Permittivity (ε):** Characterizes the influence of a medium on the forces of attraction or repulsion between the electric charges, expressed in farad per meter ($F\cdot m^{-1}$). It may contain a loss factor. The relative permittivity ($\varepsilon_r$) is the ratio of the permittivity of a dielectric to the permittivity of the vacuum ($\varepsilon_0$).

**Plane wave:** Wave whose quantities are uniform on a perpendicular plan to a fixed direction. For an electromagnetic wave, the electric and magnetic field vectors lie in a plane perpendicular to the direction of propagation, and the product of the magnetic field strength by the impedance of the free space is equal to the electric field strength.

**Power density (S):** Radiant power incident perpendicular to a surface divided by the area of the surface. It is expressed in watt per square meter ($W\cdot m^{-2}$). It is also the product of the electrical field strength and the magnetic field strength for a plane wave. It is the appropriate quantity for very high frequencies, where the depth of penetration in the body is low.

**Radiofrequency (RF):** Electromagnetic energy in the frequency range from 8.3 kHz to 3,000 GHz (ITU definition). This domain is relevant for telecommunication.

**Reference level:** Practical or operational levels established in order to simplify the process of demonstrating the compliance with the basic restrictions.

**Risk:** Probability that a hazard may lead to serious consequences. Without probability, hazard or severity, there is no risk. The purpose of the security rules is to cancel a minimum of one of these factors.

**rms value:** "rms" stands for "root-mean-squared"; square root of the average value of the squared instantaneous values.

**RSS value:** "RSS" stands for "root sum square"; square root of the sum of the squared values.

**Spatial peak value:** Term used to describe the highest level of a particular quantity averaged over a small mass or area in the human body.

**Specific absorption (SA):** Absorbed energy per unit mass of biological tissue during a high-frequency pulse. It is expressed in joule per kilogram ($J{\cdot}kg^{-1}$). SA is the time integral of the specific energy rate (SAR) during a pulse.

**Specific absorption rate (SAR):** Rate at which the energy is absorbed in body tissue expressed in watts per kilogram ($W{\cdot}kg^{-1}$). Physical quantity relating adverse thermal effects to high-frequency exposure.

**Unperturbed field:** Electric, magnetic or electromagnetic field whose lines or propagation are not modified by the presence of a human body.

**Wave:** Propagation of a disturbance that changes reversibly physical properties of the medium in which it propagates. It carries energy without transporting matter.

**Wavelength (λ):** Distance between two successive points of a periodic wave in the direction of propagation, at which the oscillation has the same phase. The wavelength is the ratio of the propagation velocity (v) to the frequency of the wave (f) let $\lambda = v/f$.

# Bibliography

## Publications


[AFS 10] AFSSET, Agence française de sécurité sanitaire de l'environnement et du travail, Avis relatif à la synthèse de l'expertise internationale sur les effets sanitaires des champs électromagnétiques extrêmement basses fréquences, March 2010.

[ALA 11] ALANKO T. *et al.*, "Examples of occupational ELF electric and magnetic field exposure in Finland", *10th International Conference European Bioelectromagnetics Association*, pp. 21–24, Rome, Italy, February 2011.

[AND 01] ANDREUCCETTI D., BINI M., CHECCUCCI A., IGNESTI A., MILLANTA L., OLMI R., RUBINO N., "Protezione dai campi elettromagnetici non ionozzanti", IFAC-CNR, 3rd ed., Florence, Italy, 2001.

[AND 06] ANDREUCCETTI D., ZOPPETTI N., "Quasi-Static Electromagnetic Dosimetry: From Basic Principles to Examples of Applications", *International Journal of Occupational Safety and Ergonomics (JOSE)*, vol. 12, no. 2, pp. 201–215, 2006.

[AND 07] ANDREUCCETTI D., "Equipments for measuring induced/contact currents: functio-ning principles, characteristics, measurement methods and procedures", *PHARE Twinning Project*, Warsaw, Poland, pp. 10–13, December 2007.

[AND 13a] ANDREUCCETTI D. *et al.*, "Nota applicativa. Il metodo del picco ponderato nella valutazione dell'esposizione umana al campi electtromagnetici", IFAC-CNR, Florence, Italy, 2013.

[AND 13b] ANDREUCCETTI D., BOGI A. *et al.*, "I campi elettromagnetici in ambito sanitario: valutazione e prevenzione del rischio", G.E.C.O. Eventi, Pisa, 2013.

[ANF 15] ANFR, Agence nationale des fréquences, Protocole de mesure visant à vérifier sur site pour les stations émettrices fixes, le respect des limitations, en termes de niveaux de référence, de l'exposition du public aux champs électromagnétiques prévus par le décret no. 2002-775 du 3 mai 2002, ANFR/DR 15-3.1, version 3.1, 9 July 2015.

[ANS 13] ANSES, Agence nationale de sécurité sanitaire de l'alimentation, de l'environnement et du travail, Avis relatif à la mise à jour de l'expertise "Radiofréquences et santé", Maisons-Alfort, October 2013.

[ANS 16] ANSES, Agence nationale de sécurité sanitaire de l'alimentation, de l'environnement et du travail, Rapport d'expertise collective: compatibilité électromagnétique des dispositifs médicaux exposés à des sources radiofréquences, Maisons-Alfort, April 2016.

[APA 06] APAT, Agenzia per la protezione dell'ambiente e per i servizi tecnici, "Metodologie, tecniche e procedure per la riduzione delle emissioni dei campi elettromagnetici nell'ambiente", APAT/ISPRA, Rome, disponible à l'adresse: www.isprambiente.gov.it, 2006.

[ARP 02] ARPANSA, Australian Radiation Protection And Nuclear Safety Agency, "Maximum Exposure Levels to Radiofrequency Fields – 3 kHz to 300 GHz", available at: www.arpansa.gov.au, March 2002.

[ARS 96] D'ARSONVAL J.A., "Dispositifs pour la mesure des courants alternatifs de toutes fréquences", note du Dr J.A. d'Arsonval, *CR Soc. Biol.*, pp. 450–451, Paris, 2 May 1896.

[BER 88] BERNHARDT J.H., "The establishment of frequency dependent limits for electric and magnetic fields and evaluation of indirect effects", *Radiation and Environmental Biophys.*, no. 27, pp. 1–27, 1988.

[BMA 11] BMAS-Forschungsbericht FB 400-E, Electromagnetic fields at workplaces, 2011.

[BOL 01] BOLOMEY J.C., GARDIOL F.E., *Engineering Applications of the Modulated Scatterer Technique*, Artech House, London, 2001.

[BUL 13] BULLO M. *et al.*, "Analysis of Stray EM Fields Generated by Induction Cooktops", *IEEE Electromagnetic Compatibility Magazine*, vol. 2, 2013.

[CNR 12] CNR, Centre national de référence RFID, Technologies de l'information – Identification par radiofréquences (RFID) – Méthodologie de mesure du champ électrique et de la puissance rayonnée en RFID UHF, version 2.0, 2012.

[COP 13] COPIC, Comité opérationnel sur les ondes de téléphonie mobile, Diminution de l'exposition aux ondes électromagnétiques émises par les antennes relais de téléphonie mobile, Rapport de synthèse des expérimentations, August 2013.

[DIM 02] DIMBYLOW P.J., "Fine resolution calculations of SAR in the human body for frequencies up to 3 GHz", *Phys. Med. Biol.*, no. 16, pp. 2835–2846, August 2002.

[DUR 86] DURNEY C.H. *et al.*, Radiofrequency Radiation Dosimetry Handbook, Electrical Engineering Department, University of Utah, Salt Lake City, USA, October 1986.

[ECC 14] EEC, Electronic communications committee (CEPT), ERC Recommendation 70-03 relating to the use of Short Range Devices (SRD), 2014.

[EDF 05] EDF-RTE, Electricité de France, Réseau de transport de l'électricité, "Les champs électromagnétiques de très basse fréquence", www.rte-france.com, 2005.

[EMF 16] EMF-Portal, "Internet information platform", University Hospital RWTH Aachen, Allemagne, available at : www.emf-portal.org, June 2016.

[FEL 13] FELBLINGER J., Présentation: *hardware* pour l'IRM fonctionnelle, INSERM U947, CHU, Nancy, 2013.

[FEY 15] FEYNMAN R., *Le cours de physique de Feynman, électromagnétisme*, Dunod, Paris, 2015.

[FRE 15] FREUDENSTEIN F., WIEDEMANN P.M., VARSIER N., "Exposure knowledge and risk perception of RF EMF", *Front Public Health*, vol. 2, no. 289, www.ncbi.nlm.nih.gov, 2015.

[GAB 96a] GABRIEL C., Compilation of the dielectric properties of body tissues at RF and microwave frequencies, Report N.AL/OE-TR-1996-0037, Occupational and environmental health directorate, Radiofrequency Radiation Division, Brooks Air Force Base, Texas, USA, June 1996.

[GAB 96b] GABRIEL C., GABRIEL S., CORTHOUT E., "The dielectric properties of biological tissues: I. Literature survey", *Phys. Med. Biol.*, no. 41, pp. 2231–2249, 1996.

[GAB 96c] GABRIEL S., LAU R.W., GABRIEL C., "The dielectric properties of biological tissues: II. Measurements in the frequency range 10 Hz to 20 GHz", *Phys. Med. Biol.*, no. 41, pp. 2251–2269, 1996.

[GAB 96d] GABRIEL S., LAU R.W., GABRIEL C., "The dielectric properties of biological tissues: III. Parametric models for the dielectric spectrum of tissues", *Phys. Med. Biol.*, no. 41, pp. 2271–2293, 1996.

[GAU 06] GAUVIN D., NGAMGA DJEUTCHA E., LEVALLOIS P., "Exposition aux champs électromagnétiques: mise à jour des risques pour la santé et pertinence de la mise en œuvre du principe de précaution", INSPQ/Institut national de santé publique du Québec, Montreal, Canada, 2006.

[GAU 16] GAUTIER M., GAUVIN D., "Evaluation des effets sur la santé des champs électromagnétiques dans le domaine des radiofréquences", INSPQ/Institut national de santé publique du Québec, Montreal, Canada, 2016.

[GRA 14] GRAVELEAU S., "Les différents moyens de création d'un champ magnétique", Note technique, SREM Technologies, La Flèche, www.srem.fr 2014.

[HAN 15] HANSSON MILD K., SANDSTRÖM M., "Electromagnetic fields in working life: A guide to risk assessment", European Trade Union Institute, University of Umea, Sweden, www.etui.org, 2015.

[HIR 11] HIRATA A. *et al.*, "An electric field induced in the retina and brain at threshold magnetic flux density causing magnetophosphenes", *Phys. Med. Biol.*, no. 56, pp. 4091–4101, 2011.

[HPA 08] HPA, Health Protection Agency, RCE-6: Static Magnetic Fields, Report of the independent Advisory Group on Non-ionising Radiation, UK, www.gov.uk/government/publications, May 2008.

[HSE 08] HSE, Health and Safety Executive, Assessment uncertainties relating to electromagnetic fields (EMF) measurement and computation, Research Report 649, EM Fields Ltd, Lambourne, UK, www.hse.gov.uk/research/ rrpdf/rr649.pdf, 2008.

[IEE 02] IEEE Std C95.6™, IEEE Standard for Safety Levels with Respect to Human Exposure to Electromagnetic Fields, 0-3 kHz, 2002.

[IEE 05] IEEE Std C95.1™, IEEE Standard for Safety Levels with Respect to Human Exposure to Radio Frequency Electromagnetic Fields, 3 kHz to 300 GHz, 2005.

[IEE 10] IEEE Std C95.3.1™, IEEE Recommended Practice for Measurements and Computations of Electric, Magnetic and Electromagnetic Fields with respect to Human Exposure to such fields, 0 Hz to 300 GHz, 2010.

[IAR 13] IARC, International Agency for research on cancer, "Non-Ionizing radiation, part 2: Radiofrequency electromagnetic field", *IARC Monographs*, vol. 102, IARC, Lyon, 2013.

[ICN 98] ICNIRP, International Commission of Non-ionizing Radiation, "Guidelines for Limiting Exposure to Time-Varying Electric, Magnetic, and Electromagnetic Fields (up to 300 GHz)", *Health Physics*, vol. 74, pp. 494–522, 1998.

[ICN 03] ICNIRP, International Commission of Non-ionizing Radiation, "Guidance on determining compliance of exposure to pulsed and complex non-sinusoidal waveforms below 100 kHz with ICNIRP guidelines", www.icnirp.org, 2003.

[ICN 08] ICNIRP, International Commission of Non-ionizing Radiation, "Statement on EMF-emitting new technologies", *Health Physics Society*, vol. 94, no. 4, pp. 376–392, April 2008.

[ICN 09a] ICNIRP, International Commission of Non-ionizing Radiation, "Guidelines on limits of exposure to static magnetic fields", www.icnirp.org, 2009.

[ICN 09b] ICNIRP, International Commission of Non-ionizing Radiation, "Exposure to high frequency electromagnetic fields, biological effects and health consequences (100 kHz-300 GHz): Review of the Scientific Evidence and Health Consequences", *High Frequency Review*, 2009.

[ICN 10] ICNIRP, International Commission of Non-ionizing Radiation, "Guidelines for limiting exposure to time-varying electric and magnetic fields (1 Hz to 100 kHz)", *Health Physics*, vol. 99, no. 6, pp. 818–836, 2010.

[ICN 14] ICNIRP, International Commission of Non-ionizing Radiation, "Guidelines for limiting exposure to electric fields induced by movement of the human body in a static magnetic field and by time-varying magnetic fields below 1 Hz", www.icnirp.org, 2014.

[ITI 10] ITIS FOUNDATION, BAUTZ W., CHEN J., CHRIST A. *et al.*, "The Virtual Family – Development of Surface-based Anatomical Models of Two Adults and Two Children for Dosimetric Simulations", *Physics in Medicine and Biology*, vol. 55, no. 2, p. N23-N38, January 2010.

[ITI 14] ITIS FOUNDATION, CHERUBINI E., DI GENNARO F., FARCITO S. *et al.*, "Development of a New Generation of High-Resolution Anatomical Models for Medical Device Evaluation: The Virtual Population 3.0", *Physics in Medicine and Biology*, vol. 59, no. 18, pp. 5287–5303, June 2014.

[JAR 12] JARRIGE P. *et al.*, "Mesure dosimétrique par capteur électro-optique fibré (*Pigtailed electro-optic sensor for microwave dosimetric assessment*)", *JS'12*, CNAM, pp. 3–4, Paris, April 2012.

[JCG 10] JCGM 100:2008, "Evaluation of measurement data, Guide to the expression of uncertainty in measurement", revised version, www.bipm.org/utils/common/documents/jcgm/JCGM_100_2008_F.pdf, 2010.

[JOK 00] JOKELA K., "Restricting exposure to pulsed and broadband magnetic fields", *Health Physics*, vol. 79, no. 4, 2000.

[JOU 02] Décret no. 2002-775 du 3 mai 2002 pris en application du 12° de l'article L. 32 du Code des postes et télécommunications et relatif aux valeurs limites d'exposition du public aux champs électromagnétiques émis par les équipements utilisés dans les réseaux de télécommunication ou par les installations radioélectriques, Revised version, 28 January 2008.

[KAS 11] KASLER B., VETTER D., *Comprendre l'IRM: manuel d'autoapprentissage*, Elsevier Masson, Paris, 2011.

[LAG 16] LAGROYE I., "Champs magnétiques, champs électromagnétiques et santé", *Year Book Santé et Environnement*, John Libbey Eurotext, Montrouge, 2016.

[LAM 13] LAMBROZO J. *et al.*, "Les médecins généralistes français face aux champs électromagnétiques", *Presse Med.*, 2013.

[LAPI 1907] LAPICQUE L., "Recherches quantitatives sur l'excitation électrique des nerfs traitée comme une polarisation", *J. Physiol.*, no. 9, pp. 620–635, 1907.

[LAPI 1909] LAPICQUE L., "Définition expérimentale de l'excitabilité", *Soc. Biologic*, no. 77, pp. 280–283, 1909.

[LEG 15] LEGROS A. *et al.*, "Human magnetophosphene perception and EEG response to 50 and 60 Hz magnetic stimuli up to 50 mT", *Conference Brain Stimulation*, 2015.

[LET 13] LETERTE T., Méthodologie pour l'évaluation des signaux émis par les technologies émergentes: applications à la compatibilité électromagnétique des systèmes et à l'exposition des personnes, PhD Thesis, Supélec, Paris, December 2013.

[MAR 10] MARTINU K., "Cours RAD6005: introduction à l'IRMf", Centre de recherche, Institut universitaire de gériatrie de Montréal, University of Montreal, www.unf-montreal.ca, 2010.

[MIL 15] HANSSON MILD K., SANDSTRÖM M., "Electromagnetic fields in working life: A guide to risk assessment", *European Trade Union Institute*, Umea University, Sweden, available at : www.etui.org, 2015.

[MIS 93] MISAKIAN M., "Coil Probe Dimension and Uncertainties During Measurements of Nonuniform ELF Magnetic Fields", *Journal of Research of the National Institute of Standards and Technology*, vol. 98, no. 3, May-June 1993.

[NEU 03] NEUBAUER G., HAIDER H., LAMEDSCHWANDNER K. *et al.*, "Measurement Methods and Legal Requirements for Exposure Assessment next to GSM Base Stations", *15th Int. Zürich Symposium and Technical Exhibition on Electromagnetic Compatibility*, Seibersdorf Laboratories, Austria/OFCOM, Biel-Bienne, Switzerland, 18-20 February 2003.

[PER 04] PERSON C., Développement de sondes de mesures de SAR-Méthodologie, URSI France, available at: www.ursi-france.telecom-paristech.fr/evenements/journees-scientifiques/2004.html, 2004.

[PER 10] PERRIN A., SOUQUES M., *Champs électromagnétiques, environnements et santé*, Springer, Paris, 2010.

[PIC 12] PICARD D., BERRAHMA K., "Dosimètres radiofréquences et diffraction par le corps de l'usager", *JS'12*, CNAM, Paris, 3-4 April 2012.

[ROA 09] ROACH W.P., *Radiofrequency Radiation Dosimetry Handbook*, July 2009.

[SCE 15] SCENIHR, Scientific Committee on Emerging and Newly Identified Health Risks, "Opinion on Potential health effects of exposure to electromagnetic fields (EMF)", available at: http://ec.europa.eu/health/scientific_committees/consultations/public_consultations/scenihr_consultation_19_en.htm, January 2015.

[SLO 15] SLOBOZHANYUK A.P., PODDUBNY A.N., RAAIJMAKERS A.J.E. *et al.*, "Enhancement of Magnetic Resonance Imaging with Metasurfaces", Itmo University, Saint-Petersbourg, available at: www.arxiv.org/ pdf/1507. 01411.pdf, 2015.

[TAK 01] TAKI M., "Characteristics, Dosimetry and Measurement of EMF", *WHO meeting on EMF Biological Effects and Standards Harmonization in Asia and Oceania*, Seoul, South Korea, pp. 22–24, October 2001.

[TAN 95] TANDLER D., "Measurement of Low Frequency Electrical and magnetic Fields", *EMF Congress*, March 1995.

[TAR 13] TARAO H. *et al.*, "Numerical Evaluation of Currents Induced in a Worker by ELF Non-Uniform Electric Fields in High Voltage Substations and Comparison with Experimental Results", *Bioelectromagnetics*, no. 34, pp. 61–73, 2013.

[TEN 91] TENFORDE T.S., "Biological interactions of extremely low frequency electric and magnetic fields", *Bioelectrochem. Bioenerg.*, no. 24, pp. 1–17, 1991.

[THU 10] THUROCZY G., BAKOS J., RAVAZZANI P., "Report on the level of exposure (frequency, patterns and modulation) in the European Union – part 2: Extremely low frequency (ELF)", *European Health Risk Assessment Network on Electromagnetic Fields Exposure (EHFRAN) Project*, www.efhran.polimi.it/docs, 2010.

[WER 79] WERTHEIMER N., LEEPER E., "Electrical wiring configurations and childhood cancer", *Am. J. Epidemiol.*, no. 109, pp. 273–284, 1979.

[WHO 84] WHO, World Health Organization, "Environmental Health Criteria 35: Extremely Low Frequency (ELF) fields", www.who.int/peh-emf, 1984.

[WHO 93] WHO, World Health Organization, "Environmental Health Criteria 137: Electromagnetic fields (300 Hz to 300 GHz)", www.inchem.org/documents/ehc/ehc/ehc137.htm, 1993.

[WHO 04] WHO, World Health Organization, "Electromagnetic Hypersensitivity", *Proceedings International Workshop on EMF Hypersensitivity*, Prague, Czech Republic, 25-27 October 2004.

[WHO 05] WHO, World Health Organization, "Champs électromagnétiques et santé publique: hypersensibilité électromagnétique", www.who.int/peh-emf/publications/facts/fs296_fr/en/, 2005.

[WHO 07] WHO, World Health Organization, "Environmental Health Criteria 238: Extremely low frequency fields", www.who.int/peh-emf/publications/Complet_DEC_2007.pdf, 2007.

[WIA 13] WIART J., "LEXNET", Project presentation, http://whist.mines-telecom.fr, December 2013.

[WOU 13] WOUT J. *et al.*, "Determination of the duty cycle of WLAN for realistic radio frequency electromagnetic field exposure assessment", *Prog. in Biophys. Molecular Biol.*, Elsevier, vol. 111, pp. 30–36, 2013.

[ZOP 09] ZOPPETTI N., ANDREUCCETTI D., "Review of open problems in assessing compliance with 2004/40/EC Directive exposure limit values for low-frequency current density by means of numerical techniques", *Radiation Protection Dosimetry*, vol. 137, no. 3-4, pp. 247–251, 2009.

[ZWE 09] ZWEIACKER P., *Vivre dans les champs électromagnétiques*, Presses polytechniques et universitaires romandes, Lausanne, 2009.

## European Union documents

[EC 001] Directive 2013/35/EU of the European Parliament and of the Council on the minimum health and safety requirements regarding the exposure of workers to the risks arising from physical agents (electromagnetic fields), (20th individual Directive within the meaning of Article 16(1) of Directive 89/391/EEC) and repealing Directive 2004/40/EC, *Official Journal of the European Union*, no. L 179, p. 21, 2013.

[EC 002] Directive 89/391/EEC-OSH, Framework Directive of 12 June 1989 on the introduction of measures to encourage improvements in the safety and health of workers at work, 1989.

[EC 003] Council of the European union, Legislative acts and other instruments: Corrigendum/ Rectificatif, subject: see [EC 001], 13359/2/13 REV 2, 17 October 2013.

[EC 004] European Commission, Non-binding guide to good practice for implementing Directive 2013/35/EU, "Electromagnetic Fields", *Practical Guide*, vol. 1, 2014.

[EC 005] European Commission, Non-binding guide to good practice for implementing Directive 2013/35/EU, "Electromagnetic Fields", *Case Studies*, vol. 2, 2014.

[EC 006] European Commission: SEC(2011) 750 final, Commission Staff working paper impact assessment Accompanying document to the proposal for a directive of the European Parliament and of the Council on the minimum health and safety requirements regarding the exposure of workers to the risks arising from physical agents "Electromagnetic fields", 20th individual Directive within the meaning of Article 16(1) of Directive 89/391/EEC, 2011.

[EC 007] Directive 90/385/EEC, of the Council of the European Communities of 20 June 1990 on the approximation of the laws of the Member States relating to active implantable medical devices, 1990.

[EC 008] Directive 1996/29/EURATOM of the Council of the European Union laying down basic safety standards for the protection of the health of workers and the general public against the dangers arising from ionizing radiation, *Official Journal of the European Union*, no. L 159, p. 114, 1996.

[EC 009] Directive 2014/53/EU of the European Parliament and of the Council on the harmonisation of the laws of the Member States relating to the making available on the market of radio equipment and repealing Directive 1999/5/EC, no. L 153, p. 45, 2014.

[EC 010] Directive 2006/25/EC of the European Parliament and of the Council on the minimum health and safety requirements regarding the exposure of workers to risks arising from physical agents (artificial optical radiation) (19th individual Directive within the meaning of Article 16(1) of Directive 89/391/EEC), OJ L 114, p. 28, 2006.

[EC 011] Directive 89/391/EEC of the Council of the European Communities on the introduction of measures to encourage improvements in the safety and health of workers at work, no. L 183/1, p. 8, 1989.

[EC 99] European Communities, "Council Recommendation of 12 July 1999 on the limitation of exposure of the general public to electromagnetic fields (0 Hz to 300 GHz) (1999/519/EC)", *Official Journal of the European Communities (OJEC)*, no. L 199, pp. 59–70, July 1999.

## Standards

[STD 01] IEC 61786-1: Measurement of DC magnetic, AC magnetic and AC electric fields from 1 Hz to 100 kHz with regard to exposure of human beings – Part 1: Requirements for measuring instruments – Proposed Horizontal Standards, 2013.

[STD 02] IEC 61786-2: Measurement of DC magnetic, AC magnetic and AC electrical fields from 1 Hz to 100 kHz with regard to exposure of human beings – Part 2: Basic standard for measurements – Proposed Horizontal Standards, 2015.

[STD 03] EN 62311: Assessment of electronic and electrical equipment related to human exposure restrictions for electromagnetic fields (0 Hz to 300 GHz), 2008.

[STD 04] IEC 62233: Measurement methods for electromagnetic fields of household appliances and similar apparatus with regard to human exposure, 2013.

[STD 05] EN 62110: Electric and magnetic field levels generated by AC power systems – Measurement procedures with regard to public exposure, 2010.

[STD 06] IEC 62822-1 – Edition 1.0 2016-03: Electric welding equipment – Assessment of restrictions related to human exposure to electromagnetic fields (0 Hz to 300 GHz) – Part 1: Product family standard, 2016.

[STD 07] IEC 62822-2 – Edition 1.0 2016-03: Electric welding equipment – Assessment of restrictions related to human exposure to electromagnetic fields (0 Hz to 300 GHz) – Part 2: Arc welding equipment, 2016.

[STD 08] IEC 62822-3 – Pr 2016-03: Electric welding equipment – Assessment of restrictions related to human exposure to electromagnetic fields (0 Hz to 300 GHz) – Part 3: Resistance welding equipment, 2016.

[STD 09] IEC 62209-1: Human exposure to radio frequency fields from hand-held and body-mounted wireless communication devices – Human models, instrumentation, and procedures – Part 1: Procedure to determine the specific absorption rate (SAR) for handheld devices used in close proximity of the ear (frequency range of 30 MHz to 6 GHz), 2005.

[STD 10] IEC 62209-2: Human exposure to radio frequency fields from hand-held and body-mounted wireless communication devices – Human models, instrumentation, and procedures – Part 2: Procedure to determine the specific absorption rate (SAR) for wireless communication devices used in close proximity to the human body (frequency range of 30 MHz to 6 GHz), 2010.

[STD 11] EN 50505: Basic standard for the evaluation of human exposure to electromagnetic fields from equipment for resistance welding and allied processes, 2008 (obsolete standard).

[STD 12] IEC/EN 62110, Electric and magnetic field levels generated by AC power systems – measurement procedures with to public exposure, 2009.

[STD 13] UTE C99-132: Protocol for 50 Hz magnetic fields generated by electricity transportation facilities *in situ* measurement, 2010.

[STD 14] IEC 60974-9: Arc welding equipment – Part 9: Installation and use, 2010.

[STD 15] IEC 62369-1: Evaluation of human exposure to electromagnetic fields from Short Range Devices (SRDs) in various applications over the frequency range 0-300 GHz. Part 1: Fields produced by devices used for Electronic Article Surveillance, Radio Frequency Identification and similar systems, 2009.

[STD 16] IEC 17025: General requirements for the competence of testing and calibration laboratories, 2005.

[STD 17] EN 55011: Industrial, scientific and medical equipment – Radio-frequency disturbance characteristics – Limits and methods of measurement, 2010.

[STD 19] IEC 62226-1: Exposure to electric or magnetic fields in the low and intermediate frequency range – Methods for calculating the current density and internal electric field induced in the human body – Part 1: General, 2005.

[STD 20] IEC 62226-2-1: Exposure to electric or magnetic fields in the low and intermediate frequency range – Methods for calculating the current density and internal electric field induced in the human body – Part 2-1: Exposure to magnetic fields – 2D models, 2005.

[STD 21] EN 62226-3-1: Exposure to electric or magnetic fields in the low and intermediate frequency range – Methods for calculating the current density and internal electric field induced in the human body – Part 3-1: Exposure to electric fields – Analytical and 2D numerical models, 2008.

[STD 23] EN 50519: Assessment of workers' exposure to electric and magnetic fields of industrial induction heating equipment, 2010.

[STD 24] EN 50413: Basic standard on measurement and calculation procedures for human exposure to electric, magnetic and electromagnetic fields (0 Hz to 300 GHz), 2008.

[STD 25] EN 50527-1: Procedure for the assessment of the exposure to electromagnetic fields of workers bearing active implantable medical devices – Part 1: General, 2010.

[STD 26] EN 50527-2-1: Procedure for the assessment of the exposure to electromagnetic fields of workers bearing active implantable medical devices – Part 2-1: Specific assessment for workers with cardiac pacemakers, 2011.

[STD 27] EN 60335-2-25: Household and similar electrical appliances – Safety – Part 2-25: Particular requirements for microwave ovens, including combination microwave ovens – 2010.

[STD 28] EN 50492: Basic standard for the in-situ measurement of electromagnetic filed strength related to human exposure in the vicinity of base stations, 2009.

[STD 30] IEC – CISPR 16-1: Specification for radio disturbance and immunity measuring apparatus and methods: Radio disturbance and immunity measuring apparatus, 2002.

[STD 31] IEC 62764-1. Measurement procedures of magnetic field levels generated by electronic and electrical equipment in the automotive environment with respect to human exposure (in progress).

[STD 32] IEC 60050-161: International Electrotechnical Vocabulary. Chapter 161: Electro-magnetic compatibility, 1990.

[STD 33] IEC 62232: Determination of RF field strength and SAR in the vicinity of radio-communication base stations for the purpose of evaluating human exposure.

[STD 34] EN 60601-1-2: Medical electrical equipment – Part 1-2: General requirements for basic safety and essential performance – Collateral standard: Electromagnetic compatibility – Requirements and tests, 2007.

[STD 35] ANSI C63.19 - 2011: American National Standard Methods of Measurement of Compatibility between Wireless Communications Devices and Hearing Aids.

# Index

## F, I



## M, P



## R, S, T



## V, W

Other titles from

iSTE

in

Waves

## 2017

PARET Dominque, SIBONY Serge
*Musical Techniques: Frequencies and Harmony*

## 2016

ANSELMET Fabien, MATTEI Pierre-Olivier
*Acoustics, Aeroacoustics and Vibrations*

BAUDRAND Henri, TITAOUINE Mohammed, RAVEU Nathalie
*The Wave Concept in Electromagnetism and Circuits: Theory and Applications*

PARET Dominique
*Antennas Designs for NFC Devices*

PARET Dominique
*Design Constraints for NFC Devices*

WIART Joe
*Radio-Frequency Human Exposure Assessment*

## 2015

PICART Pascal
*New Techniques in Digital Holography*

## 2014

APPRIOU Alain
*Uncertainty Theories and Multisensor Data Fusion*

JARRY Pierre, BENEAT Jacques N.
*RF and Microwave Electromagnetism*

LAHEURTE Jean-Marc
*UHF RFID Technologies for Identification and Traceability*

SAVAUX Vincent, LOUËT Yves
*MMSE-based Algorithm for Joint Signal Detection, Channel and Noise Variance Estimation for OFDM Systems*

THOMAS Jean-Hugh, YAAKOUBI Nourdin
*New Sensors and Processing Chain*

TING Michael
*Molecular Imaging in Nano MRI*

VALIÈRE Jean-Christophe
*Acoustic Particle Velocity Measurements using Laser: Principles, Signal Processing and Applications*

VANBÉSIEN Olivier, CENTENO Emmanuel
*Dispersion Engineering for Integrated Nanophotonics*

## 2013

BENMAMMAR Badr, AMRAOUI Asma
*Radio Resource Allocation and Dynamic Spectrum Access*

BOURLIER Christophe, PINEL Nicolas, KUBICKÉ Gildas
*Method of Moments for 2D Scattering Problems: Basic Concepts and Applications*

GOURE Jean-Pierre
*Optics in Instruments: Applications in Biology and Medicine*

LAZAROV Andon, KOSTADINOV Todor Pavlov
*Bistatic SAR/GISAR/FISAR Theory Algorithms and Program Implementation*

LHEURETTE Eric
*Metamaterials and Wave Control*

PINEL Nicolas, BOURLIER Christophe
*Electromagnetic Wave Scattering from Random Rough Surfaces: Asymptotic Models*

SHINOHARA Naoki
*Wireless Power Transfer via Radiowaves*

TERRE Michel, PISCHELLA Mylène, VIVIER Emmanuelle
*Wireless Telecommunication Systems*

## 2012

LALAUZE René
*Chemical Sensors and Biosensors*

LE MENN Marc
*Instrumentation and Metrology in Oceanography*

LI Jun-chang, PICART Pascal
*Digital Holography*

## 2011

BECHERRAWY Tamer
*Mechanical and Electromagnetic Vibrations and Waves*

GOURE Jean-Pierre
*Optics in Instruments*

GROUS Ammar
*Applied Metrology for Manufacturing Engineering*

LE CHEVALIER François, LESSELIER Dominique, STARAJ Robert
*Non-standard Antennas*

## 2010

BEGAUD Xavier
*Ultra Wide Band Antennas*

MARAGE Jean-Paul, MORI Yvon
*Sonar and Underwater Acoustics*

## 2009

BOUDRIOUA Azzedine
*Photonic Waveguides*

BRUNEAU Michel, POTEL Catherine
*Materials and Acoustics Handbook*

DE FORNEL Frederique, FAVENNEC Pierre-Noël
*Measurements using Optic and RF Waves*

FRENCH COLLEGE OF METROLOGY
*Transverse Disciplines in Metrology*

## 2008

FILIPPI Paul J.T.
*Vibrations and Acoustic Radiation of Thin Structures*

LALAUZE René
*Physical Chemistry of Solid-Gas Interfaces*

## 2007

KUNDU Tribikram
*Advanced Ultrasonic Methods for Material and Structure Inspection*

PLACKO Dominique
*Fundamentals of Instrumentation and Measurement*

RIPKA Pavel, TIPEK Alois
*Modern Sensors Handbook*

## 2006

BALAGEAS Daniel *et al.*
*Structural Health Monitoring*

BOUCHET Olivier *et al.*
*Free-Space Optics*

BRUNEAU Michel, SCELO Thomas
*Fundamentals of Acoustics*

FRENCH COLLEGE OF METROLOGY
*Metrology in Industry*

GUILLAUME Philippe
*Music and Acoustics*

GUYADER Jean-Louis
*Vibration in Continuous Media*

Printed and bound by CPI Group (UK) Ltd, Croydon, CR0 4YY
16/09/2026
14963072-0008